# Optics Manufac[illegible]

## Optical Sciences and Applications of Light

*Series Editor*
James C. Wyant
University of Arizona

Optics Manufacturing: Components and Systems, *Christoph Gerhard*

Femtosecond Laser Shaping: From Laboratory to Industry, *Marcos Dantus*

Photonics Modelling and Design, *Slawomir Sujecki*

Handbook of Optomechanical Engineering, Second Edition, *Anees Ahmad*

Lens Design: A Practical Guide, *Haiyin Sun*

Nanofabrication: Principles to Laboratory Practice, *Andrew Sarangan*

Blackbody Radiation: A History of Thermal Radiation Computational Aids and Numerical Methods, *Sean M. Stewart and R. Barry Johnson*

High-Speed 3D Imaging with Digital Fringe Projection Techniques, *Song Zhang*

Introduction to Optical Metrology, *Rajpal S. Sirohi*

Charged Particle Optics Theory: An Introduction, *Timothy R. Groves*

Nonlinear Optics: Principles and Applications, *Karsten Rottwitt and Peter Tidemand-Lichtenberg*

Numerical Methods in Photonics, *Andrei V. Lavrinenko, Jesper Lægsgaard, Niels Gregersen, Frank Schmidt, and Thomas Søndergaard*

# Optics Manufacturing
## Components and Systems

Christoph Gerhard

CRC Press is an imprint of the
Taylor & Francis Group, an **informa** business

CRC Press
Taylor & Francis Group
6000 Broken Sound Parkway NW, Suite 300
Boca Raton, FL 33487-2742

First issued in paperback 2019



ISBN-13: 978-1-4987-6459-9 (hbk)
ISBN-13: 978-1-138-74678-7 (pbk)



**Library of Congress Cataloging-in-Publication Data**

Names: Gerhard, Christoph, 1977- author.
Title: Optics manufacturing : components and systems / Christoph Gerhard.
Description: Boca Raton : Taylor & Francis, a CRC title, part of the Taylor & Francis imprint, a member of the Taylor & Francis Group, the academic division of T&F Informa, plc, [2017] | Series: Optical sciences and applications of light | Includes bibliographical references and index.
Identifiers: LCCN 2017035810 | ISBN 9781498764599 (hardback : acid-free paper) | ISBN 9781498764612 (ebook)
Subjects: LCSH: Optical instruments--Design and construction.
Classification: LCC TS513 .G47 2017 | DDC 681/.4--dc23
LC record available at https://lccn.loc.gov/2017035810

**Visit the Taylor & Francis Web site at**
**http://www.taylorandfrancis.com**

**and the CRC Press Web site at**
**http://www.crcpress.com**

# Dedication

*To Martina, Benjamin, and Philipp.*

# Contents

# Preface

In many cases, it is daring to disagree with a Nobel Prize Laureate. However, in this particular case, the author would like to emphasize that the statement "Writing textbooks on optics seems to bring nothing but trouble," taken from Max Born's autobiography, does not apply at all. Writing the present textbook was a challenging but interesting and instructive task due to the complexity of the subject, optics manufacturing. Even though the production of optical components is a traditional craft, the methods and processes are continuously improved, broadened, and even not fully understood. The aim of the present work is to give a complete overview on classical and the newest methods and approaches for the manufacturing of optical components and systems.

Such devices represent essential key components in modern engineering and everyday life. In 2006, the current chancellor of the University of Cambridge, David John Sainsbury, prognosticated, "There is good reason to believe that the impact of photonics in the twenty-first century will be as significant as electronics was in the twentieth or steam in the nineteenth." Actually, the latest market analyses agree with this statement. The education of skilled personnel and specialists in the fields of theoretical and practical optics manufacturing is thus of essential importance for the next generation technologies.

This book is thus intended to contribute to the education of optical engineers and skilled workers as a reference work. For this purpose, most chapters end with a short summary of the most important aspects, followed by a formulary of the relevant equations including the used symbols and abbreviations. Moreover, exercises on the covered basic principles of optics and approaches and techniques of optics manufacturing including detailed solutions are found in the Appendix.

The author thanks Stephanie Henniges, Geoff Adams, Ashley Gasque, and Marc Gutierrez for their help and his friends and family for support and encouragement during the preparation of the manuscript.

**Christoph Gerhard**
*Göttingen, Germany, June 2017*

# Author

After his apprenticeship as optics technician, Prof. Dr. Christoph Gerhard worked as a skilled worker and instructor in optical manufacturing. He then completed his diploma study in precision manufacturing technology in Göttingen, Germany, and Orsay, France, and a subsequent master study in optical engineering and photonics in Göttingen, Germany, and Bremen, Germany. After working in industry as a product manager for precision optics and optical design software for 3 years, he worked as a research associate in the fields of laser and plasma processing of glasses and as a lecturer for optics manufacturing and optical system design. In 2014, he earned a Doctoral degree in natural sciences and physical technologies at Clausthal University of Technology in Clausthal-Zellerfeld, Germany, and subsequently worked as an adjunct Professor for physical technologies at the University of Applied Sciences in Göttingen, Germany. Currently, he is a Professor for laser and plasma technology at the Technical University of Applied Sciences in Wildau, Germany, and a visiting Professor for computer-assisted optical system design at the Polytechnic University of Milan, Italy.

For his work on the development of laser sources for materials processing applications and novel approaches in laser-based structuring of optical glasses, he was awarded the Georg-Simon-Ohm Award by the German Physical Society in 2009 and the Young Talent Award Green Photonics of the Fraunhofer Society in 2015. He is a member of the German Physical Society and the German Society for Plasma Technology.

Dr. Gerhard has published over 30 articles on optical fabrication, testing, and system design, and he is the chief editor of a textbook on laser ablation. He also has authored two textbooks on the principles of optics and plasma-assisted laser ablation of glasses.

# 1 Introduction

Most likely, the first optical component produced by mankind was a simple mirror made of polished stone or metal; the goal was to imitate the reflection of smooth water surfaces. It is now known that about 5000 years ago, bronze mirrors were used in Mesopotamia, and the master builders of the ancient Egyptian civilization employed such mirrors for the illumination of burial sites in the Valley of the Kings. Another recorded example of mirrors in former times is their use as defensive weapons during the Siege of Syracuse in the third century BCE. According to legend, the attacking Roman fleet was inflamed and destroyed by a setup of mirrors invented by the Greek mathematician and physicist *Archimedes of Syracuse* (c. 287–c. 212 BCE). However, there is no evidence for this historical tradition.

Optics manufacturing with glass seems to be a very old and traditional craft (Temple, 2000). The oldest presently known lens, a quartz lens called the Nimrud lens in literature, is dated around 1000 BCE and was found in the ancient Assyrian city Nimrud in northern Mesopotamia (Barker, 1930). Some centuries later, the production of glass was described in detail by a cuneiform inscription in the library of the ancient Assyrian king *Ashurbanipal* (668 BCE–c. 627 BCE) in Nineveh, Upper Mesopotamia. It can be assumed that the Nimrud lens was used as magnifying glass. Maybe the inventor or maker of this lens was inspired by a raindrop on a leaf, which allowed the very first "microscopic" observation of surfaces by accident.

In the course of the centuries, the manufacture of glasses and optical components was improved iteratively. In medieval times, lens grinders and spectacle makers produced simple eyeglasses that were mainly used by monks and copyists for the duplication of ecclesiastical documents and bibles. Based on this handcraft, the first telescopes were realized at the beginning of the seventeenth century, and it is told that the famous Italian astronomer and physicist *Galileo Galilei* (1564–1642) learned how to manufacture optical lenses in order to improve the image quality of the telescopes he used for his astronomical discoveries.

Even though now modern machines, working materials, and equipment are in hand, the essential production steps of optics manufacturing as summarized in Figure 1.1 have not changed since that time.

The initial point of modern optics manufacturing is the analysis of a given imaging task; it thus starts with basic theoretical considerations. As a result of such analysis, the required optical component or system in terms of its type and focal length is determined. After the evaluation of the impact of deviations in contour accuracy and glass bulk defects on the imaging quality, the manufacturing tolerances are identified and documented in the form of a manufacturing drawing. This drawing represents the actual basis of any optics manufacturing process. As shown in Figure 1.2, optical

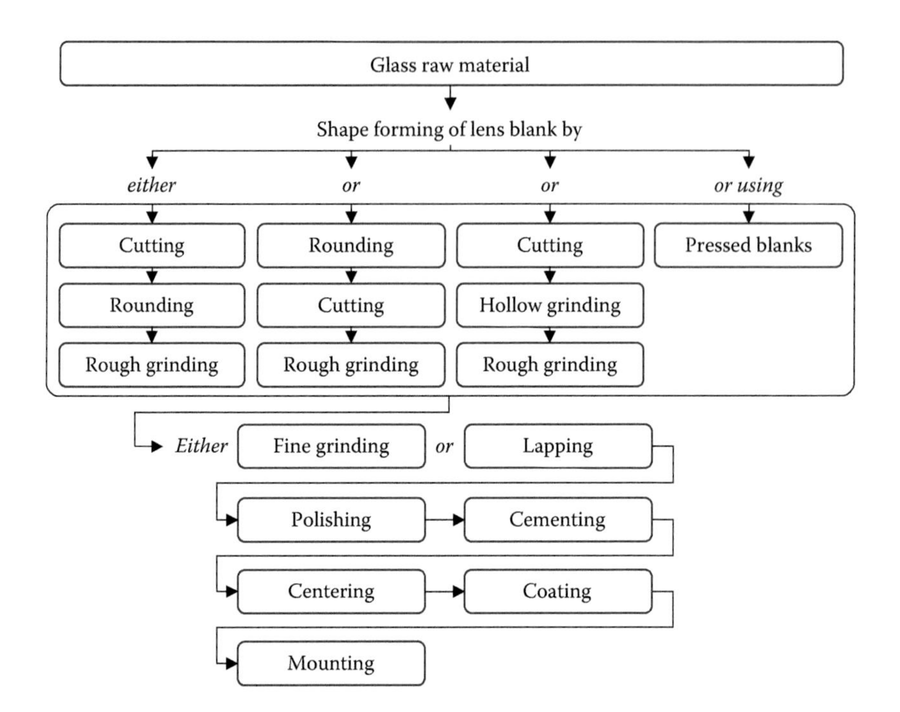


**FIGURE 1.1** Flow diagram of optics manufacturing summarizing the essential stages of production.

elements are then produced from the raw material, mainly glass, by different methods and techniques and a large variety of tools, machines, and apparatuses.

First, the preform of an optical component is produced by compression molding or cutting, rounding, and hollow drilling. The optically active surfaces of this component are then rough ground, fine ground, and finish ground; either bound abrasives or loose abrasives are used. The final surface shaping process is polishing, where the target surface accuracy and cleanliness are realized. Since the optical axis of a lens is usually tilted with respect to its border cylinder after the abovementioned manufacturing steps, centering is performed subsequently; the goal is to minimize such tilt, the lens centering error. In most cases, optically active surfaces of optical elements such as lenses, plates, or prisms are coated with antireflective coatings, mirror coatings, or thin film polarizers, depending on the final use and application of the particular component. Single components may further be cemented to groups such as achromatic lenses, triplets, or prism groups and finally, optomechanical setups are produced by mounting optics in mounts or on holders by different techniques.

All these manufacturing steps require specific strategies and tools, machines, and working materials. Further, suitable tolerances and test procedures are necessary. These aspects and the materials and methods used in optics manufacturing are introduced in the present book, in which classical and established techniques as well as unconventional and novel approaches are considered. Finally, an overview of the basics and manufacturing methods of microoptics is given.

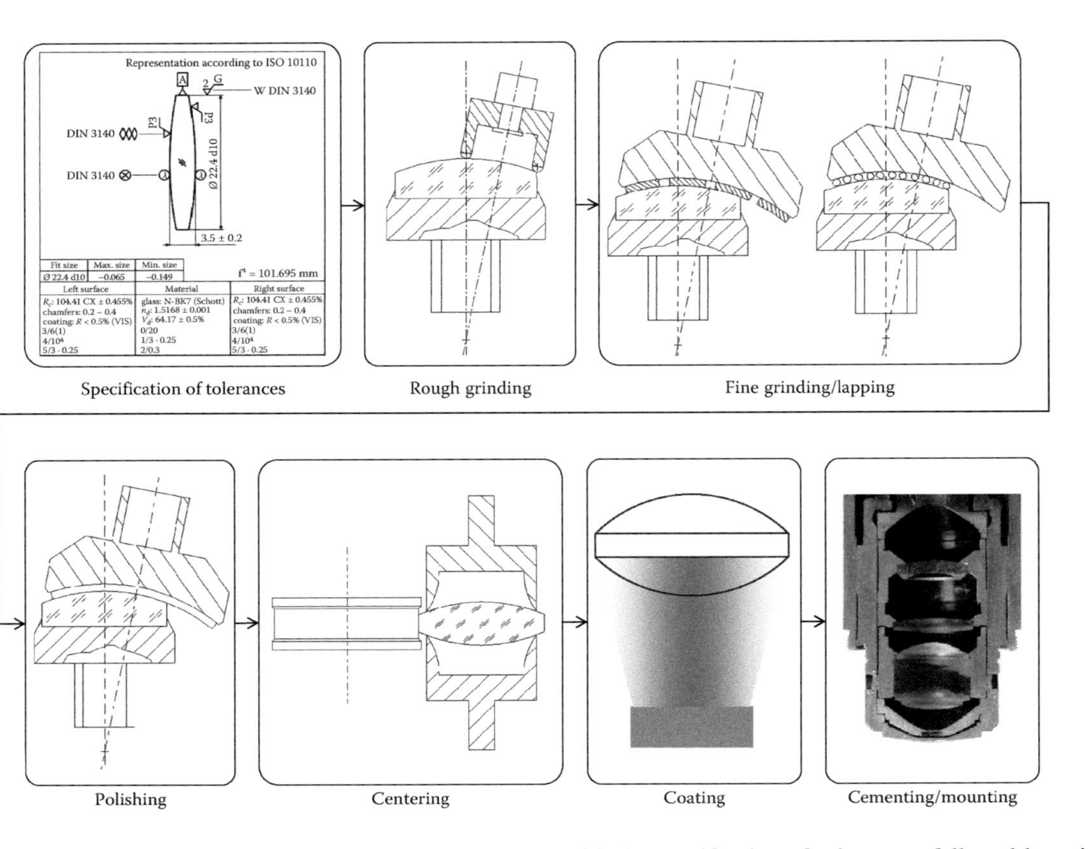


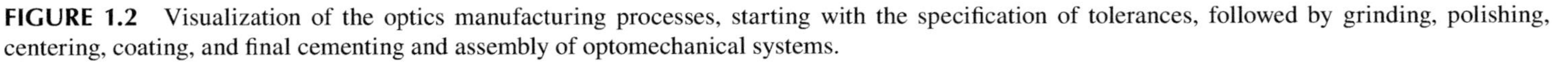

**FIGURE 1.2** Visualization of the optics manufacturing processes, starting with the specification of tolerances, followed by grinding, polishing, centering, coating, and final cementing and assembly of optomechanical systems.

## REFERENCES

Barker, W.B. 1930. Lens work of the ancients II: The Nineveh lens. *British Journal of Physiological Optics* 4:4–6.

Temple, R. 2000. *The Crystal Sun—Rediscovering a Lost Technology of the Ancient World.* London: Century/Souvenir Press Ltd.

# 2 Basics of Light Propagation

## 2.1 INTRODUCTION

Even though light propagates as a transversal wave, its propagation is usually described and characterized in the form of light rays. Such light rays are given by the perpendicular on the wave front and are the basis of geometrical optics. However, any change in the propagation direction of a light ray, due to refraction or reflection, can be attributed to the Huygens-Fresnel principle, named after the Dutch physicist *Christiaan Huygens* (1629–1695) and the French physicist *Augustin-Jean Fresnel* (1788–1827). Geometrical optics thus indirectly takes the wave properties or rather the shape of the wave front of light into account.

In this chapter, the basics of refraction and reflection of light rays are presented. These effects are the underlying principles for the function of optical components and systems. Further, the effect of dispersion is introduced and the phenomena of transmission and absorption are discussed.

## 2.2 REFRACTION

Classical optical elements, such as lenses, prisms, or plates, are based on a simple but fundamental phenomenon that occurs at optical interfaces: *refraction*. This mechanism was first observed and described by the Greek mathematician *Claudius Ptolemy* (c. 100–170) in the second century; the Persian optical engineer *Abu Sad al-Ala ibn Sahl* (c. 940–1000) formulated the law of refraction in the tenth century. However, this basic law is now known as *Snell's law*, named after the Dutch astronomer *Willebrord van Roijen Snell* (1580–1626) who worked in the field of geodesy in the seventeenth century. Snell's law is given by

$$n \cdot \sin\varepsilon = n' \cdot \sin\varepsilon', \tag{2.1}$$

where $n$ is the index of refraction (for its definition see Section 3.2.3.1) in front of an optical interface, $\varepsilon$ the *angle of incidence* of an incoming light beam,[1] $n'$ is the index of refraction behind the optical interface, and $\varepsilon'$ is the *angle of refraction* as shown in Figure 2.1.

As a result, an incident light ray is deviated as expressed by the angle of refraction and the deviation from the original direction of propagation, respectively. The angle of refraction can be calculated according to

$$\varepsilon' = \arcsin\left(\frac{n \cdot \sin\varepsilon}{n'}\right). \tag{2.2}$$

[1] The angle of incidence is sometimes abbreviated *AOI*.

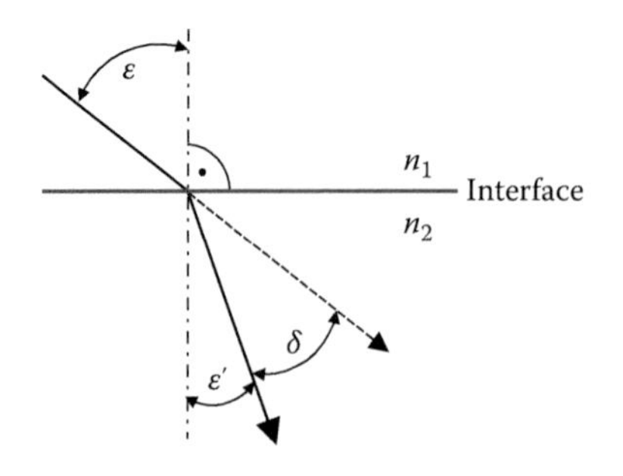


**FIGURE 2.1** Visualization of Snell's law describing refraction at an interface of two different optical media.

The deviation from the original direction of propagation $\delta$ simply follows from

$$\delta = \varepsilon - \varepsilon'. \tag{2.3}$$

One has to consider that optical media feature certain dispersion characteristics. This means that a particular index of refraction $n(\lambda)$ results for each wavelength of light. Consequently, incident polychromatic light is dispersed into its spectral components when passing through a lens. This effect gives rise to chromatic aberration (i.e., the formation of different positions of the focal point for each wavelength as shown in Figure 2.2). The description and quantification of the dispersion characteristics of optical media are presented in more detail in Section 3.2.3.2.

## 2.3 REFLECTION

The total *reflectance* $R_{tot}$ at any optical interface or surface of optical components is composed of two specific reflectances, $R_s$ and $R_p$, which represent the particular reflectance of s-polarized light and p-polarized light, respectively.[2] $R_{tot}$ is then given by the arithmetic mean of these two components,

$$R_{tot} = \frac{R_s + R_p}{2}. \tag{2.4}$$

Equation 2.4 implies that the surface reflectance depends on the polarization of incident light. This behavior is described by the *Fresnel equations*, named after the

[2] The denominations "s" and "p" describe the orientation of polarization of light with respect to its plane of incidence. The orientation is defined as follows: Light can be described as an electromagnetic wave featuring both an electric and a magnetic field where the electric field strength is significantly higher than the magnetic field strength. The orientation and polarization direction of such a light wave thus refers to the direction of the vector of the electric field. If this vector direction is perpendicular to the plane of incidence, the light is referred to as perpendicularly polarized or shortly s-polarized (German: "senkrecht" → abbreviation "s"). Light waves with a vector direction parallel to the plane of incidence are called parallel-polarized or p-polarized (German: "parallel" → abbreviation "p").

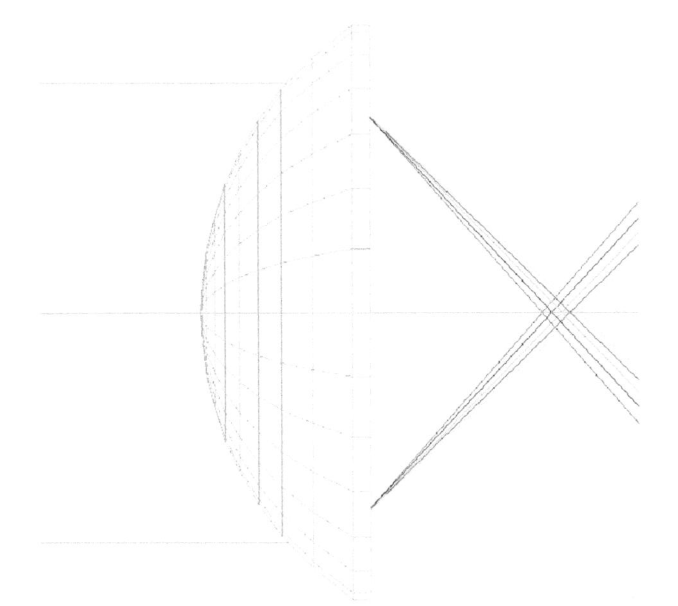

**FIGURE 2.2** Visualization of chromatic aberration by a single lens (Figure was generated using the software "WinLens3D Basic" from Qioptiq Photonics GmbH & Co. KG.)

French physicist *Augustin-Jean Fresnel* (1788–1827). According to *Fresnel*, the reflectance for s-polarized light is given by

$$R_s = \left| \frac{n \cdot \cos\varepsilon - n' \cdot \cos\varepsilon'}{n \cdot \cos\varepsilon + n' \cdot \cos\varepsilon'} \right|^2, \tag{2.5}$$

whereas the reflectance of p-polarized light follows from

$$R_p = \left| \frac{n' \cdot \cos\varepsilon - n \cdot \cos\varepsilon'}{n' \cdot \cos\varepsilon + n \cdot \cos\varepsilon'} \right|^2. \tag{2.6}$$

Equations 2.5 and 2.6 can also be expressed as a function of the angle of incidence, written as

$$R_s = \left| \frac{n \cdot \cos\varepsilon - n' \cdot \sqrt{1 - \left( \frac{n}{n'} \cdot \sin\varepsilon \right)^2}}{n \cdot \cos\varepsilon + n' \cdot \sqrt{1 - \left( \frac{n}{n'} \cdot \sin\varepsilon \right)^2}} \right|^2, \tag{2.7}$$

and

$$R_p = \left| \frac{n \cdot \sqrt{1 - \left( \frac{n}{n'} \cdot \sin \varepsilon \right)^2} - n' \cdot \cos \varepsilon}{n \cdot \sqrt{1 - \left( \frac{n}{n'} \cdot \sin \varepsilon \right)^2} + n' \cdot \cos \varepsilon} \right|^2 . \tag{2.8}$$

Here, the angle of refraction is not considered directly but rather results from inserting Snell's law into Equations 2.5 and 2.6. As a direct consequence of Equations 2.7 and 2.8, the reflectance of s-polarized light at an optical interface is higher than the reflectance of p-polarized light as shown in Figure 2.3. This effect becomes of crucial importance for the design of antireflective and mirror coatings.

Two main statements can be defined on the basis of Figure 2.3: first, the reflectance of both polarization states is equal at low angles of incidence from 0° to approx. 8°, $R_s = R_p$. In this case, the total reflectance is given by the approach

$$R_{tot} = \left( \frac{n' - n}{n' + n} \right)^2 . \tag{2.9}$$

Second, the reflectance for p-polarized light reaches a minimum at a certain angle of incidence (approx. $\varepsilon = 56°$ in the example shown in Figure 2.3). This specific angle of incidence is the *Brewster's angle*, named after the Scottish physicist *Sir David Brewster* (1781–1868) (Brewster, 1815) and given by

$$\varepsilon_B = \arctan\left( \frac{n'}{n} \right). \tag{2.10}$$

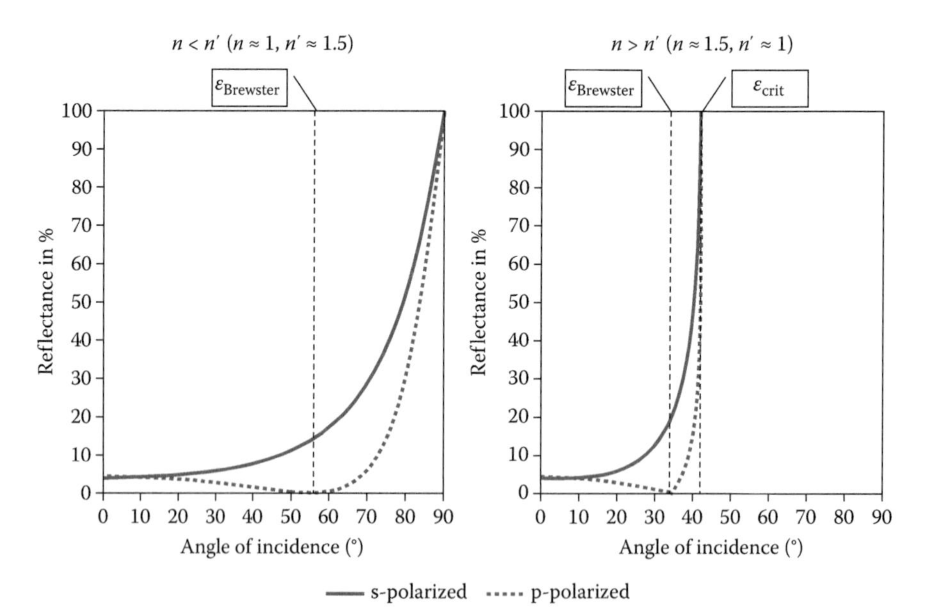


**FIGURE 2.3** Visualization of reflectance vs. angle of incidence according to Fresnel equations.

In optical setups, this effect is used in order to separate nonpolarized light into two polarized fractions.

Another significant angle in terms of reflection is the critical angle of total internal reflection $\varepsilon_{crit}$. It occurs at the interface from an optically dense to an optically thinner medium (i.e., from a medium with a comparatively high index of refraction $n'$ to a medium with a comparatively low index of refraction $n$). At the *critical angle of total internal reflection*, given by

$$\varepsilon_{\text{crit}} = \arcsin\left(\frac{n}{n'}\right), \tag{2.11}$$

or even higher angles of incidence, a light beam is totally reflected. This effect is the underlying mechanism for the realization of classical optical step index fibers and reflection prisms, as for example Porro-prisms (see Section 4.3.1).

## 2.4 TRANSMISSION AND ABSORPTION

*Transmission* and *absorption* are generally described by the Beer-Lambert law[3] (Lambert, 1760; Beer, 1852), named after the German mathematician *August Beer* (1825–1863) and the Swiss mathematician *Johann Heinrich Lambert* (1728–1777). This law relates the intensity of light $I_t$ after passing an optical medium with a certain thickness $t$ (given in meters or subunities) to the initial light intensity $I_0$ according to

$$I_t = I_0 \cdot e^{-\alpha \cdot t}. \tag{2.12}$$

Here, $\alpha$ is the material-specific *absorption coefficient*, usually given in $cm^{-1}$. A selection of absorption coefficients for optical glasses is given in Table 2.1.

**TABLE 2.1**
**Absorption Coefficient $\alpha$ of Different Glass Types at a Wavelength of $\lambda = 546$ nm**

| Glass Type | Absorption Coefficient $\alpha$ in $cm^{-1}$ |
|---|---|
| Crown glass K7 | 0.0024072 |
| Borosilicate crown glass N-BK7 | 0.0016045 |
| Barium crown glass N-BaK1 | 0.0020050 |
| Flint glass F2 | 0.0008008 |
| Heavy flint glass N-SF1 | 0.0056346 |
| Soda lime glass | 0.0487220 |

*Source:* Schott, Optical Glass Data Sheets, 2015; Rubin, *Solar Energy Materials*, 12, 275–288, 1985.

[3] The Beer-Lambert law is also known as the Bouguer-Lambert law, since it is based on the work of the French physicist *Pierre Bouguer* (1698–1758) (Bouguer, 1729).

Equation 2.12 can be rewritten as

$$\frac{I_t}{I_0} = e^{-\alpha \cdot t}. \tag{2.13}$$

This way of representing clearly shows that the exponential function in Equations 2.12 and 2.13 represents the transmission characteristics of the considered optical medium. One has to distinguish between two different ways of specifying transmission: the exponential function in Equation 2.12 is the so-called *internal transmittance* $T_i$,

$$T_i = e^{-\alpha \cdot t}, \tag{2.14}$$

which merely takes absorption within the bulk material of optical media into account. This internal absorption is given by the absorption coefficient $\alpha$ or the extinction coefficient $\kappa$[4] but can also be expressed by the absorbance $A$, given by

$$A = 1 - T_i. \tag{2.15}$$

When taking the reflection of light at the entrance and exit surface of an optical medium with limited dimensions into account, the *total transmittance* $T_{total}$ is defined. It is given by

$$T_{total} = (1 - R)^2 \cdot e^{-\alpha \cdot t}, \tag{2.16}$$

where $R$ is the reflectance at the interface of the optical medium. According to the Fresnel equations, total transmission thus directly depends on the polarization of incident light, which impacts the reflectance. Equations 2.5 and 2.6 can thus be modified in order to express the total transmission for s- and p-polarization, resulting in

$$T_{total,s} = \left| \frac{2 \cdot n \cdot \cos\varepsilon}{n \cdot \cos\varepsilon + n' \cdot \cos\varepsilon'} \right|^2, \tag{2.17}$$

and

$$T_{total,p} = \left| \frac{2 \cdot n \cdot \cos\varepsilon}{n' \cdot \cos\varepsilon + n \cdot \cos\varepsilon'} \right|^2, \tag{2.18}$$

respectively. Figure 2.4 visualizes the transmission behavior of light at optical interfaces resulting from these equations.

[4] Normally, we call $n$ the index of refraction of an optical medium. However, the index of refraction is a complex parameter, given by $N = n + i\,\kappa$. The term $n$ thus merely represents the real part of the complex index of refraction, and the imaginary part is given by the extinction coefficient $\kappa$. This constant describes absorption in optical media and is directly related to the absorption coefficient $\alpha$ according to $\alpha \cdot c = 4 \cdot \kappa \cdot \pi \cdot f$, where $c$ is the velocity of light and $f$ its frequency.

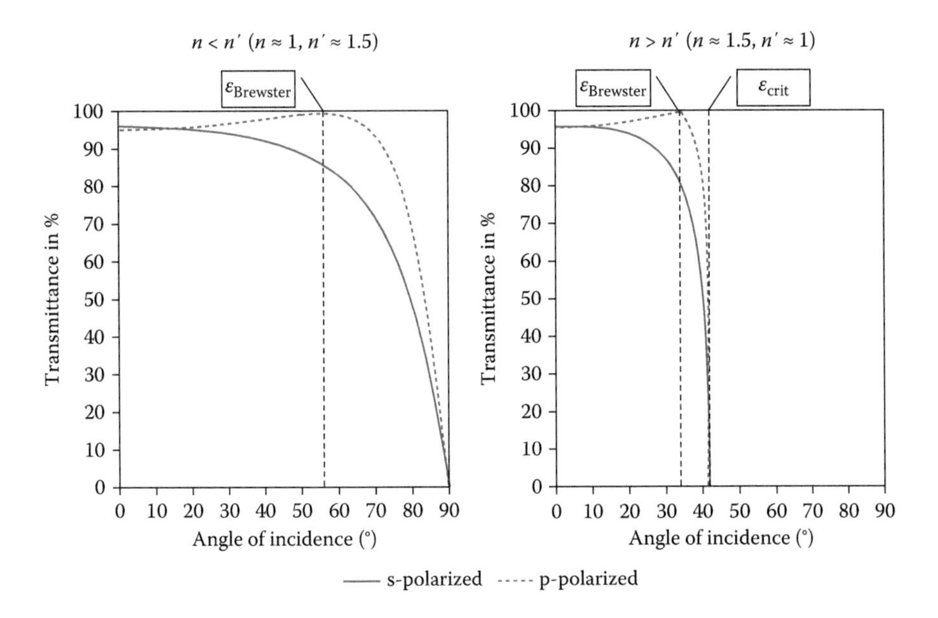


**FIGURE 2.4** Visualization of transmittance vs. angle of incidence according to Fresnel equations.

One has to consider that in addition to surface reflections and absorption within a glass bulk material scattering at impurities and inclusions has a further impact on the total transmission of an optical component as shown in Figure 2.5.

However, a mathematical description of losses in transmission resulting from scattering becomes nearly impossible due to the arbitrary distribution and size of

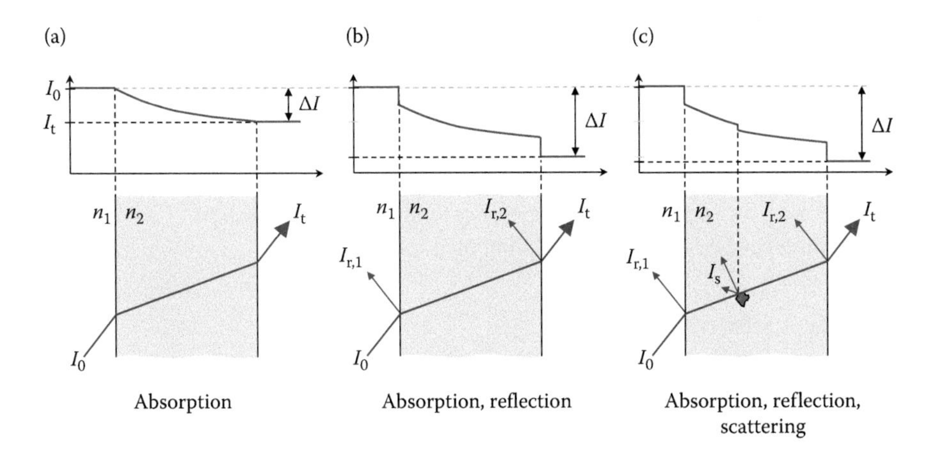


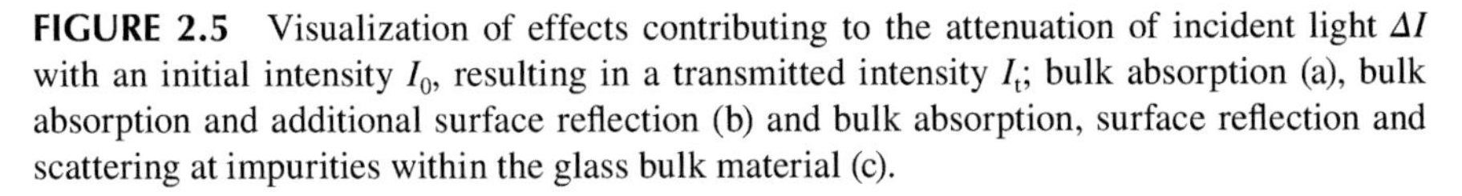

**FIGURE 2.5** Visualization of effects contributing to the attenuation of incident light $\Delta I$ with an initial intensity $I_0$, resulting in a transmitted intensity $I_t$; bulk absorption (a), bulk absorption and additional surface reflection (b) and bulk absorption, surface reflection and scattering at impurities within the glass bulk material (c).

such impurities. The purity of optical media is thus of essential importance and is discussed in more detail in Section 6.2.2.

## 2.5 SUMMARY

The deviation of an incident light ray from its original propagation direction at optical interfaces is referred to as refraction. The underlying reason for this effect is a difference in indices of refraction at an optical interface. Refraction is described by Snell's law, which links the angle of incidence to the angle of refraction. Since the index of refraction of any optical medium is wavelength-dependent, a particular angle of refraction results for each wavelength. This effect is referred to as dispersion.

The reflectance at optical interfaces depends on the polarization of the incident light as expressed by the Fresnel equations. For p-polarized light, the reflectance reaches a minimum at a specific angle of incidence, the so-called Brewster's angle. Total reflection of light occurs at interfaces from media with a high index of refraction to other media with a lower index of refraction as quantified by the critical angle of total internal reflection.

The absorption of light in optical bulk material follows from its material-specific and wavelength-dependent absorption coefficient, leading to an exponential decay of light intensity along the propagation direction. Such decay is described by the Beer-Lambert law (a.k.a. Bouguer-Lambert law). This interrelationship allows for the determination of internal transmittance and absorbance, respectively, of optical bulk material. When additionally taking the surface reflection into account, the total transmittance can be quantified. This parameter is consequently also polarization-dependent.

## 2.6 FORMULARY AND MAIN SYMBOLS AND ABBREVIATIONS

**Snell's law:**

$$n \cdot \sin\varepsilon = n' \cdot \sin\varepsilon'$$

$n$ index of refraction in front of an optical interface
$\varepsilon$ angle of incidence of an incoming light beam (a.k.a. *AOI*)
$n'$ index of refraction behind an optical interface
$\varepsilon'$ angle of refraction

**Angle of refraction $\varepsilon'$:**

$$\varepsilon' = \arcsin\left(\frac{n \cdot \sin\varepsilon}{n'}\right)$$

$n$ index of refraction in front of an optical interface
$\varepsilon$ angle of incidence of an incoming light beam (a.k.a. *AOI*)
$n'$ index of refraction behind an optical interface

**Total reflectance $R_{tot}$ at an optical interface:**

$$R_{\text{tot}} = \frac{R_s + R_p}{2}$$

$R_s$ reflectance of s-polarized light
$R_p$ reflectance of p-polarized light

**Reflectance $R_s$ of s-polarized light (Fresnel equation for reflection):**

$$R_s = \left| \frac{n \cdot \cos\varepsilon - n' \cdot \cos\varepsilon'}{n \cdot \cos\varepsilon + n' \cdot \cos\varepsilon'} \right|^2$$

$n$ index of refraction in front of an optical interface
$\varepsilon$ angle of incidence of an incoming light beam (a.k.a. *AOI*)
$n'$ index of refraction behind an optical interface
$\varepsilon'$ angle of refraction

**Reflectance $R_p$ of p-polarized light (Fresnel equation for reflection):**

$$R_p = \left| \frac{n' \cdot \cos\varepsilon - n \cdot \cos\varepsilon'}{n' \cdot \cos\varepsilon + n \cdot \cos\varepsilon'} \right|^2$$

$n$ index of refraction in front of an optical interface
$\varepsilon$ angle of incidence of an incoming light beam (a.k.a. *AOI*)
$n'$ index of refraction behind an optical interface
$\varepsilon'$ angle of refraction

**Total reflectance $R_{tot}$ (special case, valid for low angles of incidence < approx. 8°):**

$$R_{\text{tot}} = \left( \frac{n' - n}{n' + n} \right)^2$$

$n$ index of refraction in front of an optical interface
$n'$ index of refraction behind an optical interface

**Brewster's angle $\varepsilon_B$**

$$\varepsilon_B = \arctan\left( \frac{n'}{n} \right)$$

$n'$ index of refraction behind an optical interface
$n$ index of refraction in front of an optical interface

**Critical angle of total internal reflection $\varepsilon_{crit}$:**

$$\varepsilon_{\text{crit}} = \arcsin\left( \frac{n}{n'} \right)$$

$n$ index of refraction in front of an optical interface
$n'$ index of refraction behind an optical interface

**Beer-Lambert law (a.k.a. Bouguer-Lambert law):**

$$I_t = I_0 \cdot e^{-\alpha \cdot t}$$

$I_t$ transmitted intensity of light
$I_0$ initial intensity of light
$\alpha$ absorption coefficient (material-specific and wavelength-dependent)
$t$ thickness of optical medium

**Internal transmittance $T_i$:**

$$T_i = e^{-\alpha \cdot t}$$

$\alpha$ absorption coefficient (material-specific and wavelength-dependent)
$t$ thickness of optical medium

**Absorbance $A$:**

$$A = 1 - T_i$$

$T_i$ internal transmittance

**Total transmittance $T_{total}$:**

$$T_{total} = (1-R)^2 \cdot e^{-\alpha \cdot t}$$

$R$ surface reflectance
$\alpha$ absorption coefficient (material-specific and wavelength-dependent)
$t$ thickness of optical medium

**Total transmittance $T_{total,s}$ of s-polarized light (Fresnel equation for transmission):**

$$T_{total,s} = \left| \frac{2 \cdot n \cdot \cos\varepsilon}{n \cdot \cos\varepsilon + n' \cdot \cos\varepsilon'} \right|^2$$

$n$ index of refraction in front of an optical interface
$\varepsilon$ angle of incidence of an incoming light beam (a.k.a. *AOI*)
$n'$ index of refraction behind an optical interface
$\varepsilon'$ angle of refraction

**Total transmittance $T_{total,p}$ of p-polarized light (Fresnel equation for transmission):**

$$T_{total,p} = \left| \frac{2 \cdot n \cdot \cos\varepsilon}{n' \cdot \cos\varepsilon + n \cdot \cos\varepsilon'} \right|^2$$

$n$ index of refraction in front of an optical interface
$\varepsilon$ angle of incidence of an incoming light beam (a.k.a. *AOI*)
$n'$ index of refraction behind an optical interface
$\varepsilon'$ angle of refraction

## REFERENCES

Beer, A. 1852. Bestimmung der Absorption des rothen Lichts in farbigen Flüssigkeiten. *Annalen der Physik und Chemie* 86: 78–88.

Bouguer, P. 1729. *Essai d'optique, sur la gradation de la lumière*. Paris, France: Claude Jombert.

Brewster, D. 1815. On the laws which regulate the polarisation of light by reflection from transparent bodies. *Philosophical Transactions of the Royal Society of London* 105:125–159.

Lambert. J.H. 1760. *Photometria, sive de mensura et gradibus luminis, colorum et umbrae.* Augsburg, Germany: Sumptibus Vidae Eberhardi Klett.

Rubin, M. 1985. Optical properties of soda lime silica glasses. *Solar Energy Materials* 12: 275–288.

Schott, A.G. 2015. Optical Glass Data Sheets.

# 3 Optical Materials

## 3.1 INTRODUCTION

In optics manufacturing, many optical materials are used: plastics, crystals, liquids, gradient index materials, glass ceramics, and glasses. Due to the wide range of available refractive indices, the latter represent the most important and commonly used optical materials. The term "glass" originates from the old Germanic word "glasa," that is, "shining" or "sparkling," and glass is one of the oldest artificially produced raw materials. The first traditional instruction for the production of glasses was found in the library of the ancient Assyrian king *Ashurbanipal* (668 BC–c. 627 BC) in Nineveh, Upper Mesopotamia. Here, a cuneiform inscription was found, roughly saying: "Take 60 parts of sand, 180 parts of ash from sea plants, and 5 parts of chalk and you obtain glass." Since that time, the basic recipe of this type of glass, also referred to as crown glass, has not been significantly modified, and glass was mainly used for containers such as bottles, carafes, and drinking glasses or for stained glass windows.

More than 2000 years after the first mention of a glass recipe, a considerably different type of glass, so-called flint glass, was discovered in England. On the basis of this discovery, the production of different glasses with well-defined optical properties started in the early nineteenth century and was further developed by the cooperation of the German physicist and optical scientist *Ernst Karl Abbe* (1840–1905), the German chemist and glass technologist (and inventor of borosilicate glass) *Friedrich Otto Schott* (1851–1935), and the German entrepreneur *Carl Zeiss* (1816–1888) in the late nineteenth century. Based on this development and the continuously increasing demand in optical media with different properties, a large number of optical glasses are available today.

In this chapter, the chemical composition, manufacturing methods, and essential properties of such glasses are presented. Moreover, glass ceramics, gradient index materials, and crystals, as well as possible applications of these optical media, are introduced.

## 3.2 OPTICAL GLASSES

### 3.2.1 Composition of Optical Glasses

Optical glasses can be classified into two main categories: *single-component glasses* and *multicomponent glasses*. Prime examples of single-component glass are *quartz glass* and *fused silica*, which exclusively consists of silicon dioxide ($SiO_2$). Quartz glass thus represents the purest basic glass where the basic module is a tetrahedron consisting of one silicon (Si) atom and four oxygen (O) atoms. As shown in Figure 3.1a, each silicon atom is linked to four oxygen atoms. Since the oxygen

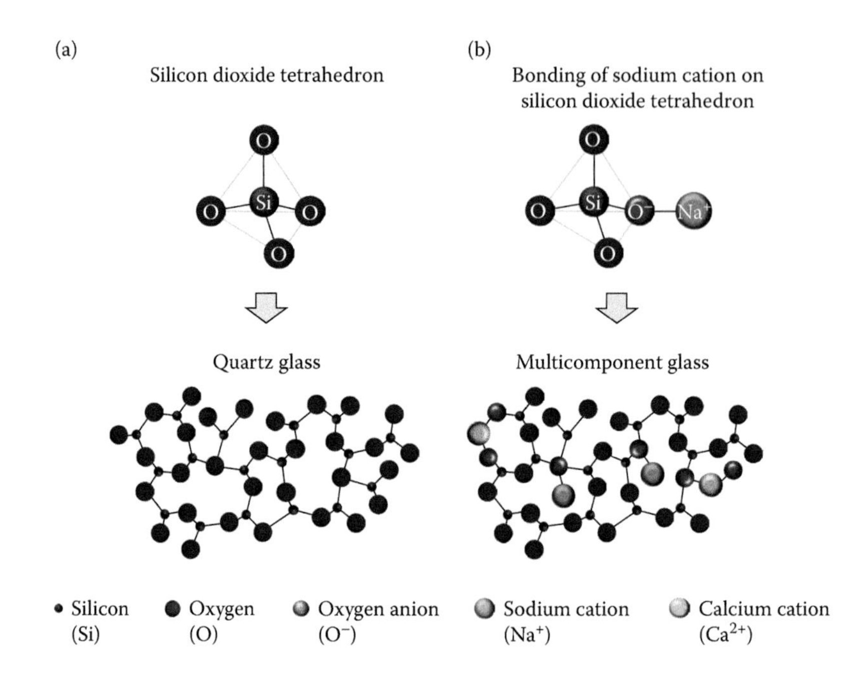


**FIGURE 3.1** Comparison of the basic structures of quartz glass (a) and multicomponent glass (b).

atoms are shared by silicon atoms, the chemical composition can be expressed by the total formula $SiO_{4/2}$, consequently resulting in $SiO_2$. In contrast to the single-component quartz glass, multicomponent glasses consist of a number of different compounds since glass-forming elements (e.g., silicon) are replaced by ions of other elements, for example, sodium ($Na^+$) or calcium ($Ca^{2+}$) as shown in Figure 3.1b. As a result, the chemical composition and structure are modified, leading to a modification of the optical properties.

The compounds of such multicomponent glasses are classified on the basis of their function where the main component is the so-called network former.

#### 3.2.1.1 Network Formers

The function of glass *network formers* is to form a cross-linked network. This basic skeletal structure is also referred to as glass matrix. The most important and well-established network formers are listed in Table 3.1. The large variety (in terms of refractive index and dispersion characteristics) of optical glasses is obtained by the addition of further compounds, the so-called network modifiers.

#### 3.2.1.2 Network Modifiers

The addition of *network modifiers* leads to an alternation of the glass matrix as constituted by the network formers and to a modification of the optical properties, respectively. They further act as flux melting agents, which reduce the melting

**TABLE 3.1**
**Overview on Well-Established Network Formers in Optical Glasses**

| Network Former | Total Formula |
|---|---|
| Silicon dioxide | $SiO_2$ |
| Boron trioxide | $B_2O_3$ |
| Germanium dioxide | $GeO_2$ |
| Phosphorus pentoxide | $P_2O_5$ |

**TABLE 3.2**
**Overview on Well-Established Network Modifiers in Optical Glasses**

| Network Modifiers | Total Formula |
|---|---|
| Barium oxide | $BaO$ |
| Calcium oxide | $CaO$ |
| Cesium oxide | $Cs_2O$ |
| Potassium oxide | $K_2O$ |
| Arsenic oxide | $As_2O_3$ |
| Lithium oxide | $Li_2O$ |
| Sodium oxide | $Na_2O$ |

temperature of a glass melt and, in some cases, even act as network formers. A selection of commonly used network modifiers is given in Table 3.2.

#### 3.2.1.3 Stabilizers

Chemical stability is achieved by adding the third important component of optical glasses, so-called *stabilizers*, a.k.a. intermediates that can serve as both network formers and modifiers. Depending on the network formers and modifiers used for a particular glass, different stabilizers can be applied as listed in Table 3.3.

These oxides have a further impact on the optical properties of the glass; especially lead oxide was used extensively in former times in order to achieve high refractive indices. However, as a result of the Restriction of Hazardous Substances Directive, which became effective in the early 2010s, hazardous compounds such as lead or arsenic are currently successively replaced by other uncritical substances except for some special cases, for example very heavy flint glasses.

#### 3.2.1.4 Glass Families and Types

Optical multicomponent glasses are generally classified into two so-called glass families, *crown glasses* and *flint glasses*, where the classification is based on the Abbe number (for definition see Section 3.2.3.2.1) of the particular glass. The terms "crown" and "flint" date back to the production process and origin of the raw materials that are used. The first term originated because in former times, windowpanes

**TABLE 3.3**
**Overview on Stabilizers in Optical Glasses**

| Stabilizers/Intermediates | Total Formula |
|---|---|
| Aluminum oxide | $Al_2O_3$ |
| Lead oxide | PbO |
| Calcium oxide | CaO |
| Polonium oxide | PoO |
| Tin oxide | SnO |
| Cadmium oxide | CdO |
| Titanium dioxide | $TiO_2$ |
| Beryllium oxide | BeO |
| Zirconium oxide | $ZrO_2$ |
| Iron oxide | FeO, $Fe_2O_3$ |
| Nickel oxide | NiO |
| Cobalt oxide | CoO |

were manufactured by glassblowers. First, a cylindrical bottle-like vessel was produced. Second, the upper and lower parts of this vessel were removed to create a hollow cylinder. This cylinder was finally cut along its axis, heated, and unfolded; the hollow cylinder was thus transformed into a plane glass pane. Because the work piece exhibited the appearance of a bifurcated crown in the course of the unfolding process, it was called "crown glass." This historic denomination for glasses (mainly based on the network former silicon dioxide) continues to survive.

The second term, "flint," originates from the raw material that was used for the production of such glasses. For the production of the first flint glasses in 1650 in England, the raw material was extracted from flint stones. This raw material featured a comparatively high share of lead oxide, resulting in a high index of refraction in comparison to crown glasses.

Today, the notations "crown" and "flint," abbreviated "K" and "F," respectively, are extended by suitable prefixes where required in order to state the particular glass type more precisely by subcategories. The prefixes used are based on specific glass properties or admixtures. For example, the abbreviation "SF" indicates a so-called "heavy flint glass" where the letter "S" originates from the German term for "heavy" (*schwer*). This naming is due to the fact that heavy flint glasses contained a high share in lead oxide, resulting not only in a high index of refraction, but also in a high weight.[1] Another example is the crown glass subcategory "BaK" where the letter "Ba" represents the symbol of the element barium. BaK-glasses thus contain a considerable portion of this element. Finally, the denomination of glasses may result from both the specific glass properties and major admixtures as shown in Table 3.4.

The denomination of any glass subcategory gives a first hint of its optical properties, such as index of refraction or dispersion characteristics. These properties

[1] Actually, the index of refraction of any glass is directly related to its weight (i.e., the mass density), which can be expressed by a linear internship. Due to this fact, the index of refraction is also sometimes referred to as "optical density."

**TABLE 3.4**
**Selection of Different Glass Types Including Denomination and Abbreviation**

| Crown Glasses | | Flint Glasses | |
|---|---|---|---|
| **Denomination** | **Abbreviation** | **Denomination** | **Abbreviation** |
| Borosilicate crown | BK | Barium flint | BaF |
| Barium crown | BaK | Barium light flint | BaLF |
| Fluorite crown | FK | Barium heavy flint | BaSF |
| Lanthanum crown | LaK | Lanthanum flint | LaF |
| Phosphate crown | PK | Lanthanum heavy flint | LaSF |
| Heavy phosphate crown[a] | PSK | Heavy flint | SF |
| Heavy crown | SK | Light flint | LF |
| Very heavy crown | SSK | Very light flint | LLF |

[a] The term "heavy" can be replaced by the term "dense" as found on some glass datasheets.

**TABLE 3.5**
**Chemical Composition of Selected Optical Multicomponent Glasses**

| Glass Type | Chemical Composition/Content of Oxides in Mass% | | | | | | |
|---|---|---|---|---|---|---|---|
| | $SiO_2$ | $B_2O_3$ | $Na_2O$ | $K_2O$ | BaO | ZnO | PbO |
| Boron crown (BK) | 60 ± 10 | 15 ± 5 | 15 ± 5 | 15 ± 5 | — | — | — |
| Barite crown (BaK) | 50 ± 10 | — | 7.5 ± 2.5 | 7.5 ± 2.5 | 22.5 ± 7.5 | 10 ± 5 | — |
| Crown flint (KF) | 60 ± 10 | — | 7.5 ± 2.5 | 12.5 ± 7.5 | — | — | 12.5 ± 7.5 |
| Barite flint (BaF) | 40 ± 10 | 7.5 ± 2.5 | 5 ± 5 | 5 ± 5 | 25 ± 15 | — | 12.5 ± 7.5 |
| Heavy flint (SF) | 37.5 ± 12.5 | — | — | — | — | — | 60 ± 10 |

*Source:* Data taken from Vogel, W. et al., *Jenaer Rundschau*, 1, 75–76, 1965 (in German).

directly follow from the chemical composition, which is realized during the glass manufacturing process. The chemical compositions of some selected glass types are listed in Table 3.5.

#### 3.2.1.5 Colored Glasses

In some cases, dyes are added to optical glasses. Coloring can be achieved by very small admixtures of positive metal ions, that is, metal cations, in the range of some parts per million (ppm). A selection of suitable elements and the resulting glass coloring is listed in Table 3.6.

Some of these cations give rise to different coloring where the actual color results from the valency stage state of the particular ion. Such *colored glasses* are used to create homogeneous bulk color filters.

**TABLE 3.6**
**Elements Used for Glass Coloring and Resulting Glass Coloring**

| Element | Coloring |
|---|---|
| Titanium (Ti) | Purple |
| Vanadium (Va) | Green |
| Chromium (Cr) | Green |
| Iron (Fe) | Blue or yellow |
| Cobalt (Co) | Blue, pink, or green |
| Nickel (Ni) | Blue or yellow |
| Copper (Cu) | Blue |

*Source:* Schaeffer, H.A., and Langfeld, R., *Werkstoff Glas*. Berlin and Heidelberg: Springer Verlag, 2014 (in German).

## 3.2.2 Manufacturing of Optical Glasses

### 3.2.2.1 Manufacturing of Fused Silica

The most basic glass, quartz glass, can be manufactured in different ways. First, it can be produced by classical melting as described in the following section. Second, it can be obtained from a flame pyrolysis process by chemical vapor deposition (see Section 11.4.1). The latter allows the production of synthetic quartz glass of high purity. This type of quartz glass is referred to as fused silica. In the course of a *flame pyrolysis* production process, gaseous silicon tetrachloride ($SiCl_4$) and oxygen ($O_2$) are inserted into a tube as shown in Figure 3.2a, where the tube is usually made of glass.

This tube is simultaneously heated by a flame, for example from a Bunsen burner, resulting in the heat-induced formation of solid (s) silicon dioxide ($SiO_2$) and volatile gaseous (g) chlorine ($Cl_2$), according to

$$SiCl_4(g) + O_2(g) \rightarrow SiO_2(s) + 2Cl_2(g), \tag{3.1}$$

as shown in Figure 3.2b.

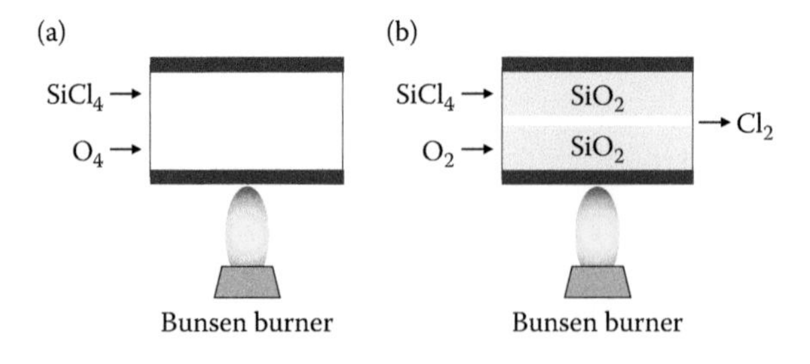


**FIGURE 3.2** Schematic of a flame pyrolysis process for the production of synthetic quartz glass of high purity (fused silica) consisting of (a) insertion of reactant gases into a tube and (b) deposition of solid silicon dioxide at the tube walls.

Depending on the process parameters, the deposited silicon dioxide (i.e., the fused silica) features high purity and homogeneity. Flame pyrolysis can also be applied for the deposition of multicomponent glasses. This approach is one of the most important manufacturing methods for the production of blanks for optical step index fibers. Here, a thin layer of fused silica is first deposited on the tube wall. Subsequently, a further process gas is used and another solid-state phase is deposited onto the previously deposited silicon dioxide layer. For example, the use of gaseous germanium tetrachloride ($GeCl_4$) results in the formation of germanium dioxide ($GeO_2$), which features a higher index of refraction than does silicon dioxide. After depositing such different layers at the inner surface of the tube, it is finally heated and drawn ("collapsed") to an optical fiber.

#### 3.2.2.2 Classical Melting of Multicomponent Glasses

Classical melting is used for the production of high volumes of multicomponent glasses as needed for the manufacture of standard optical components. Such melting is performed in crucibles or large furnaces with volumes of up to some tens of cubic meters. It consists of several subsequent essential steps, preparation, meltdown, plaining, cooling, and forming as visualized in Figure 3.3 and described in more detail in the following paragraphs.

**Step 1: Preparation**

*Preparation* is the first and dominant step for producing any glass. Here, the particular components (i.e., oxides) are selected, weighed, and mixed in powder form, the so-called mixture or *batch*. Since a number of components will volatilize in the course of the subsequent manufacturing steps but play important roles as intermediates or chemical catalysts, the composition of the batch may not correspond to the final stoichiometric composition of the solid glass. Batch mixing and preparation thus represent a sophisticated

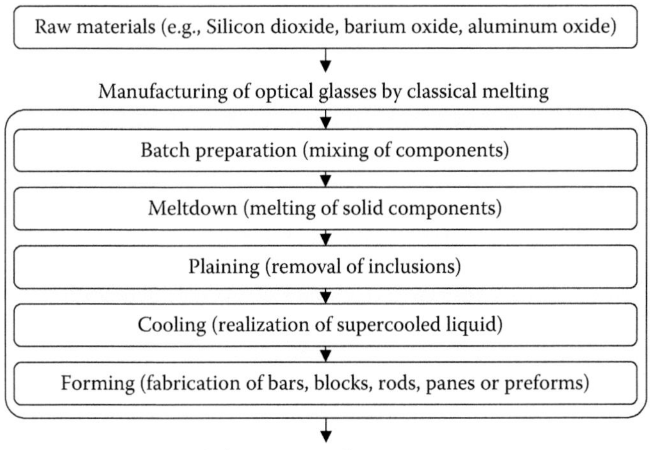


**FIGURE 3.3** Flow diagram of a classical glass melting process.

and complex process and remain one of the main secrets of glass manufacturers.[2] However, as mentioned at the beginning of this chapter, glass making is a very old craft and the most basic glass recipe was archived in *Ashurbanipal*'s library: "Take 60 parts of sand, 180 parts of ash from sea plants, and 5 parts of chalk and you obtain glass." Once these compounds are mixed and stirred, the resulting batch is filled into a crucible or furnace (which are usually made of heat-resistant material such as chamotte). Then, the next step of glass making, the meltdown, is performed.

**Step 2: Meltdown**

After filling the batch into a crucible or furnace, both the particular vessel and the batch surface are heated by burners and other auxiliary heat sources. This step is referred to as *meltdown:* the powdery batch is heated up to approximately 1000°C–1600°C, depending on its composition. In the course of this process, the batch is decomposed into liquid glass compounds, the final network formers, network modifiers and stabilizers, as well as volatile, gaseous by-products. Referring to the historic glass recipe from *Ashurbanipal*'s library, different functions can be allocated to the particular components mentioned there as shown in Figure 3.4.

The duration of the meltdown process depends on the glass composition but typically takes 10–15 h. The loss in weight of the initial batch amounts to approximately 40% due to the evaporation of volatile products in the course of the process. In some cases, the furnace may thus be successively refilled with further batch material.

| Historical: | "Sand" | "Ash from sea plants" | "Chalk" |
|---|---|---|---|
| Comment: | | Kelp/ash of seaweed | |
| Compound: | Silicon dioxide | Potassium carbonate | Calcite |
| Raw material: | $SiO_2$ | $K_2CO_3$ | $CaCO_3$ |
| | ↓ | ↓ | ↓ |
| After heating: | $SiO_2$ (l) | $K_2O$ (l) + $CO_2$(g) | CaO(l) + $CO_2$(g) |
| | ↓ | ↓ | ↓ |
| Final function: | Network former (silicon dioxide) | Network modifier (potassium oxide) | Stabiliser (calcium oxide) |

**FIGURE 3.4** Composition of glass according to the library of the ancient Assyrian king *Ashurbanipal* including the modern interpretation and allocation of function of the particular components.

[2] The preparation of glass melts is handled with the utmost discretion as shown by the fact that the famous glassmakers of Murano, an island close to Venice, Northern Italy, were not allowed to leave the Republic of Venice.

Since the total volume of the glass melt can only be heated from its surface and the interfaces with the crucible or furnace, it usually exhibits a comparatively cold center region with hot boundary regions. This difference in temperature results in an automatic stirring of the glass melt by convection currents. However, a glass melt contains a considerable number of impurities after the actual meltdown process as shown in Figure 3.5.

Such impurities include crystals, stones, or nonmolten batch material, as well as air or gas bubbles. The first may result from inappropriate heating and cooling of the glass melt or may simply be chips from the used crucible or furnace. The second come from gaseous inclusions and air in the powdery batch, unable to swell up to the melt surface due to its high viscosity. Impurities can thus be classified into heavy and light inclusions and need to be removed from the melt. Such removal is carried out in the course of a further essential manufacturing step: plaining.

**Step 3: Plaining**

*Plaining* describes the removal of inclusions from a glass melt where the goal is to obtain the best possible homogeneity. It can be performed by different approaches: (1) increasing the temperature of the melt, (2) adding refining agents to the melt, (3) stirring the glass melt, or (4) a combination of these three techniques. Since the purity and quality of glass raw material (for more information see Section 6.2) is directly impacted by the plaining process, this step is exhausted and thus takes up to 8 h where stirring takes 4 h in case of combined plaining.

In the course of plaining by increasing the temperature, the glass melt is further heated up by some hundreds of degrees centigrade; the final temperature thus amounts to approximately 2000°C–2200°C. The resulting glass melt is attenuated and thinner. Due to this decrease in viscosity and the accompanying increase in mobility of impurities, heavy inclusions (for example, stones) sink and become sediment at the bottom of the furnace,

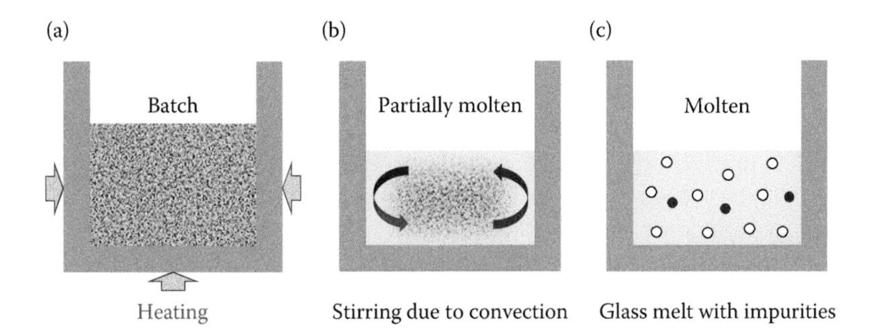


**FIGURE 3.5** Different stages during a glass meltdown process. The initial batch is heated (a), resulting in a hot, molten phase at the outer regions and a comparatively cold core of the total volume, which are automatically stirred due to convection (b). Finally, the melt features a number of heavy (dark circles) and light (light circles) inclusions (c).

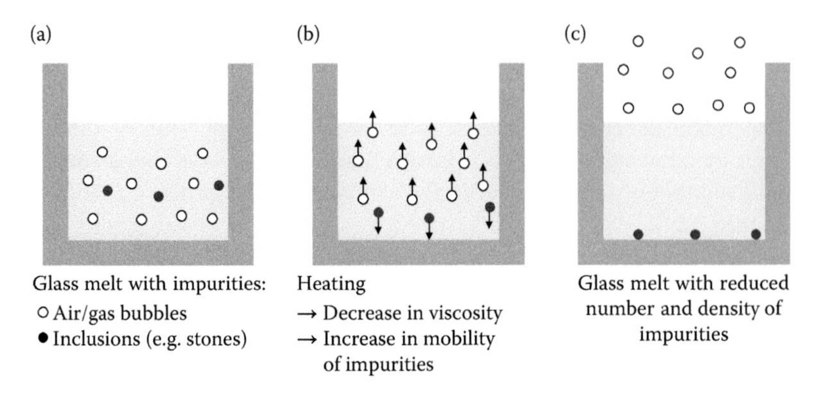


**FIGURE 3.6** Principle of plaining by increasing temperature; a glass melt with light and heavy inclusions is heated in order to decrease the viscosity of the melt (a). Heavy inclusions consequently sink to the bottom of the furnace, and light inclusions well up (b), resulting in a reduction of the number and density of impurities within the glass melt (c).

whereas light inclusions (air or gas bubbles) well up to the surface and leave the melt as shown in Figure 3.6.

Plaining can also be obtained by adding so-called *refining agents* such as sodium nitrate ($NaNO_3$) or potassium nitrate ($KNO_3$) to the glass melt. These refining agents are provided in the solid state; they sink to the bottom of the furnace and decompose due to the high temperature in this heated region. As a result of this decomposition, relatively big bubbles with high buoyant force are formed. These bubbles well up and collect smaller bubbles with low buoyant force on the way to the surface of the glass melt.

Another method for plaining is stirring the glass melt applying stirring staffs made of materials of high temperature stability (for example chamotte or clay). Such stirring gives rise to a mechanically induced motion of the glass melt. This motion is further supported by the differences in temperatures of different regions within the melt volume.

### Step 4: Cooling

After plaining, the fourth and last essential step of glass manufacturing, *cooling*, is performed. Generally, the cooling procedure of a melt directly impacts the nature of the resulting solid. During the cooling of crystalline solids (for example, metals), a well-defined crystallization temperature $T_c$ is found as shown in Figure 3.7. This temperature indicates the beginning of crystallization and thus the transition from the liquid to the solid state. In contrast, glasses feature not a crystallization temperature but rather a certain temperature range where the transition from the liquid to the plastic state occurs during the cooling of the glass melt.[3] Glasses can thus be referred to as supercooled liquids. The transition from the liquid to the

[3] For multicomponent glasses, the transition from the liquid to the plastic state is found within a range of viscosity from $10^4$ to $10^{13}$ dPa·s.

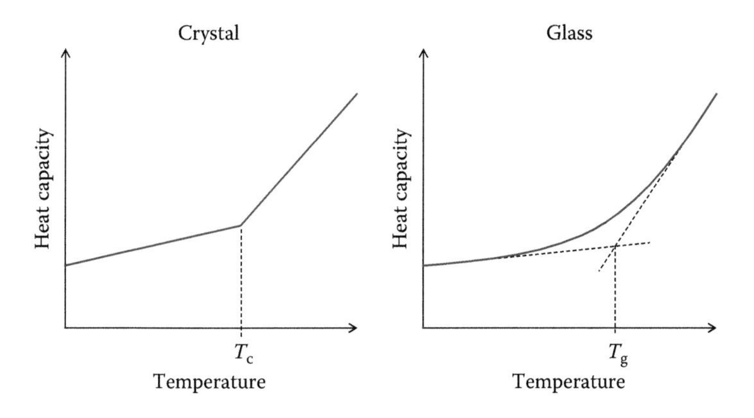


**FIGURE 3.7** Heat capacity vs. temperature during cooling of a crystal (left) and a glass (right). For crystals, a characteristic crystallization temperature $T_c$ is found whereas glasses feature a certain temperature range of transition from the liquid to the plastic state, indicated by the glass transition temperature $T_g$ as determined by extrapolation.

plastic state is described or quantified by the *glass transition temperature* $T_g$ (glass transition point). This characteristic temperature is determined by the linear extrapolation of the cooling curve as shown in Figure 3.7.

As a result of the smooth transition, glasses and glassy materials feature an amorphous structure since no phase transition occurs. This is achieved by well-controlled cooling procedures: a glass melt is cooled down from its initial temperature to a temperature above the glass transition temperature as listed in Table 3.7.

Further cooling is then performed in appropriate cooling steps; defined decreases in temperature per time unit are made below the so-called lower strain temperature. Finally, the glass is cooled down to ambient temperature. In the course of this process, further characteristic glass temperatures can be found as listed in Table 3.8.

The lower strain temperature represents an important limit temperature. Below this temperature, original stress relief of glass is inhibited until the melt temperature reaches the annealing point. Another characteristic

**TABLE 3.7**
**Overview on Glass Transition Temperatures/Temperature Ranges of Different Multicomponent Glasses**

| Glass Type | Glass Transition Temperature in Centigrade |
|---|---|
| Borosilicate glass | ≈500 |
| Soda-lime glass | ≈520°C–600°C |
| Lead glasses | ≈400 |
| Aluminosilicate glasses | ≈800 |
| Quartz glass | ≈1200 |

**TABLE 3.8**
**Characteristic Glass Temperatures during Cooling, Expressed by the Common Logarithm $l_g$ of the Temperature-Dependent Glass Viscosity $\eta$**

| Denomination of Characteristic Glass Temperature | Temperature, Expressed by $l_g(\eta)$ in Pa·s |
|---|---|
| Lower strain temperature | 14.5 |
| Annealing temperature | 13 |
| Softening temperature | 7.6 |
| Processing temperature | 4.0 |

*Source:* Pfaender, H.G., *Schott Glaslexikon*, mgv-verlag, Landsberg, Germany, 1997 (in German).

temperature is the softening point, where glass flows and deformation of a glass component due to its proper weight occurs. This fact may be taken into account during some high-temperature coating processes. Finally, the processing temperature indicates the point at which glass can be shaped by molding or hot embossing (see Section 7.2.1).

The cooling process and strategy of a glass melt not only impacts the glass formation in terms of its amorphous structure, but it also helps with the formation of internal stress. Inappropriate cooling may give rise to local tensile or compressive stress, which leads to birefringence as described in more detail in Section 6.2.1. In order to avoid this disturbing effect, the glass cooling process needs to be controlled and performed in well-defined steps as mentioned above. It may thus be a time-consuming procedure. Simple glasses (for example, windowpanes) can be cooled down quite quickly, that is, within some hours. However, the cooling time can also amount to several years, depending on the particular glass as well as the dimension and final application of the optical component, which is produced from the glass volume. As an example, the cooling time of the primary mirror of the very large telescope (VLT) was 3 months. This mirror was produced by the German glass manufacturer Schott in the 1990s and has a diameter of 8.2 m and a weight of 40 tons. During cooling from an initial temperature of 1000°C to ambient temperature, the melt was filled into a cylindrical crucible that was constantly rotated. As shown in Figure 3.8, the mirror surface was preshaped by the formation of a curved melt surface due to centrifugal forces in the course of the cooling process.

In this example, the comparatively long cooling time was thus of notable use in order to reduce the time and effort of subsequent manufacturing steps on the one hand and to save material on the other hand.

**Step 5: Forming**

The example of preshaping the VLT mirror during cooling as mentioned above represents an exceptional case of forming glass melts. However, the final use and application are also considered during classical glass forming. As shown in Figure 3.9, the viscous glass frit, i.e., the molten glass mass, may be formed into blocks, bars, panes, rods, or blanks after cooling.

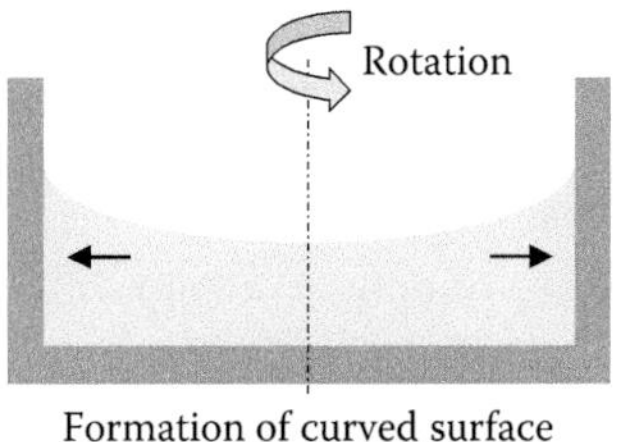


**FIGURE 3.8** Strategy of cooling of the glass melt for the very large telescope (VLT); the melt was poured into a cylindrical vessel and rotated during cooling for 3 months, resulting in a preshaping of the optically active mirror surface due to centrifugal forces.

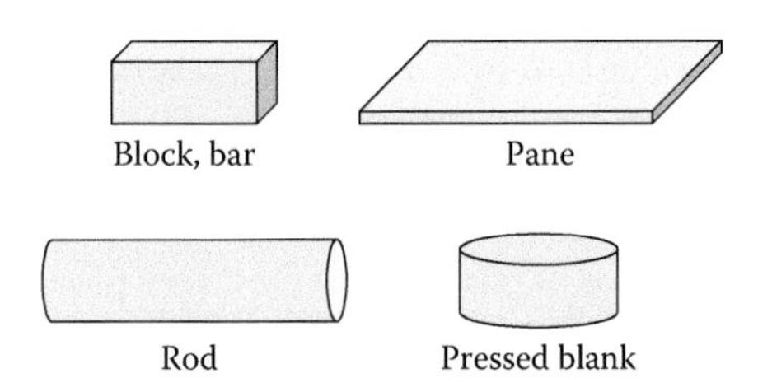


**FIGURE 3.9** Examples of raw glass geometries as provided by glass manufacturers.

Blocks and bars are typically produced by founding in casts, whereas glass panes are usually manufactured by the so-called float process. In the latter case, the hot glass melt is cast on molten tin, commonly referred to as a tin bath. Since the surface of molten tin features high flatness, the glass is automatically smoothed at the contact surface where it floats on the molten tin. The rear side of the glass is simultaneously smoothed by flame polishing. Finally, blanks are produced by molding, hot embossing, or deep drawing. This allows the production of preforms (for example, preshaped lenses or prisms). However, the raw glass material is usually provided as blocks or bars for classical optical manufacturing in order to reduce costs that could result from the special design of forming dies for molding, which is uneconomic for low lot sizes.

### 3.2.3 Characterization of Optical Glasses

After the production process, optical glasses are tested in order to quantify a number of characteristics of glass properties and to specify the glass quality. Mechanical properties include the hardness or grindability and thermal properties (for example, the coefficient of thermal expansion and chemical properties such as acid or alkali resistance; see Section 6.2.4). In addition, optical properties (for example, stress birefringence, bubbles, and inclusions as well as striae) are determined as discussed in more detail in Sections 6.2.1 through 6.2.3. Finally, the essential and basic optical properties, that is, the internal transmission (see Section 2.4), the index of refraction, and dispersion characteristics are measured. The latter two parameters represent the most important glass properties and are introduced in Sections 3.2.3.1 and 3.2.3.2.

#### 3.2.3.1 Index of Refraction

For the design and manufacture of optical components, the most important parameters of optical glasses are the *index of refraction* and the *dispersion*. Generally, the index of refraction $n$ of a transparent medium is defined as the ratio of the speed of light in vacuum $c$ and the speed of light within the medium $v$:

$$n = \frac{c}{v}. \tag{3.2}$$

One has to consider that $n$ depends on the wavelength of light as shown schematically in Figure 3.10.

This behavior is characterized by the medium's dispersion (see Section 3.2.3.2). In practice, $n$ is given for defined wavelengths (i.e., Fraunhofer lines, discrete laser wavelengths, etc.) in manufacturer's data sheets. The nominal index of refraction of any glass is usually referred to a wavelength of 546.07 nm (i.e., Fraunhofer line e).[4] As an example, the nominal index of refraction of the crown glass N-BK7 from Schott is $n_e = 1.51872$. In the past, the reference wavelength 587.56 nm (i.e., Fraunhofer line d) was used; it is additionally given in glass manufacturer's data sheets. The index of refraction is usually understood as a material constant. However, it features several dependencies. First, it is directly related to the absorption coefficient (see Section 2.4). This interrelation is considered by expressing the index of refraction as a complex number,

$$N = n + i\kappa, \tag{3.3}$$

where the real part $n$ is the index of refraction and the imaginary part $\kappa$ is the *extinction coefficient*. The extinction coefficient finally gives the absorption coefficient $\alpha$ according to

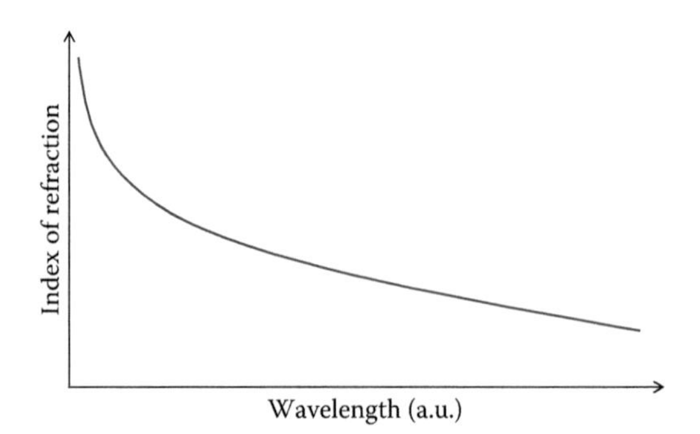


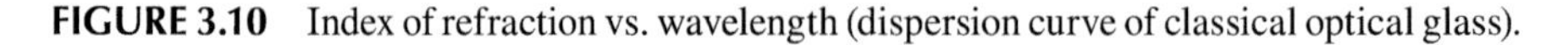

**FIGURE 3.10** Index of refraction vs. wavelength (dispersion curve of classical optical glass).

[4] This wavelength was chosen because it is well adapted to the human eye, which features a maximum sensitivity at a wavelength of 555 nm at daylight.

$$\alpha = \frac{2 \cdot \kappa \cdot \omega}{c_0}, \tag{3.4}$$

with $\omega$ being the angular frequency of light given by

$$\omega = 2 \cdot \pi \cdot f = \frac{2 \cdot \pi \cdot c_0}{\lambda}, \tag{3.5}$$

and $c_0$ being the speed of light in vacuum.[5]

Second, the index of refraction may depend on the intensity of light[6] and its polarization. The latter dependency is of special interest for the manufacture of optical glasses. External effects such as mechanical stress may cause anisotropy of a glass. As a result, the index of refraction becomes dependent on the propagation direction and polarization of light. This is expressed by particular indices of refraction, the ordinary for s-polarized light and the extraordinary for p-polarized light. The ordinary index of refraction $n_o$ is given by

$$n_o = \frac{c_0}{c_s}, \tag{3.6}$$

whereas the extraordinary index of refraction $n_{eo}$ follows from

$$n_{eo} = \frac{c_0}{c_p}. \tag{3.7}$$

Here, $c_s$ and $c_p$ are the speed of light for s-polarized light and p-polarized light, respectively, within the optical medium. This behavior is referred to as *birefringence* $\Delta n$. It is given by

$$\Delta n = n_{eo} - n_o \tag{3.8}$$

and has to be taken into account and specified during the design of optical components or systems as well as the manufacture of glasses and optics (see Section 6.2.1).

Finally, the index of refraction of optical media depends on temperature. For optical glasses, this dependency is expressed by the absolute temperature coefficient of index of refraction,

$$\frac{dn_{absolute}}{dT} = \frac{dn_{relative}}{dT} + n \cdot \frac{dn_{air}}{dT}. \tag{3.9}$$

Here, $dn_{relative}/dT$ is the relative temperature coefficient of the index of refraction[7] of the glass and $dn_{air}/dT$ is the temperature coefficient of the index of refraction of air. The latter can be calculated on the basis of the so-called Edlén equation, named

[5] The speed of light in vacuum amounts to $c_0 = 299{,}792{,}458\ \text{m/s} \approx 3 \cdot 10^8\ \text{m/s}$.

[6] The dependency of the index of refraction on the intensity may lead to the self-focusing of light when passing through an optical medium, mainly occurring in crystals.

[7] The relative temperature coefficient of index of refraction refers to air at standard conditions whereas the absolute temperature coefficient of index of refraction refers to vacuum.

after the Swedish astrophysicist *Bengt Edlén* (1906–1993). This equation is basically given by

$$(n-1)=\frac{p\cdot(n_s-1)}{96095.43}\cdot\frac{\left[1+10^{-8}\cdot(0.613-0.00998\cdot p)\cdot T\right]}{1+0.03661\cdot T}, \quad (3.10)$$

where $p$ is the air pressure (given in Pa), $T$ is the air temperature (given in °C), and $n_s$ is the index of refraction of air at standard conditions ($p$=1013.25 hPa, $T$=15°C) (Edlén, 1966).[8]

The absolute temperature coefficient of the index of refraction can also be calculated on the basis of the glass-specific temperature coefficients[9] $D_0$, $D_1$, $D_2$, $E_0$, $E_1$, and $\lambda_{TK}$ according to

$$\frac{dn_{absolute}}{dT}=\frac{n^2-1}{2\cdot n}\cdot\left(D_0+2\cdot D_1\cdot\Delta T+3\cdot D_2\cdot\Delta T^2+\frac{E_0+2\cdot E_1\cdot\Delta T}{\lambda^2-\lambda_{TK}^2}\right). \quad (3.11)$$

Here, $n$ is the (wavelength-dependent) index of refraction of the glass at the initial temperature $T_0$ and $\Delta T$ is the change in temperature.

#### 3.2.3.2 Dispersion Characteristics

The dependency of the index of refraction on the wavelength of the incident light is described by the *dispersion characteristics* of an optical medium. Dispersion can be quantified by different models or values such as the Abbe number, the Cauchy equation, the Sellmeier equation, or the partial dispersion as described in the Sections 3.2.3.2.1 through 3.2.3.2.4.

##### *3.2.3.2.1 Abbe Number*

The dispersion of any optical glass can be described by its *Abbe number*, also referred to as *V-number*. This characteristic value is given by

$$V_e=\frac{n_e-1}{n_{F'}-n_{C'}}. \quad (3.12)$$

Here, the indices refer to the Fraunhofer lines as listed in Table 3.9. $V_e$ thus refers to a wavelength of 546.07 nm and results from the refractive index $n_e$ at this wavelength and the denominator in Equation 3.12. This denominator denotes the *principal dispersion*.

Sometimes, another definition of the Abbe number,

$$V_d=\frac{n_d-1}{n_F-n_C}, \quad (3.13)$$

[8] The Edlén equation was updated several times in the past due to the fact that more precise data on basic parameters of air were found after *Edlén*'s fundamental work in the mid-1960s (Birch and Downs, 1993; Ciddor, 1996).

[9] These coefficients as well as the absolute temperature coefficients of index of refraction can be found in glass manufacturer's data sheets, classified into temperature ranges, and given for selected wavelengths.

**TABLE 3.9**
**Fraunhofer Lines Used for the Calculation of the Abbe Numbers $V_e$ and $V_d$, Respectively**

| Fraunhofer Line | Wavelength in nm |
|---|---|
| *e* | 546.07 |
| *F′* | 479.99 |
| *C′* | 643.85 |
| *d* | 587.56 |
| *F* | 486.13 |
| *C* | 656.27 |

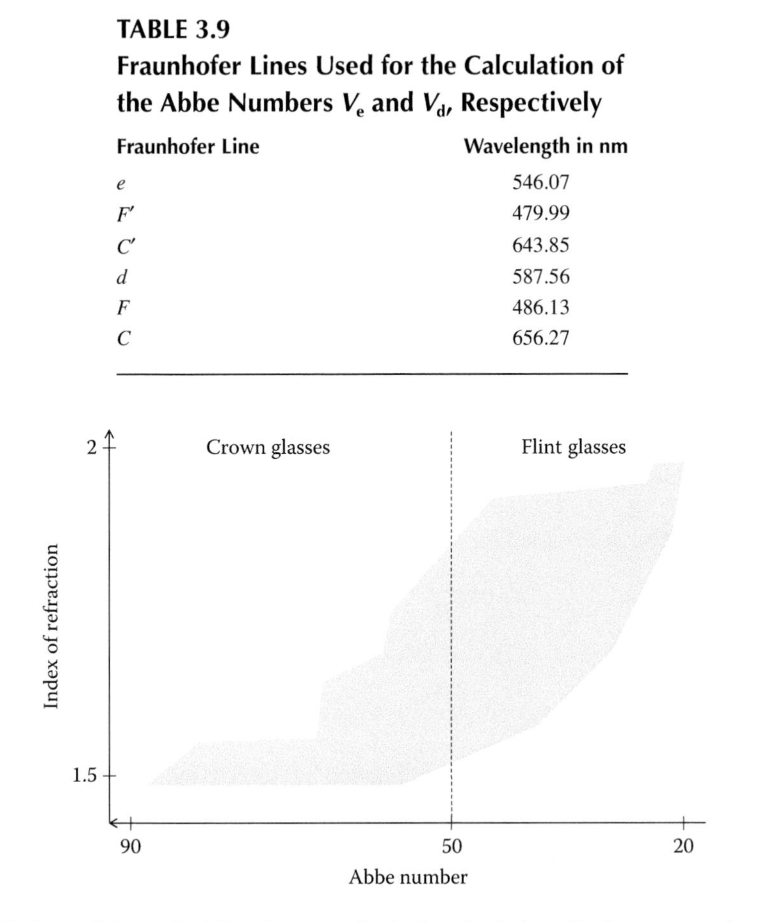


**FIGURE 3.11** Schematic Abbe diagram, displaying the index of refraction vs. the Abbe number, including the main glass families: crown and flint glass.

is found in literature. Officially, this Abbe number $V_d$ was recently replaced by $V_e$, but is still used in practice.

As already mentioned above, there are two main types of optical glasses, crown glass and flint glass. The classification is based on the Abbe number. Crown glasses feature an Abbe number higher than 50 ($V_e > 50 \rightarrow$ crown glasses) whereas the Abbe number of flint glasses is lower than 50 ($V_e < 50 \rightarrow$ flint glasses). The classification of optical glasses is usually visualized and identified by the so-called *Abbe diagram* (*glass map* or "*n* vs. *V* diagram"). Here, the index of refraction ($n_d$ or $n_e$, respectively) is plotted vs. the Abbe number ($V_d$ or $V_e$, respectively) as shown in Figure 3.11.

#### *3.2.3.2.2 Cauchy Equation*

The Cauchy equation, named after the French mathematician *Augustin-Louis Cauchy* (1789–1857) and also known as *dispersion formula*, is a parametric description of the dispersion characteristics of an optical medium. In its easiest version, it is given by

$$n(\lambda) = A + \frac{B}{\lambda^2} + \cdots, \tag{3.14}$$

where $A$ and $B$ (and $C$ etc., not shown in Equation 3.14) are the material-specific so-called Cauchy parameters. As an example, the parameter $A$ amounts to 1.45800 and $B$ to 0.00354 for the boron crown glass N-BK7 from the glass manufacturer Schott (Schott 2015). Knowing the Cauchy parameters of an optical medium thus allows the calculation of its index of refraction at any wavelength of interest. However, one has to consider that the Cauchy equation is only valid for a limited wavelength range, that is, roughly visible light.

#### *3.2.3.2.3 Sellmeier Equation*

Another parametric description of the dispersion characteristics of optical media is the Sellmeier equation, which is also referred to as Sellmeier dispersion formula. It is named after the German physicist *Wolfgang von Sellmeier* (1871). Similar to the Cauchy equation, it is based on material-specific coefficients, the Sellmeier coefficients $B_1$, $B_2$, $B_3$, $C_1$, $C_2$, and $C_3$. For known Sellmeier coefficients, the index of refraction can be determined for any wavelength according to

$$n(\lambda) = \sqrt{1 + \frac{B_1 \cdot \lambda^2}{\lambda^2 - C_1} + \frac{B_2 \cdot \lambda^2}{\lambda^2 - C_2} + \frac{B_3 \cdot \lambda^2}{\lambda^2 - C_3}}. \tag{3.15}$$

An example for Sellmeier coefficients of a particular glass is given by the data listed in Table 3.10.

As already mentioned, the Cauchy equation has limited validity in terms of range of wavelength. In the case of optical standard glasses, it gives reliable results for the visible wavelength range. In comparison, the Sellmeier equation can be applied in the near ultraviolet (approximately 250–400 nm) and infrared (approximately 780–1600 nm) wavelength ranges.

**TABLE 3.10**
**Sellmeier Coefficients for the Boron Crown Glass N-BK7 from the Glass Manufacturer Schott**

| Sellmeier Coefficient | Value |
|---|---|
| $B_1$ (unitless) | 1.039 |
| $B_2$ (unitless) | 0.232 |
| $B_3$ (unitless) | 1.011 |
| $C_1$ (given in μm²) | 0.006 |
| $C_2$ (given in μm²) | 0.020 |
| $C_3$ (given in μm²) | 103.561 |

*Source:* Schott, A.G., Optical Glass Data sheets, 2015. http://www.schott.com/d/advanced_optics/ac85c64c-60a0-4113-a9df-23ee1be20428/1.1/schott-optical-glass-collection-datasheets-english-17012017.pdf.

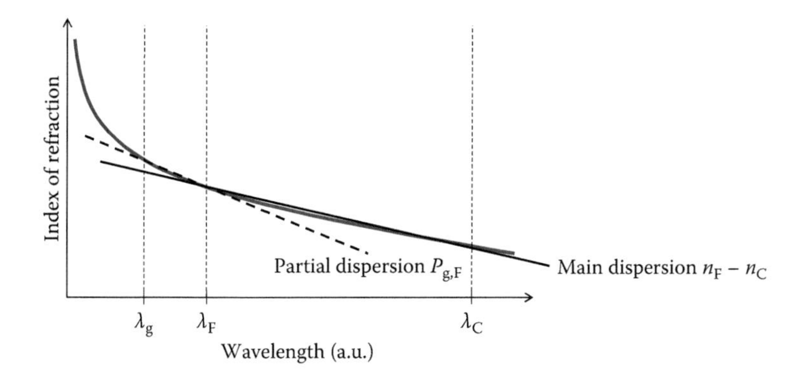


**FIGURE 3.12** Definition and comparison of main dispersion and partial dispersion, the latter being defined for the Fraunhofer lines g and F in the shown example.

#### 3.2.3.2.4 Partial Dispersion

The *partial dispersion* $P$ and $P_{x,y}$, respectively, allow the description of dispersion characteristics for a certain wavelength range of interest between the arbitrarily-chosen wavelengths x and y. It is given by

$$P_{x,y} = \frac{n_x - n_y}{n_F - n_C} \tag{3.16}$$

and thus relates the dispersion for the chosen wavelengths to the main dispersion $n_F - n_C$, which is also found in the obsolete definition of the Abbe number. From a mathematical point of view, the partial dispersion represents the slope of the regression line between the chosen wavelengths and consequently indicates the bending of the dispersion curve as shown in Figure 3.12.

Partial dispersion is thus a helpful parameter for the design of achromatic or apochromatic lenses since the remaining, so-called secondary, spectrum of such systems directly results from the different bendings in dispersion curves of the used optical glasses. However, the variety of optical glasses in terms of refractive index and dispersion characteristics allows the realization of complex optical systems and the minimization of optical aberrations such as chromatic aberration. The approach for the correction of this unwanted effect by combining two lenses made of different glasses is discussed in more detail in Section 5.3.2.

## 3.3 GLASS CERAMICS

In addition to optical glasses, *glass ceramics* are important materials in optical manufacturing. Glass ceramics can be described as partially crystalized glasses where polycrystals with a maximum size of approximately 1 $\mu$m are embedded in an amorphous glass matrix. As a result, these media feature high mechanical stability and low coefficients of thermal expansion *CTE* as well as low transmission. Glass ceramics are thus mainly used as material for the production of large mirror substrates.

As an example, the main components of glass ceramics are lithium oxide ($Li_2O$), aluminum oxide ($Al_2O_3$), and silicon dioxide ($SiO_2$). Such a glass ceramic represents the most important type; it is referred to as the LAS system (lithium–aluminum–silicon system). In order to induce epitaxic crystal growth,[10] zirconium(IV) oxide and titanium(IV) oxide are added to the LAS matrix and act as nucleation agents. The main crystal phases within LAS glass ceramics are a high-quartz solid solution and a keatite solid solution (Hummel, 1951; Smoke, 1951).

## 3.4 GRADIENT INDEX MATERIALS

The use of *gradient index materials*, usually abbreviated GRIN materials, allows realizing both converging and diverging lenses without any classical surface shaping. Here, focusing or defocusing is achieved by a gradient in refractive index within the lens bulk material. This gradient can be either positive or negative. Starting at the center of a GRIN lens, the refractive index thus increases or decreases with rising distance to the lens center, found at its optical axis as shown in Figure 3.13.

This behavior is obtained by an ion exchange process. For this purpose, a glass rod with a given refractive index $n_0$ is placed within a salt solution. As a result of diffusion, ions are exchanged between the glass bulk material and the salt solution. For example, lithium ions ($Li^+$) from the glass material are replaced by sodium ions ($Na^+$) from the salt solution that enter the glass and take the place of removed $Li^+$. Depending on the diffusion coefficient (for definition see Section 13.4.7) of the involved media (glass and solution), a radial gradient of ion concentration is formed. As a consequence, a radial gradient of refractive index is resulting since the index of refraction of any glass is directly related to its chemical composition.

The resulting radial distribution of the refractive index $n(r)$ within a GRIN lens is described by a hyperbolic secant distribution (abbreviated by sec $h$ in Equation 3.17) and is given by

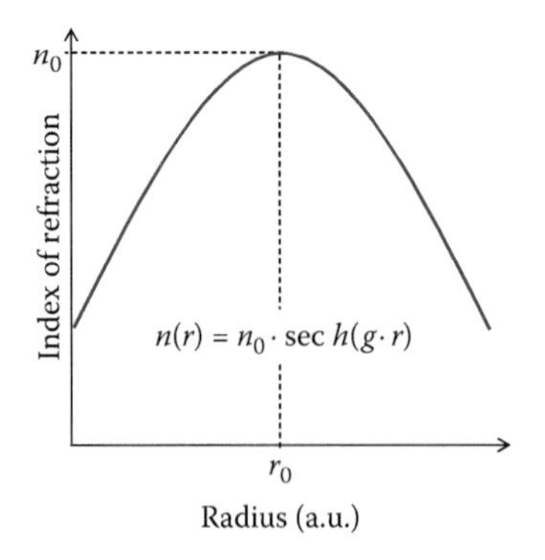


**FIGURE 3.13** Example for the hyperbolic secant distribution of the index of refraction within a GRIN material with a central index of refraction $r_0$.

[10] Epitaxy describes the growth of crystal layers on a substrate crystal where the structure and orientation of the grown crystals correspond to the structure and orientation of the substrate crystal.

$$n(r) = n_0 \cdot \sec h(g \cdot r). \tag{3.17}$$

Here, $n_0$ is the index of refraction at the center of the GRIN lens, and $g$ is the *geometrical gradient constant* that can be either positive or negative. The working principle of such GRIN lenses is presented in more detail in Section 13.3.1.

## 3.5 CRYSTALS

A number of different crystals are used as raw material for optical components. *Crystals* are versatile media and feature a variety of technically usable properties. For instance, some crystals show a high transmission in the ultraviolet and infrared wavelength range (see Table 3.11); they are therefore used for producing lenses and windows for UV and IR laser sources.

The natural birefringence of some crystals such as calcite or ammonium dihydrogen phosphate allows realizing polarization prisms due to the fact that such media feature different refractive indices for s-polarized light and p-polarized light, respectively. Further, crystals are suitable carrier media for laser active dopants. Especially in laser physics, the interaction of some crystals with externally applied acoustic waves, electric or magnetic fields, or high light intensity is of significant relevance. These interactions are generally described by the acousto-optic effect,[11] the

**TABLE 3.11**
**Refractive Index *n*, Transmission Range *T* and Possible Applications of Selected Crystals**

| Crystal | *n* @ 633 nm | *T* in nm | Applications |
|---|---|---|---|
| Lithium fluoride (LiF) | 1.39 | 120–6,500 | UV optics |
| Calcium fluoride ($CaF_2$) | 1.43 | 150–9,000 | UV optics |
| Magnesium fluoride ($MgF_2$) | 1.39 | 130–7,000 | UV and IR optics |
| Sapphire ($Al_2O_3$) | 1.76 | 170–5,000 | UV laser windows |
| Zinc selenide (ZnSe) | 2.60 | 550–18,000 | $CO_2$-laser optics |
| Gallium arsenide (GaAs) | 3.83 | 1,000–15,000 | $CO_2$-laser optics |
| Germanium (Ge) | 4.10 | 1,800–23,000 | IR optics (e.g., night vision, thermal imaging) optical fiber core material |
| Silicon (Si) | 3.35 | 1,200–7,000 | Mirrors IR optics |
| Yttrium aluminum garnet ($Y_3Al_2Al_3O_{12}$) a.k.a. YAG | 1.83 | 210–5,500 | Carrier medium for laser active dopants |
| Calcite ($CaCO_3$) | 1.66 | 210–5,200 | Polarization prisms |

[11] The acousto-optic effect describes the periodical variation in index of refraction within an optical medium due to mechanical strain induced by sound waves. It can thus be used for the realization of dynamic diffractive optical gratings.

**TABLE 3.12**
**Selection of Crystals Suitable for Realizing Optically Active Devices on the Basis of Nonlinear Optical Phenomena as for Example the Acousto-, Magneto-, and Electro-Optical Effect**

| Effect | Suitable Crystal | Crystal Acronym |
|---|---|---|
| Acousto-optic effect | Tellurite ($TeO_2$) | |
| Magneto-optic effect | Terbium gallium garnet ($Tb_3Ga_5O_{12}$) | TGG |
| Electro-optic effect | Monopotassium phosphate ($KH_2PO_4$) | KDP |
| | Potassium dideuterium phosphate ($KD_2PO_4$) | KD*P |
| | Barium borate ($BaB_2O_4$) | BBO |
| Sum-frequency generation | Potassium dideuterium phosphate ($KD_2PO_4$) | KD*P |
| | Potassium titanyl phosphate ($KTiOPO_4$) | KTP |

magneto-optic effect,[12] the electro-optic effect,[13] and sum-frequency generation.[14] A selection of crystals subject to such mechanisms is found in Table 3.12.

Finally, yttrium aluminum garnet ($Y_3Al_5O_{12}$, acronym: YAG) and yttrium orthovanadate ($YVO_4$) shall be mentioned as examples of well-established carrier media for laser active dopants.

## 3.6 SUMMARY

Multicomponent glasses that are mainly used in optics manufacturing consist of network formers, network modifiers, and stabilizers. The basic cross-linked glass matrix is formed by the network formers whereas network modifiers alter the glass composition and its optical properties, respectively, and additionally reduce the melting temperature. Chemical stabilization is achieved by stabilizers that may also act as network formers. Multicomponent glasses are classified into two main

[12] The magneto-optic effect is also known as Faraday effect, named after the English scientist *Michael Faraday* (1791–1867). It occurs in optical media that are exposed to strong external magnetic fields. This leads to a rotation of the initial polarization direction of light when passing through such a medium. The magneto-optic effect is used for realizing rotators or optical isolators.

[13] A number of different electro-optic effects occur in suitable optical media when exposed to strong external electric fields. Mainly, these effects can be classified into (1) a change in absorption and (2) a change in index of refraction. In practice, it is used for optical modulators and switches where the most famous example is the Pockels cell, named after the German physicist *Friedrich Carl Pockels* (1865–1913).

[14] Sum-frequency generation describes the multiplication of frequency of light of high intensity within nonlinear optical media. The special case of frequency doubling is also known as second harmonic generation (SHG) and allows the conversion of laser light of a fundamental wavelength. As an example, the fundamental laser light of a Nd:YAG-laser with a wavelength of 1064 nm can be converted into laser irradiation at a wavelength of 532 nm by SHG.

glass families: crown glass and flint glass, with a number of subcategories or glass types. The denomination of these glass types is based on essential components or properties.

Optical glasses can be manufactured by flame pyrolysis or by classical melting; the latter method is usually applied for high volumes. Here, a powdery batch is prepared in the first step. This batch is then molten down in crucibles or furnaces where the melting temperature amounts to approximately 1000°C–1600°C. The glass melt is homogenized in the course of the plaining process, which can be performed by increasing the temperature of the melt, adding refining agents, or mechanical stirring. In the course of the subsequent cooling process, the glass melt is cooled down to form a supercooled liquid with low internal stress. Finally, glass bars, blocks, rods, panes, or preforms are fabricated, tested, and characterized.

Apart from further essential parameters, the nominal index of refraction, given for a reference wavelength, and the dispersion characteristics of glasses are determined. Dispersion can be specified by the Abbe number, the partial dispersion, the Cauchy equation, or the Sellmeier equation including the particular glass-specific Cauchy and Sellmeier coefficients.

Optical glasses play a major role in optics manufacturing. However, other materials are used for special applications and devices. In this context, glass ceramics stand out due to a low coefficient of thermal expansion and a high mechanical stability. This optical medium consists of single crystallites, which are embedded in an amorphous glass matrix. Further special optical materials are gradient index materials where the distribution of the index of refraction is described by a radial gradient with respect to a center axis. Such distribution is achieved by ion exchange processes where ions from the initial glass network are replaced by other ions, consequently resulting in an alteration of the index of refraction in a locally selective manner. Finally, optical crystals offer a wide field of applications in laser technology or optical communication due to nonlinear optical active effects that arise from interactions of suitable crystals with external influences such as electric or magnetic fields. Most crystals further feature higher transmission in the ultraviolet and infrared wavelength range than optical glasses.

## 3.7 FORMULARY AND MAIN SYMBOLS AND ABBREVIATIONS

**Index of refraction *n* of a transparent optical medium:**

$$n = \frac{c}{v}$$

$c$ speed of light in vacuum
$v$ speed of light within optical medium

**Complex index of refraction *N*:**

$$N = n + i\kappa$$

$n$ index of refraction (real part)
$\kappa$ extinction coefficient (imaginary part)

**Absorption coefficient $\alpha$:**

$$\alpha = \frac{2 \cdot \kappa \cdot \omega}{c_0}$$

$\kappa$ extinction coefficient
$\omega$ angular frequency of light
$c_0$ speed of light in vacuum

**Angular frequency of light $\omega$:**

$$\omega = 2 \cdot \pi \cdot f = \frac{2 \cdot \pi \cdot c_0}{\lambda}$$

$f$ frequency of light
$c_0$ speed of light in vacuum
$\lambda$ wavelength of light

**Ordinary index of refraction $n_o$:**

$$n_o = \frac{c_0}{c_s}$$

$c_0$ speed of light in vacuum
$c_s$ speed of s-polarized light within optical medium

**Extraordinary index of refraction $n_{eo}$:**

$$n_{eo} = \frac{c_0}{c_p}$$

$c_0$ speed of light in vacuum
$c_p$ speed of p-polarized light within optical medium

**Birefringence $\Delta n$:**

$$\Delta n = n_{eo} - n_o$$

$n_{eo}$ extraordinary index of refraction
$n_o$ ordinary index of refraction

**Absolute temperature coefficient of index of refraction $dn_{absolute}/dT$:**

$$\frac{dn_{absolute}}{dT} = \frac{dn_{relative}}{dT} + n \cdot \frac{dn_{air}}{dT}$$

$dn_{relative}/dT$ relative temperature coefficient of index of refraction
$dn_{air}/dT$ temperature coefficient of index of refraction of air
or

$$\frac{dn_{\text{absolute}}}{dT} = \frac{n^2 - 1}{2 \cdot n} \cdot \left( D_0 + 2 \cdot D_1 \cdot \Delta T + 3 \cdot D_2 \cdot \Delta T^2 + \frac{E_0 + 2 \cdot E_1 \cdot \Delta T}{\lambda^2 - \lambda_{\text{TK}}^2} \right)$$

$n$ index of refraction of the glass at the initial temperature
$\Delta T$ change in temperature
*Note:* $D_0$, $D_1$, $D_2$, $E_0$, $E_1$, and $\lambda_{\text{TK}}$ are glass-specific temperature coefficients.

**Edlén equation (for calculation of index of refraction *n* of air):**

$$(n-1) = \frac{p \cdot (n_s - 1)}{96095.43} \cdot \frac{\left[1 + 10^{-8} \cdot (0.613 - 0.00998 \cdot p) \cdot T\right]}{1 + 0.03661 \cdot T}$$

$p$ air pressure
$n_s$ index of refraction of air at standard conditions
$T$ air temperature

**Abbe number $V_e$ relating to a center wavelength of 546.07 nm (new definition):**

$$V_e = \frac{n_e - 1}{n_{F'} - n_{C'}}$$

$n_e$ index of refraction at a wavelength of 546.07 nm
$n_F'$ index of refraction at a wavelength of 479.99 nm
$n_C'$ index of refraction at a wavelength of 643.85 nm

**Abbe number $V_d$ relating to a center wavelength of 587.56 nm (old definition):**

$$V_d = \frac{n_d - 1}{n_F - n_C}$$

$n_d$ index of refraction at a wavelength of 587.56 nm
$n_F$ index of refraction at a wavelength of 486.13 nm
$n_C$ index of refraction at a wavelength of 656.27 nm

**Cauchy equation (dispersion formula):**

$$n(\lambda) = A + \frac{B}{\lambda^2} + \cdots$$

$A$, $B$ material-specific Cauchy parameters
$\lambda$ wavelength of interest

**Sellmeier equation (Sellmeier dispersion formula):**

$$n(\lambda) = \sqrt{1 + \frac{B_1 \cdot \lambda^2}{\lambda^2 - C_1} + \frac{B_2 \cdot \lambda^2}{\lambda^2 - C_2} + \frac{B_3 \cdot \lambda^2}{\lambda^2 - C_3}}$$

$B$, $C$ material-specific Sellmeier coefficients
$\lambda$ wavelength of interest

**Partial dispersion $P_{x,y}$:**

$$P_{x,y} = \frac{n_x - n_y}{n_F - n_C}$$

$n_x$ index of refraction at first wavelength of interest
$n_y$ index of refraction at second wavelength of interest
$n_F$ index of refraction at a wavelength of 486.13 nm
$n_C$ index of refraction at a wavelength of 656.27 nm

**Radial distribution of refractive index $n(r)$ within gradient index material:**

$$n(r) = n_0 \cdot \sec h(g \cdot r)$$

$n_0$ index of refraction at center
$g$ geometrical gradient constant
$r$ radius

## REFERENCES


Birch, K.P., and Downs, M.J. 1993. An updated Edlén equation for the refractive index of air. *Metrologia* 30:155–162.

Ciddor, P.E. 1996. Refractive index of air: new equations for the visible and near infrared. *Applied Optics* 35:1566–1573.

Edlén, B. 1966. The refractive index of air. *Metrologia* 2:71–80.

Hummel, F.A. 1951. Thermal expansion properties of some synthetic lithia minerals. *Journal of the American Ceramic Society* 34:235–239.

Pfaender, H.G. 1997. *Schott Glaslexikon*. Landsberg, Germany: mgv-verlag (in German).

Schaeffer, H.A., and Langfeld, R. 2014. *Werkstoff Glas*. Berlin and Heidelberg: Springer-Verlag (in German).

Schott, A.G. 2015. Optical Glass Data sheets. http://www.schott.com/d/advanced_optics/ac85c64c-60a0-4113-a9df-23ee1be20428/1.1/schott-optical-glass-collection-datasheets-english-17012017.pdf.

Smoke, E.J. 1951. Ceramic compositions having negative linear thermal expansion. *Journal of the American Ceramic Society* 34:87–90.

Vogel, W., Gerth, K., and Heindorf, W. 1965. Zur Entwicklung der optischen Gläser. *Jenaer Rundschau* 1:75–76 (in German).

von Sellmeier, W. 1871. Zur Erklärung der abnormen Farbenfolge im Spectrum einiger Substanzen. *Annalen der Physik* 219:272–282 (in German).

# 4 Optical Components

## 4.1 INTRODUCTION

Single optical components are the keys to and basis of any optical system; the most famous optical components are most likely lenses. Lenses were already produced and used 3000 years ago in Mesopotamia, where the so-called Nimrud lens (a.k.a. Layard lens) was found in 1850 at the Assyrian palace of Nimrud. It is assumed that this lens was used for igniting fires and as a magnifying glass. In medieval times, simple eyeglasses made by lens grinders and spectacle makers were available. In 1608, the German-Dutch spectacle maker *Hans Lippershey* (c. 1570–1619) discovered the telescopic effect of magnification by the combination of two lenses. This discovery can be referred to as the first optical system.

In addition to lenses, a number of different optical components such as prisms, wedges, plates, and mirrors are used. These components, as well as the functional principles and main parameters, are introduced in this chapter.

## 4.2 LENSES

### 4.2.1 Spherical Lenses

*Lenses* represent the most important optical components of complex optical systems. Depending on the geometrical shape or function, lenses can be classified into *converging* and *diverging* ones. Assuming a bundle of light rays propagates parallel to the optical axis of a lens, the first type focuses the light rays into a real *focal point*[1] due to refraction at the curved lens surfaces whereas diverging lenses feature a virtual focal point. The particular behavior directly results from the lens shape and the orientation of the curved surfaces. This orientation is expressed by the nominations "*convex*" (from Latin "convexus" = "curved") and "*concave*" (from Latin "concavus" = "hollowed"). Usually, a convex lens surface is indicated by the abbreviation "CX" and a concave by "CC" in manufacturing drawings, specifying the basic orientation of curvature. Further, the particular *radii of curvature* $R$ or $R_c$, respectively, are given as principally shown at the example of a symmetric biconvex lens in Figure 4.1.

A complete definition of a lens further requires information about its *center thickness* $t_c$, its *diameter* $D$, and the material (i.e., the glass, in terms of its nominal index of refraction at a predefined reference wavelength, its dispersion characteristics, and material quality regarding acceptable striae, inclusions, birefringence, etc., compare Section 6.2). Finally, the surface quality of the optically active surfaces has to be specified in terms of contour accuracy, cleanliness, surface roughness, etc., as

[1] In practice, there is no infinitely small focal point due to aberrations such as spherical aberration and additional chromatic aberration in polychromatic light. The nominal focal length of any lens thus refers to a reference wavelength, usually 546.07 nm (i.e., Fraunhofer line e) or 587.56 nm (i.e., Fraunhofer line d).

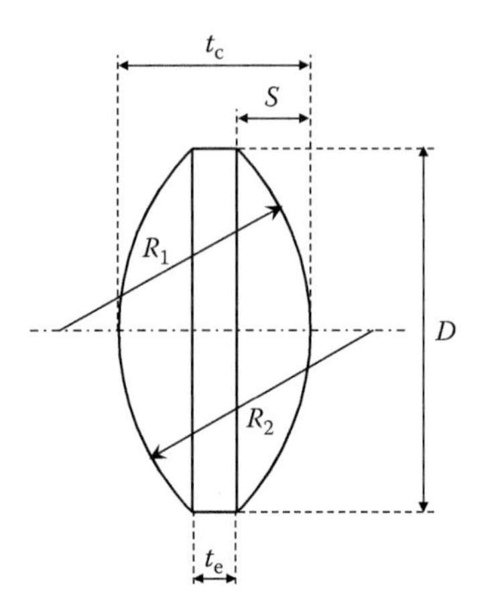


**FIGURE 4.1** Definition of essential lens parameters at the example of a biconvex spherical lens with the radii of curvature $R_1$ and $R_2$, the center thickness $t_c$, the diameter $D$, the sagitta $S$, and the edge thickness $t_e$.

presented in more detail in Section 6.3. In addition to this mandatory information, related lens parameters such as the *sagitta S* of a lens surface and the *edge thickness* $t_e$ can be derived from the lens geometry via its diameter, radii of curvature, and center thickness. An overview on different lens types and the particular relation between the radii of curvature and the center and edge thickness is given in Table 4.1.

**TABLE 4.1**
**Different Types of Lenses Including a Short Description on the Basis of the Radii of Curvature $R_1$ and $R_2$ and the Center Thickness $t_c$ and Edge Thickness $t_e$**

| Lens Type | a.k.a. | Short Description |
|---|---|---|
| | **Converging Lenses** | |
| Plano-convex | | $R_1 = \infty$ and $t_c > t_e$ |
| Symmetric biconvex | Double convex, convex-concave | $R_1 = R_2$ and $t_c > t_e$ |
| Asymmetric biconvex | Best form converging lens | $R_1 \neq R_2$ and $t_c > t_e$ |
| Concave-convex | Positive meniscus | $R_{cc} > R_{cx}$ and $t_c > t_e$ |
| | **Diverging Lenses** | |
| Plano-concave[a] | | $R_1 = \infty$ and $t_c < t_e$ |
| Symmetric biconcave | Double concave, concave-convex | $R_1 = R_2$ and $t_c < t_e$ |
| Asymmetric biconcave | Best form diverging lens | $R_1 \neq R_2$ and $t_c < t_e$ |
| Convex-concave | Negative meniscus | $R_{cx} > R_{cc}$ and $t_c < t_e$ |

[a] Plano-concave lenses are also used as substrate for concave mirrors where reflective coatings are applied to the curved front side of the lens. The focal length of such a concave mirror is then half its radius of curvature, $f = R/2$.

It turns out that the center thickness of diverging lenses is smaller than the edge thickness and vice versa, which becomes obvious when looking at the comparison of the above-listed different lens types in Figure 4.2.

Lenses are generally classified into so-called thin and thick lenses. *Thin lenses* feature significantly lower center thicknesses in comparison to the radius of curvature ($t_c \ll R$); the center thickness can thus be neglected. The *effective focal length* (EFL) of a thin lens is given by

$$EFL = \frac{1}{n-1} \cdot \left( \frac{R_1 \cdot R_2}{R_2 - R_1} \right). \tag{4.1}$$

For equal radii of curvature, $R = |R_1| = |R_2|$, this interrelationship can be rewritten as

$$EFL = \frac{1}{n-1} \cdot \frac{R}{2}. \tag{4.2}$$

The EFL of thin plano-convex lenses, where one radius of curvature is infinite, is then given by

$$EFL = \frac{R}{n-1}. \tag{4.3}$$

For *thick lenses,* the center thickness has to be taken into account; the EFL thus follows from

$$EFL = \frac{1}{n-1} \cdot \frac{n \cdot R_1 \cdot R_2}{(n-1) \cdot t_c + n \cdot (R_2 - R_1)} \tag{4.4}$$

and

$$EFL = \frac{1}{n-1} \cdot \frac{n \cdot R^2}{(n-1) \cdot t_c} \tag{4.5}$$

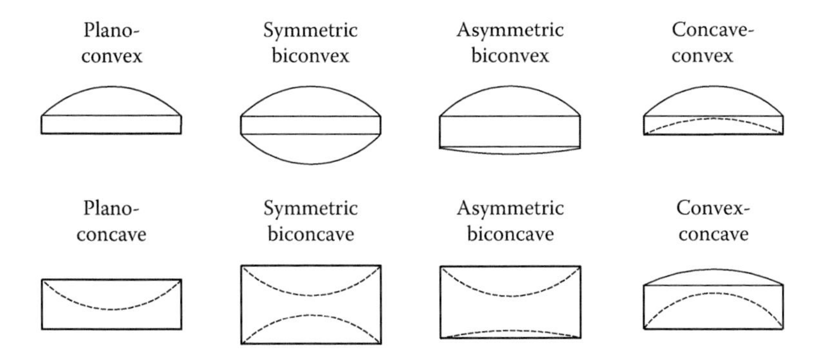


**FIGURE 4.2** Different types of spherical converging (top) and diverging (bottom) lenses.

if $R = |R_1| = |R_2|$ and

$$EFL = \frac{R}{n-1} - \frac{t_c}{n}, \tag{4.6}$$

in case of thick plano-convex or plano-concave lenses. The EFL of thick lenses thus generally results from the radii of curvature, the used material (as defined by its refractive index), and the center thickness.

### 4.2.2 Nonspherical Lenses

Up to now, we have considered lenses with spherically shaped surfaces. However, this type of surface gives rise to several aberrations, for example, spherical aberration due to surface sphericity. One approach to reducing this type of aberration is the use of *aspherical surfaces.* As shown at the example in Figure 4.3, such lenses feature complex surface shapes, such as paraboloids or even free forms.

Aspherical surfaces are described by a base sphere with the radius $R$, a coefficient of asphericity $A$, and the position-dependent sagitta $S(z)$, according to

$$S(z) = \frac{\frac{z^2}{R}}{1 + \sqrt{1 - e\left(\frac{z}{R}\right)^2}} + \sum_{i=2}^{i_{max}} A_{2i} \cdot z^{2i}. \tag{4.7}$$

The coordinate $z$ corresponds to half the lens diameter in rotational-symmetric aspheres (compare Figure 4.3) and is thus equivalent to the *ray entrance height* of incoming light rays, which propagate parallel to the optical axis of the aspheric lens. The parameter $e$ in Equation 4.7 is referred to as *conus constant* and defines the geometric basic shape of an aspheric surface as listed in Table 4.2.

In addition to spherical and aspherical lenses, *cylindrical lenses* play an essential role for the realization of high-quality optical systems but also daily-used convenience goods, for example, scanners. Such lenses feature two perpendicular sections on the surface, the meridional and the sagittal. Only one of those sections features a

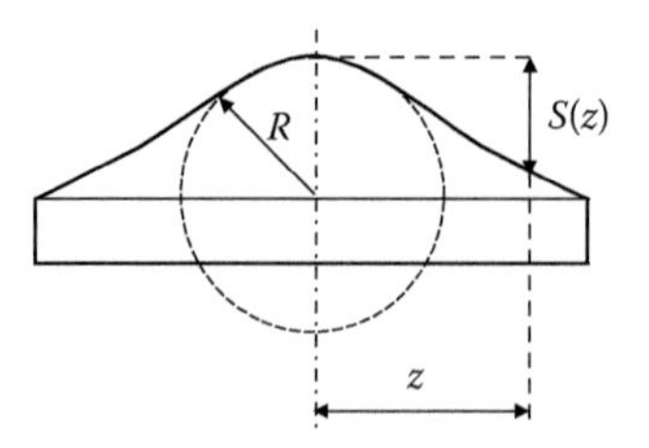


**FIGURE 4.3** Definition of essential lens parameters at the example of a convex aspherical lens with the base sphere radius of curvature $R$ and the position-dependent sagitta $S(z)$.

**TABLE 4.2**
**Conus Constant e and Resulting Geometric Basic Shape of Aspherical Lens Surfaces**

| Conus Constant | Geometric Basic Shape |
|---|---|
| $e < 0$ | Hyperbola |
| $e = 0$ | Parabola |
| $e > 0$ | Ellipse |
| $e = 1$ | Circle |

certain radius of curvature; the other section is represented by a plane. Its radius of curvature is consequently $R = 0$ mm. As a result, cylindrical lenses realize focal lines instead of focal points.

If both the meridional and sagittal sections feature different radii of curvature higher than 0 mm, the resulting optical component is referred to as a *toric lens* as shown in Figure 4.4.

Due to the different radii of curvature of the meridional and sagittal planes, this type of lens produces astigmatism. It is thus a typical lens shape for eye glasses, since astigmatism of the eye lens can be corrected by an appropriate toric surface.

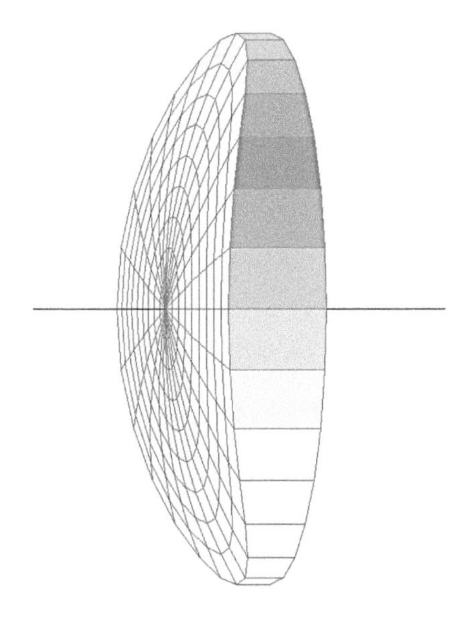

**FIGURE 4.4** Example for a toric lens with two different radii of curvature in the meridional and sagittal surface section. (Figure was generated using the software WinLens3D Basic from Qioptiq Photonics GmbH & Co. KG.)

## 4.3 PRISMS AND WEDGES

Prisms can be classified into three main types: deflection prisms, dispersion prisms, and polarization prisms. The denomination of prisms is usually based on geometric properties, mostly the prism wedge angle, or on the name of the inventors of a particular prism type. A selection of the most important and commonly used prisms is shown in Figure 4.5.

### 4.3.1 Deflection Prisms

*Deflection prisms* are used for the manipulation of light in terms of changing either its orientation or direction of propagation. A change in orientation is performed for image reversal, for example, as realized by Porro-prisms[2] in Kepler telescopes,[3] binoculars, or microscopes. Here, image reversal is due to total internal reflection (see Section 2.3) within the prism.

Further, deflection prisms are employed in order to generate a *deviation* $\delta$ of a light beam or bundle of light rays from its original propagation direction. This deviation is given by

$$\delta = \varepsilon_1 + \varepsilon_2' - \alpha. \tag{4.8}$$

Here, $\varepsilon_1$ is the angle of incidence of a light ray on the entrance surface of a deflection prism, $\varepsilon_2'$ is the exit angle of the light ray at the prism's exit surface, and $\alpha$ is the *prism wedge angle*. The exit angle can be calculated according to

$$\varepsilon_2' = \arcsin\left(n \cdot \sin\varepsilon_2\right), \tag{4.9}$$

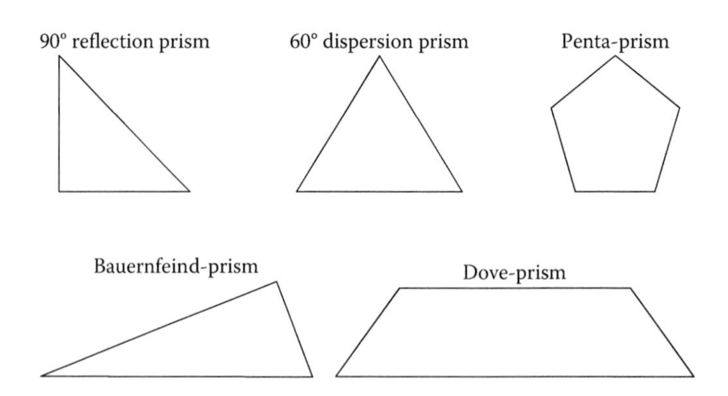


**FIGURE 4.5** Selection of different types of prisms.

[2] Named after the Italian engineer *Ignazio Porro* (1801–1875).

[3] A Kepler telescope, named after the German mathematician and astronomer *Johannes Kepler* (1571–1630), principally consists of two converging lenses, where the distance between the lenses is given by the sum of the particular focal lengths. This setup produces a magnified, but inverted, image, leading to the necessity of a further optical element for image reversal.

where $\varepsilon_2$ is the angle of incidence of light at the interface glass-surrounding medium (e.g., air) within the prism, given by

$$\varepsilon_2 = \alpha - \arcsin\left(\frac{\sin\varepsilon_1}{n}\right). \tag{4.10}$$

For instance, the effect of deflection by a prism can be used for the determination of the index of refraction $n$ of the prism material according to

$$n = \frac{\sin\left(\frac{\delta_{\min} + \alpha}{2}\right)}{\sin\left(\frac{\alpha}{2}\right)}. \tag{4.11}$$

Here, $\delta_{\min}$ is the minimum deviation caused by the prism, given by

$$\delta_{\min} = 2 \cdot \arcsin\left[n \cdot \sin\left(\frac{\alpha}{2}\right)\right] - \alpha, \tag{4.12}$$

which is found for the special case of so-called symmetric pass where the angle of incidence of light is equal to the exit angle.

### 4.3.2 Dispersion Prisms

Prisms can also be used for the segmentation of white light into its spectral fractions, i.e., wavelengths or colors, respectively. Since this effect is due to the dispersion characteristics of the prism material as presented in more detail in Section 3.2.3.2, this type of prism is referred to as *dispersion prism*[4] and applied for the setup of spectrometers or for monochromatization. This is made possible by the wavelength dependency of the index of refraction and the deviation, respectively. The *dispersion angle* $\delta_d$ (i.e., the difference in deviation $\delta$ between two considered wavelengths of interest $\lambda_1$ and $\lambda_2$) is given by

$$\delta_d = \delta(\lambda_2) - \delta(\lambda_1), \tag{4.13}$$

where $\lambda_2 < \lambda_1$ and $n_2 > n_1$, respectively.

There are a number of different types of dispersion prisms. The most common is a single prism with a prism wedge angle of 60°, made of an optical glass with a low Abbe number and a high dispersion, respectively. This prism generates both segmentation of incident white light into its spectral fractions and a general

[4] A dispersion prism was also used by the German optician and physicist *Joseph von Fraunhofer* (1787–1826) during his famous experiment in 1814 where he observed dark lines in the solar spectrum. The underlying mechanism for his observation is self-absorption within the sun due to the high electron densities in the sun plasma. These lines are now known as Fraunhofer lines, representing the reference wavelengths for a number of specifications in optics such as the index of refraction or the Abbe number.

deviation of the propagation direction of light. Segmentation without such deviation is achieved by special prism assemblies, for example, Amici prisms,[5] (i.e., a prism group consisting of several cemented prisms). As shown in Figure 4.6, the combination of such a prism group and a single lens allows setting up a simple but usable and stable spectrometer.

### 4.3.3 Polarization Prisms

The combination of two or more prisms also allows realizing *polarization prisms*, made of birefringent material such as Calcite ($CaCO_3$); see Section 3.5. As a result, non-polarized incident light is divided up into two light rays of different polarization, s-polarized and p-polarized with a certain angular distance. This angular distance can range from some degrees to more than 90° depending on the type of polarization prism, as shown by the selection of commonly used polarization prisms listed in Table 4.3.

Such polarization prisms are used in laser technology or for the analysis of liquid samples in food technology.

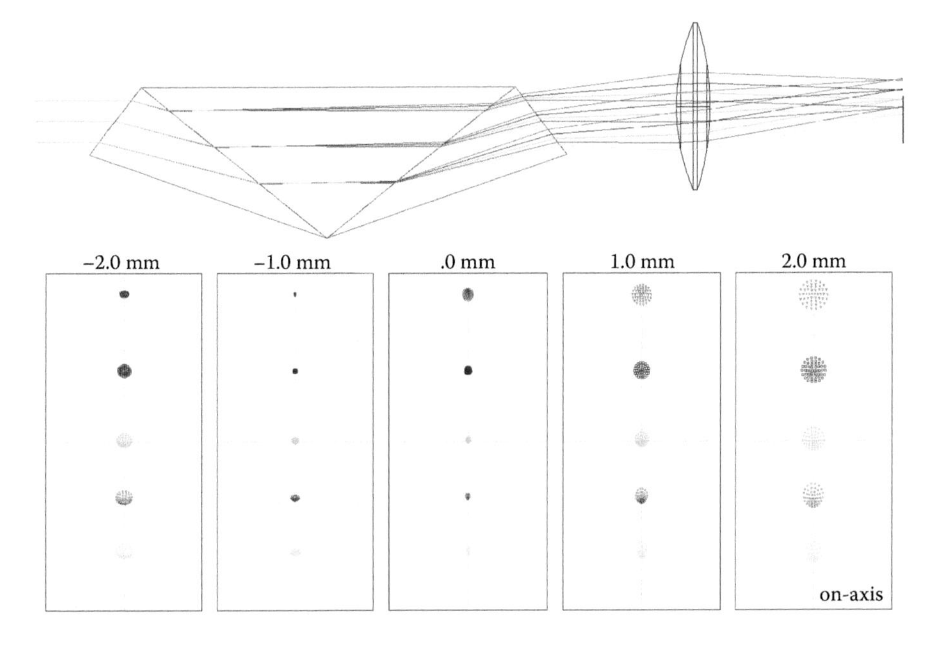


**FIGURE 4.6** Spectrometer setup consisting of a dispersive double Amici prism and a converging lens (top) including the resulting spot diagram for five wavelengths (bottom), visualizing the separation of wavelengths and spectral resolution, respectively. (Figure was generated using the software WinLens3D Basic from Qioptiq Photonics GmbH & Co. KG.)

[5] Named after the Italian astronomer *Giovanni Battista Amici* (1786–1863).

**TABLE 4.3**
**Selection of Polarization Prisms Including the Particular Setup, Characteristics, and Angular Distance of the Polarized Fractions**

| Polarization Prism Type | Setup | Paths of Rays | Angular Distance of Polarized Fractions |
|---|---|---|---|
| Nicol prism[a] | Two cemented prisms, pointed angles 68° and 22° | Transmission of one, but reflection of the other polarization direction due to total internal reflection | ≈20° |
| Rochon prism,[b] Wollaston prism[c] | Two cemented equiangularly prisms | Transmission of both polarization directions | ≈15°–45° |
| Glan-Thompson prism[d] | Two cemented equiangularly prisms | Transmission of one, but reflection of the other polarization direction due to total internal reflection | ≈45° |
| Glan-Foucault prism[e] | Two prisms, pointed angles 90° and 45°, separated by air gap | Transmission of one, but reflection of the other polarization direction at entrance interface of second prism | <90° |

[a] Named after the Scottish physicist *William Nicol* (c. 1768–1851) (Nicol, 1828).
[b] Named after the French astronomer *Alexis-Marie de Rochon* (1741–1817).
[c] Named after the British physicist *William Hyde Wollaston* (1766–1828).
[d] Named after the German physicist *Paul Glan* (1846–1898) (Glan, 1880) and the English physicist *Silvanus Phillips Thompson* (1851–1916) (Thompson, 1881).
[e] Named after the German physicist *Paul Glan* (1846–1898) and the French physicist *Jean Bernard Léon Foucault* (1819–1868); this type of prism is also known as a Glan-Taylor prism (Archard and Taylor, 1948).

### 4.3.4 Wedges

*Wedges* can be referred to as deflection prisms with small wedge angles. Normally, the entrance surface is perpendicular to the incident light, and deflection of light is due to refraction at the exit surface. The wedge angle $\alpha$ of wedges is defined to be smaller than 10°, resulting in moderate deviation $\delta$ of incoming light according to

$$\delta = \alpha \cdot (n - 1). \tag{4.14}$$

These optical components are thus used in order to realize slight deviations of propagation direction of light (e.g., for the compensation of lateral offsets).

## 4.4 PLATES

Even though *plates* or blocks feature the simplest possible geometry, this type of optical component has a large number of applications. *Plane-parallel plates* are used as windows, for example, as protection windows for laser cavities or windows of vacuum chambers, which allow the optical access to the process zone for industrial vision purposes. Another important and widespread application for plates is their use as substrates for mirror or filter coatings or polarization layers. When arranged at the Brewster's angle (see Section 2.3), a pure plane plate can act as polarizer, even without any coating. Moreover, plane plates can be applied as retarders or wave plates[6] when made of a birefringent optical medium such as calcite or glimmer. As shown in Figure 4.7, a glass plate can cause *parallel offset* $O_p$ and *longitudinal offset* $O_l$ of light rays as they pass through.

Both offsets principally depend on the thickness $t$ and the index of refraction $n$ of the plate material, where the amount of parallel offset is further related to the angle of incidence $\varepsilon$ of an incident light ray at the plate's front surface according to

$$O_p = t \cdot \sin\varepsilon \cdot \left(1 - \frac{\cos\varepsilon}{\sqrt{n^2 - \sin^2\varepsilon}}\right). \tag{4.15}$$

The longitudinal offset is given by

$$O_l = t \cdot \frac{n-1}{n}. \tag{4.16}$$

This behavior can be used for optical compensators or longitudinal or lateral displacers, for example, for shifting the image plane and realizing a defocus, respectively, in an optical system by benefiting from longitudinal offset as shown in Figure 4.7 (right).

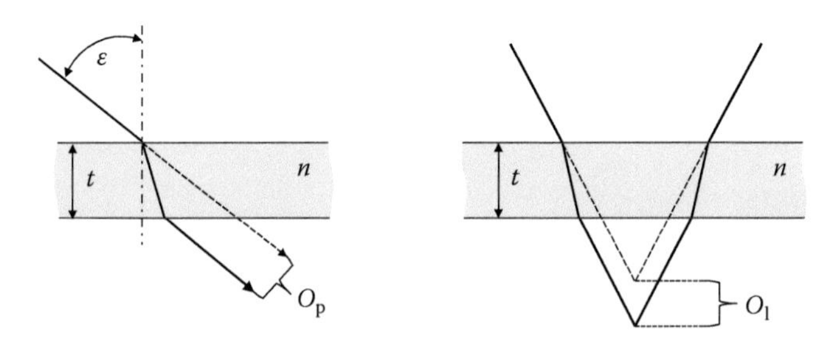


**FIGURE 4.7** Visualization of parallel offset $O_p$ (left) and longitudinal offset $O_l$ (right) of light rays passing through a plane parallel plate with the thickness $t$ and the index of refraction $n$.

[6] Wave plates are classified into two main types: half-wave plates and quarter-wave plates. The polarization vector of linearly polarized light is mirrored when passing through a half-wave plate; the orientation of the polarization is thus changed by a certain angle. In contrast, linearly polarized light is converted into elliptically polarized light when passing through a quarter-wave plate.

## 4.5 SUMMARY

Optical components are the basis of any optical system and can generally be classified into lenses, prisms, and plates. Lenses are the most important and commonly used optical components and are characterized by the basic surface shape of the optically active surfaces, spherical, aspherical, toric, or cylindrical. All types of lenses are further specified by the lens material and the center thickness, where thin lenses are distinguished from thick lenses. Spherical lenses are further defined by the radii of curvature of the optically active surface whereas aspherical lens surfaces are described by appropriate mathematical functions. Toric lens surfaces feature two different radii of curvature perpendicular to each other, and cylindrical lenses can be described as a segment of cylinder walls.

The second most important optical components are prisms, which are classified on the basis of their final functionality and use. Deflection prisms allow deflecting light from its initial direction of propagation, whereas segmentation of white light into its spectral fractions is achieved by the use of dispersion prisms. In the latter case, the effect of dispersion within optical media is utilized. Nonpolarized light can be divided into two light rays with different polarizations when applying polarization prisms made of birefringent materials. Deflection prisms with small wedge angles are referred to as wedges, and plane-parallel plates can be applied for realizing parallel and longitudinal offsets of light rays or as substrates for mirror, polarization, or filter coatings.

## 4.6 FORMULARY AND MAIN SYMBOLS AND ABBREVIATIONS

***EFL* of a thin asymmetric biconvex lens:**

$$EFL = \frac{1}{n-1} \cdot \left( \frac{R_1 \cdot R_2}{R_2 - R_1} \right)$$

$n$ index of refraction of the lens material
$R_1$ radius of curvature of the first lens surface
$R_2$ radius of curvature of the second lens surface

***EFL* of a thin symmetric biconvex lens:**

$$EFL = \frac{1}{n-1} \cdot \frac{R}{2}$$

$n$ index of refraction of the lens material
$R$ radius of curvature of both lens surfaces

***EFL* of a thin plano-convex lens:**

$$EFL = \frac{R}{n-1}$$

$n$ index of refraction of the lens material
$R$ radius of curvature of the curved lens surface

***EFL* of a thick asymmetric biconvex lens:**

$$EFL = \frac{1}{n-1} \cdot \frac{n \cdot R_1 \cdot R_2}{(n-1) \cdot t_c + n \cdot (R_2 - R_1)}$$

$n$ index of refraction of the lens material
$R_1$ radius of curvature of the first lens surface
$R_2$ radius of curvature of the second lens surface
$t_c$ lens center thickness

***EFL* of a thick asymmetric biconvex lens:**

$$EFL = \frac{1}{n-1} \cdot \frac{n \cdot R^2}{(n-1) \cdot t_c}$$

$n$ index of refraction of the lens material
$R$ radius of curvature of both lens surfaces
$t_c$ lens center thickness

**EFL of a thick plano-convex lens:**

$$EFL = \frac{R}{n-1} - \frac{t_c}{n}$$

$R$ radius of curvature of both lens surfaces
$t_c$ lens center thickness
$n$ index of refraction of the lens material

**Position-dependent sagitta *S*(*z*) of aspheric lens surfaces:**

$$S(z) = \frac{\frac{z^2}{R}}{1 + \sqrt{1 - e\left(\frac{z}{R}\right)^2}} + \sum_{i=2}^{i_{\max}} A_{2i} \cdot z^{2i}$$

$z$ coordinate along half lens diameter
$R$ base sphere radius
$e$ conus constant
$A$ coefficient of asphericity

**Deviation $\delta$ of light by deflection prisms:**

$$\delta = \varepsilon_1 + \varepsilon_2' - \alpha$$

$\varepsilon_1$ angle of incidence of light at prism entrance surface
$\varepsilon_2'$ exit angle of light at prism exit surface
$\alpha$ prism wedge angle

**Exit angle of light $\varepsilon_2'$ at prism exit surface:**

$$\varepsilon_2' = \arcsin\left(n \cdot \sin\varepsilon_2\right)$$

$n$ index of refraction of prism material
$\varepsilon_2$ angle of incidence at interface glass-surrounding medium within prism

**Angle of incidence $\varepsilon_2$ at interface glass-surrounding medium within prism:**

$$\varepsilon_2 = \alpha - \arcsin\left(\frac{\sin\varepsilon_1}{n}\right)$$

$\alpha$ prism wedge angle
$\varepsilon_1$ angle of incidence of light at prism entrance surface
$n$ index of refraction of prism material

**Index of refraction $n$ of prism material:**

$$n = \frac{\sin\left(\dfrac{\delta_{\min} + \alpha}{2}\right)}{\sin\left(\dfrac{\alpha}{2}\right)}$$

$\delta_{\min}$ minimum deviation
$\alpha$ prism wedge angle

**Minimum deviation $\delta_{\min}$ of light by deflection prisms (so-called symmetric pass):**

$$\delta_{\min} = 2 \cdot \arcsin\left[n \cdot \sin\left(\frac{\alpha}{2}\right)\right] - \alpha$$

$n$ index of refraction of prism material
$\alpha$ prism wedge angle

**Dispersion angle $\delta_d$ of dispersion prisms:**

$$\delta_d = \delta(\lambda_2) - \delta(\lambda_1)$$

$\delta(\lambda_1)$ deviation for first wavelength of interest
$\delta(\lambda_2)$ deviation for second wavelength of interest
*Note:* $\lambda_2 < \lambda_1$

**Deviation $\delta$ of light by wedges:**

$$\delta = \alpha \cdot (n - 1)$$

$\alpha$ wedge angle (<10° by definition)
$n$ index of refraction of wedge material

**Parallel offset $O_p$ by plane-parallel plates:**

$$O_p = t \cdot \sin\varepsilon \cdot \left(1 - \frac{\cos\varepsilon}{\sqrt{n^2 - \sin^2\varepsilon}}\right)$$

$t$ plate thickness
$\varepsilon$ angle of incidence of light at plate entrance surface
$n$ index of refraction of plate material

**Longitudinal offset $O_l$ by plane-parallel plates:**

$$O_l = t \cdot \frac{n-1}{n}$$

$t$ plate thickness
$n$ index of refraction of plate material

## REFERENCES

Archard, J.F., and Taylor, A.M. 1948. Improved Glan-Foucault prism. *Journal of Scientific Instruments* 25:407–409.

Glan, P. 1880. Ueber einen Polarisator. *Carl's Repertorium* 16:570.

Nicol, W. 1828. On a method of so far increasing the divergence of the two rays in calcareous-spar that only one image may be seen at a time. *Edinburgh New Philosophical Journal* 6:83–94.

Thompson, S.P. 1881. On a new polarising prism. *Philosophical Magazine* 5:349.

# 5 Design of Optical Components

## 5.1 INTRODUCTION

Although it is not a physical production step, the design of an optical component or system can be referred to as the first essential element of manufacturing, since dimensions such as radii or center thickness and the optical material of a component are defined and acceptable manufacturing tolerances are specified here. The design of optical components thus represents the basis of any manufacturing drawing and provides information on the required manufacturing accuracy.

The goal of the design of optical components, usually called optical system design, is to define and specify a suitable optical component, group, or system for a given imaging task. This involves (1) the determination of so-called conjugated parameters that result directly from the known or given parameters and (2) the choice of an appropriate optical system that allows reducing optical aberrations and performing imaging at high quality. Hence, basic considerations regarding the interrelations of optical imaging, underlying mechanisms for the formation of aberrations as well as the impact of manufacturing tolerances on the imaging performance, have to be taken into account during optical system design. These aspects are presented in the present chapter.

## 5.2 DETERMINATION OF OPTICAL COMPONENTS AND SYSTEMS

Any optical imaging can be described by characteristic parameters in the so-called *object space* (i.e., in front of an imaging system) and corresponding parameters in the *image space* found behind an imaging system. The relevant and essential parameters are shown in Figure 5.1 and listed in Table 5.1.

The parameters in object space are directly linked to the parameters in image space by a constant (e.g., the focal length of the imaging system or the magnification). They are thus referred to as *conjugated parameters*. For example, the interrelation of the object distance $a$, the image distance $a'$, and the effective focal length (EFL) is given by the general *imaging equation*,

$$\frac{1}{EFL} = \frac{1}{a} + \frac{1}{a'}. \tag{5.1}$$

Moreover, the image height $u'$ and the object height $u$ (and image distance $a'$ and object distance $a$, respectively) are linked by the *magnification* $\beta$[1] according to

[1] The magnification is also commonly abbreviated by the letter $m$.

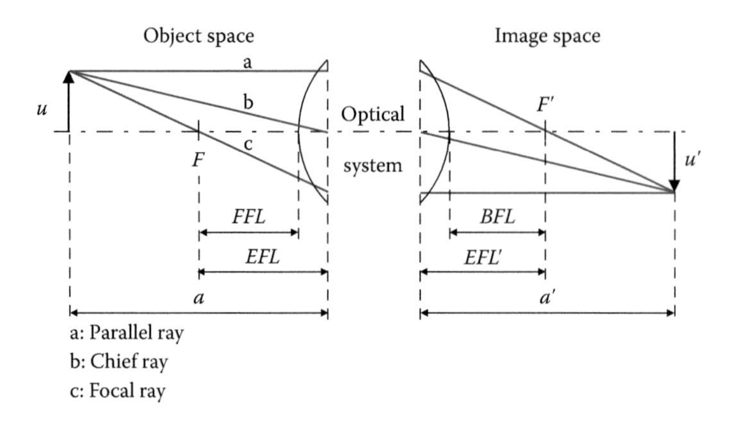


**FIGURE 5.1** Relevant conjugated parameters of optical imaging in object and image space (for explanation see Table 5.1) including construction rays.

**TABLE 5.1**
**Essential Parameters of Optical Imaging in Object and Image Space**

| Object Space | | Image Space | |
|---|---|---|---|
| **Symbol** | **Parameter** | **Symbol** | **Parameter** |
| $u$ | Object height | $u'$ | Image height |
| $a$ | Object distance | $a'$ | Image distance |
| $FFL$ | Front focal length | $BFL$ | Back focal length |
| $EFL$ | Effective (front) focal length | $EFL'$ | Effective (back) focal length |
| $F$ | Focal point, focus | $F'$ | Focal point, focus |

$$\beta = \frac{u'}{u} = \frac{a'}{a}. \tag{5.2}$$

As a consequence, nearly all wanted parameters can be determined if a sufficient number of conjugated parameters is given. Usually, the object height, the object distance, the aperture diameter of an optical system, and the image height (given by the size of the used detector) are known. This allows the calculation of the image distance, the object and image angle, the numerical aperture, and finally the required focal length of an optical component or system. Such calculation is now usually carried out via computer-assisted optical system design (Thöniß et al., 2009) employing appropriate software tools as shown in the example in Figure 5.2.

In this example, the given parameters were the object distance $a=-500\,\text{mm}$, the object height $u=5\,\text{mm}$, the image height $u'=2\,\text{mm}$,[2] and the stop radius of 12.7 mm; the free diameter of the entrance pupil of the optical system was thus one inch. The

[2] This image height corresponds to the size of a 1/4-in. CCD-chip. This chip features lateral dimensions of $3.2 \cdot 2.4\,\text{mm}^2$. Its size or image height in terms of optical system design is given by half the diameter of the resulting diagonal since the image and object height are always considered to be the maximum dimension, rotational-symmetric to the optical axis.

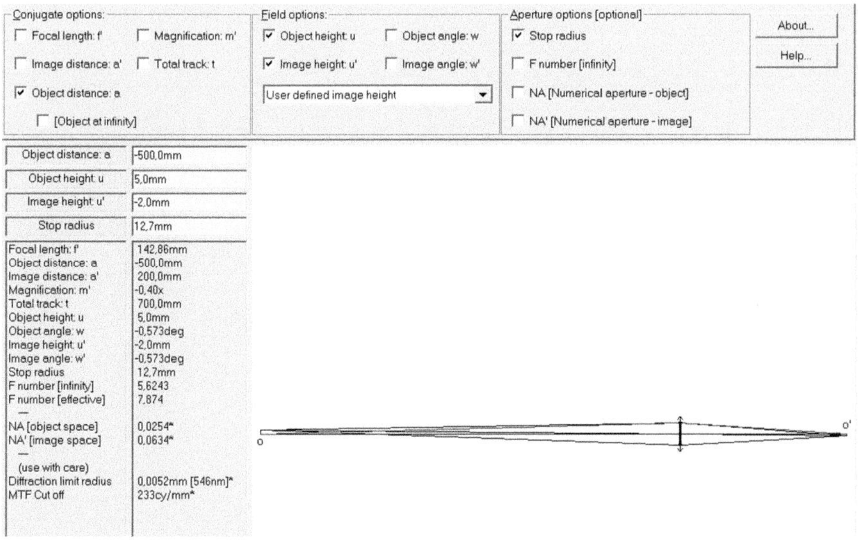

**FIGURE 5.2** Determination of conjugated parameters on the basis of a given imaging task via optical design software (here: software tool PreDesigner from Qioptiq Photonics GmbH & Co. KG).

required EFL for realizing this imaging task is then $EFL = 142.86\,\text{mm}$ according to Equation 5.1. Moreover, this value and the given object distance give the object angle, here $w = 0.573°$.

Computer-assisted optical system design further facilitates the choice of an appropriate optical system (i.e., an optical system that features the minimum optical aberrations for a given imaging task). This choice is based on theoretical calculations but also on experience and can be performed on the basis of the object angle $w$ and the F-number[3] of the optical component or system as suggested by the American optical engineer *Warren J. Smith* (1922–2008) (Smith, 2004) and shown in Figure 5.3.

For the imaging task presented above, where the object angle is $w = 0.573°$ and the F-number is 5.6243, an achromatic doublet is identified as an appropriate optical system for realizing the given imaging task with high imaging quality.

Once the required EFL and the type of optical system are determined theoretically, its design and specification have to be carried out. In the case of a single lens, this involves the determination of the radii of curvature $R_1$ and $R_2$, the center thickness $t_c$ and the index of refraction $n$ of the used optical material. These parameters can be calculated by the so-called lensmaker's equations (see also Section 4.2.1). For thin lenses, where the center thickness is negligible in comparison to the radii of curvature, the *lensmaker's equation* is given by

$$\frac{1}{EFL} = (n-1) \cdot \left( \frac{1}{R_1} - \frac{1}{R_2} \right). \tag{5.3}$$

[3] The F-number $N$ is given by the ratio of the $EFL$ of an optical system and the free diameter of its entrance pupil $D$ according to $N = EFL/D$.

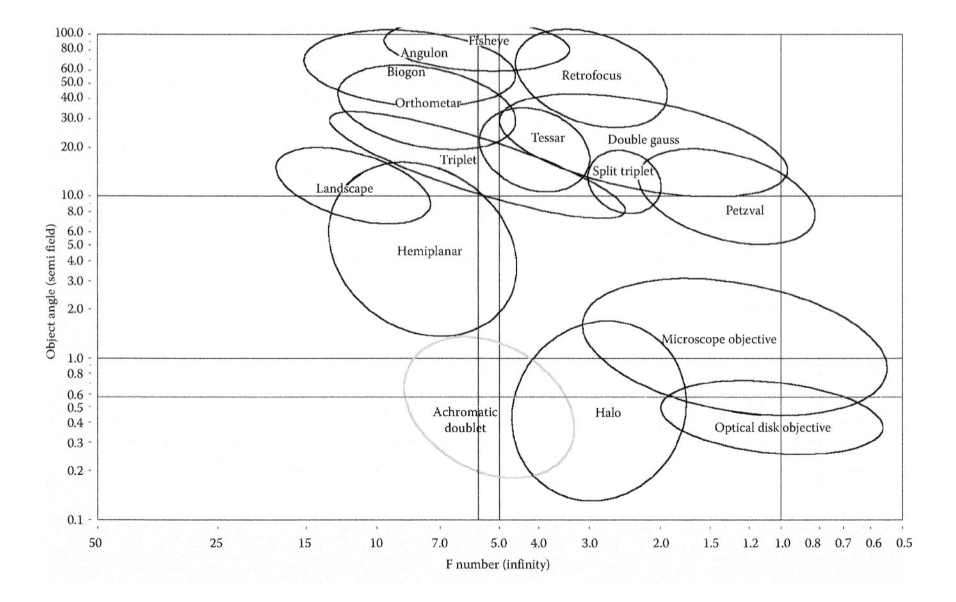


**FIGURE 5.3** Choice of an appropriate optical system for the imaging task given above (see Figure 5.2) on the basis of the object angle and the F-number via optical design software (Smith, 2004) (here: software tool PreDesigner from Qioptiq Photonics GmbH & Co. KG).

For a thick lens, the center thickness has to be taken into account according to

$$\frac{1}{EFL} = (n-1) \cdot \left( \frac{1}{R_1} - \frac{1}{R_2} \right) + \frac{(n-1)^2 \cdot t_c}{n \cdot R_1 \cdot R_2}. \tag{5.4}$$

However, as presented in more detail in Section 5.3, single lenses feature high optical aberrations. In practice, optical systems or lens groups and modules are thus applied for imaging tasks, making the calculation much more complex as described in detail in the literature (van Albada, 1955; Kingslake, 1983; Welford, 1986; Smith, 1990; Smith, 2004; Kingslake and Johnson, 2009). Hence, optical system design is now realized using appropriate design software, which allows easy and fast analysis and evaluation of the formation of optical aberrations (Harendt and Gerhard, 2008; Adams et al., 2013; Gerhard and Adams, 2015).

## 5.3 OPTICAL ABERRATIONS

### 5.3.1 Spherical Aberration

The disturbing effect of *spherical aberration* occurs due to the sphericity of classical optical components such as mirrors or lenses, where the surface shape is given by a spherical segment with the radius of curvature $R$. Hence, a particular angle of incidence results for each ray entrance height $h$ (a.k.a., height of incidence) with respect to the optical axis. According to Snell's law, a particular refraction angle consequently occurs for each light ray, depending on its entrance height.

The nominal back focal length (BFL) at a single optical interface can be calculated using the so-called vergence equation, given by

$$BFL = R \cdot \frac{n}{(n-1)}. \tag{5.5}$$

However, for high ray entrance heights $h$ on a spherically curved lens surface, this height has to be taken into account according to

$$BFL = R + \frac{h}{n \cdot \sin\left[\arcsin\left(\frac{h}{R}\right) - \arcsin\left(\frac{h}{n \cdot R}\right)\right]}. \tag{5.6}$$

As a consequence, a particular BFL is found for each ray entrance height, resulting in a difference in BFL, $\Delta BFL$, which is given by

$$\Delta BFL = BFL(h_{\min}) - BFL(h_{\max}). \tag{5.7}$$

This difference describes the variation of the focal point along the optical axis in propagation direction of light; it is referred to as *longitudinal spherical aberration.* When placing a detector at a fixed positon, usually in the Gaussian image plane,[4] this effect becomes visible by different image heights for light rays that enter the interface at different entrance heights, leading to a certain diameter of the image spot formed on the detector. This effect is known as *lateral spherical aberration*, which directly follows from longitudinal spherical aberration.

Spherical aberration of a single converging optical component can be reduced by the choice of its basic shape: plano-convex lenses feature high spherical aberration, whereas the use of so-called best form lenses (biconvex lenses with different radii of curvature) and aspherical lenses allows reducing spherical aberration. Another possibility is the use of lens groups, for example, lens doublets instead of single components. The total EFL, $EFL_{tot}$, of such a lens doublet can be calculated according to

$$\frac{1}{EFL_{tot}} = \frac{1}{EFL_1} + \frac{1}{EFL_2}. \tag{5.8}$$

Here, $EFL_1$ and $EFL_2$ are the EFLs of the two single lenses, which form the lens doublet.[5] As an example, such combination of two lenses with a particular EFL of

[4] The Gaussian image plane is given by the plane perpendicular to the optical axis at the position of the nominal focal point. This position follows from the so-called paraxial imaging model. This model is based on different assumptions, (1) merely small object and field angles $<5°$ are considered, (2) the imaging optical system is assumed to be a thin lens where the lens or system thickness can be neglected in comparison to its radii of curvature, and (3) the optical system is described by a single principal plane. In practice, this model is rarely valid, leading to a defocus of the image plane and an increase in lateral spherical aberration, respectively.

[5] One has to consider that Equation 5.8 applies for lens doublets without any spacing between the two single lenses.

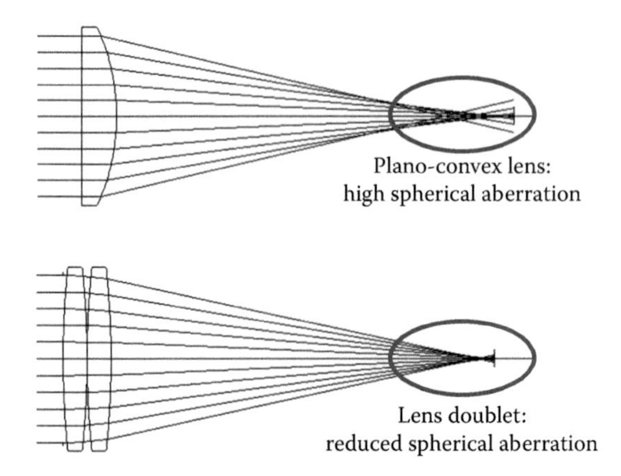


**FIGURE 5.4** Comparison of a single converging lens (top) and a converging lens doublet (bottom) with the same focal length as the single lens visualizing spherical aberration. (Figure was generated using the software WinLens3D Basic from Qioptiq Photonics GmbH & Co. KG.)

100 mm results in a doublet with a total EFL of 50 mm. In comparison to a single lens with the same EFL, the radii of curvature of the single lenses are approximately two times higher. As shown in Figure 5.4, spherical aberration is consequently reduced.

This reduction in spherical aberration is due to the fact that lower radii of curvature lead to a decrease in angle of incidence and angle of refraction, respectively, of incident light rays at high ray entrance heights on the particular optical interfaces.

### 5.3.2 Chromatic Aberration

*Chromatic aberration* occurs due to the dispersion characteristics of optical media as described in more detail in Section 3.2.3.2. According to Snell's law, a particular refraction angle results for each wavelength where the refraction angle for shorter wavelengths ("blue light") is higher than for longer wavelengths ("red light"). This effect gives rise to the formation of wavelength-dependent foci for incident white light or light with a broad wave band in general. Taking the wavelength-dependency of the index of refraction into account, Equation 5.6 can be rewritten as follows:

$$BFL(\lambda) = R + \frac{h}{n(\lambda) \cdot \sin\left[\arcsin\left(\frac{h}{R}\right) - \arcsin\left(\frac{h}{n(\lambda) \cdot R}\right)\right]}. \tag{5.9}$$

This rewritten equation shows that, in practice, the position of foci depends on both the ray entrance height $h$ and the wavelength $\lambda$. For a constant ray entrance height, the difference in effective BFL caused by dispersion is given by

$$\Delta BFL = BFL(\lambda_{\min}) - BFL(\lambda_{\max}). \tag{5.10}$$

This effect, the formation of different foci for each wavelength along the optical axis in propagation direction of light, is referred to as *longitudinal chromatic aberration*. It can also be expressed as a function of the nominal EFL in image space $EFL'$ and the particular Abbe number $V$ according to

$$\Delta EFL' = -\frac{EFL'}{V}. \tag{5.11}$$

This description reveals that the higher the Abbe number, the lower the amount of longitudinal chromatic aberration.

Longitudinal chromatic aberration further gives rise to chromatic variation of the wavelength-dependent image height. This effect is referred to as *lateral chromatic aberration*. It results from the chromatic variation of the exit pupil of an optical system due to a chromatic division of the chief ray, giving rise to the formation of color fringes around image structures.

Minimization of chromatic aberration for two selected wavelengths can be achieved by the use of *achromatic doublets*, usually consisting of a converging lens and a diverging lens that are cemented as shown in Figure 5.5.

For this purpose, the *condition for achromatism* has to be satisfied: the product of the EFL and the Abbe number $V$ of the converging lens must correspond to the

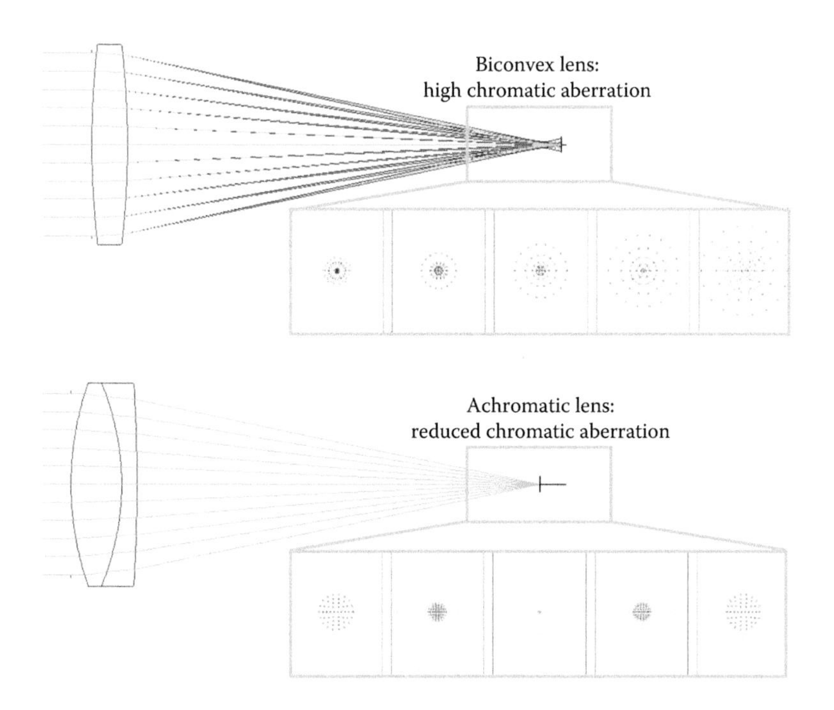


**FIGURE 5.5** Comparison of spot diagrams of a single converging lens (top) and a converging achromatic lens (bottom) with the same focal length as the single lens visualizing chromatic aberration. (Figure was generated using the software WinLens3D Basic from Qioptiq Photonics GmbH & Co. KG.)

product of the EFL and the Abbe number of the diverging lens, the latter given as a negative value according to

$$EFL_1 \cdot V_1 = -EFL_2 \cdot V_2. \tag{5.12}$$

The indices represent the converging lens (1) and the diverging lens (2), respectively. In practice, the converging lens is made of crown glass and features a high Abbe number (i.e., a low dispersion), whereas the diverging lens consists of flint glass with a low Abbe number (i.e., a high dispersion). The EFL of the diverging lens can be calculated when solving Equation 5.12,

$$EFL_2 = \frac{EFL_1 \cdot V_1}{V_2}. \tag{5.13}$$

Further, the EFL, $EFL_{A,}$ of an achromatic doublet can be calculated from the EFLs of both involved lenses according to

$$\frac{1}{EFL_A} = \frac{1}{EFL_1} + \frac{1}{EFL_2}. \tag{5.14}$$

Combining Equations 5.13 and 5.14 finally results in

$$\frac{1}{EFL_A} = \frac{1}{EFL_1} \cdot \frac{V_1 - V_2}{V_1}. \tag{5.15}$$

This interrelationship can be solved for the EFL of the converging lens,

$$EFL_1 = EFL_A \cdot \frac{V_1 - V_2}{V_1}, \tag{5.16}$$

or for the EFL of the diverging lens,

$$EFL_2 = EFL_A \cdot \frac{V_1 - V_2}{V_2}. \tag{5.17}$$

For a known or default value for the EFL of the achromatic doublet the EFLs of both lenses can thus be calculated on the basis of the Abbe numbers of the used optical materials (Gerhard and Adams, 2010; Gerhard et al., 2010).

The choice of Abbe numbers further specifies the two wavelengths where chromatic aberration is corrected. For example, when calculating the condition for achromatism, EFLs and radii of curvature of the two involved lenses on the basis of the Abbe number $V_e$, its denominator (the main dispersion $n_F' - n_C'$) represents the corrected wavelengths ($F' = 479.99$ nm and $C' = 643.85$ nm). The nominal EFL of such an achromatic lens then refers to the center wavelength given by the Abbe number $V_e$, i.e., $e = 546.07$ nm.

The general approach for the design of an achromatic doublet is to select an optical crown glass with a high Abbe number and low dispersion, respectively, for the converging lens and an optical flint glass with a low Abbe number and high dispersion, respectively, for the diverging lens. The larger the difference in Abbe number of both glasses, the longer the focal length of each lens. This leads to high radii of curvature of each particular lens surface and consequently a decrease in particular angles of incidence and refraction at the surfaces. Hence, both chromatic and spherical aberration is reduced.

After the correction of chromatic aberration for two wavelengths, a residual chromatic error, the so-called *secondary spectrum*, remains. This secondary spectrum is given by the chromatic aberration of the corrected wavelengths with respect to the center wavelength, which is used for the specification of the nominal focal length. The underlying reason for the formation of secondary spectra is the difference in bending of the dispersion curves of the used crown and flint glasses as expressed by the partial dispersion (see Section 3.2.3.2.4). The secondary spectrum becomes minimal, if the bending of the particular dispersion curves of both glasses is identical, which is not possible when using standard glasses. For the correction of this residual chromatic aberration, complex systems such as apochromatic lenses, consisting of at least three lenses, or the use of optical media with abnormal dispersion characteristics are required.

### 5.3.3 Coma

Up to now, only aberrations that occur for light rays propagating parallel to the optical axis have been discussed. In practice, inclined rays have to be considered in most imaging cases, resulting in a certain field of view and field angle, respectively. These result in the formation of further aberrations such as *coma*. Because of an inclined incidence of light rays on an optical interface, an asymmetric distribution of angles of incidence with respect to the center axis of the incident light bundle occurs; image points are shifted laterally and create a difference in image point positions. The resulting image has the appearance of a comet tail; it is referred to as *coma*. It can be overcome by Steinheil or aplanatic lenses or the appropriate choice of the position of the aperture stop where the incident light bundle is symmetric before and after refraction. This is the so-called Gleichen's stop position. The correction of coma is realized through the optical components and mechanical elements of an optical system.

### 5.3.4 Astigmatism and Petzval Field Curvature

Another aberration that occurs for inclined incidence of light rays on optical interfaces is *astigmatism*. Here, different cross sections in the meridional and the sagittal plane (perpendicular to each other) on the lens surface develop for an incident light bundle. Consequently, two different focal planes are formed where the so-called *circle of least confusion* is found between the focal planes. In this medial focus, a blurred image is formed. The focal planes are usually curved and feature different bendings for the meridional and the sagittal plane. This effect can be reduced by anastigmatic or apochromatic lenses.

After correction of astigmatism, another optical aberration, *Petzval field curvature*,[6] can emerge. This defect can be described as a special case of astigmatism since it is characterized by a rotational-symmetrically curved image plane with the same radius of curvature in the meridional and the sagittal plane; it can be minimized with Petzval lenses, protars, or curved detectors.

### 5.3.5 Distortion

Inclined incidence of light rays can further give rise to *distortion*, which usually occurs in aberration-corrected systems and depends on the position of the aperture stop. This aberration emerges if the magnification $\beta$ is not constant but decreases or increases with increasing image height. This results in a lateral deviation of an image point from its target position, which theoretically follows from the magnification valid for the paraxial case. As a consequence, a distorted image of an object's geometry is formed.

The distortion $D$ can be quantified on the basis of the actual image point coordinate $u_a'$ and the theoretical target image point coordinate $u_t'$ according to

$$D = u_a' - u_t'. \tag{5.18}$$

In optical system design, the percentage distortion $D_{per}$, given by

$$D_{per} = \frac{u_a' - u_t'}{u_t'} \cdot 100\%, \tag{5.19}$$

is usually used. Both the distortion $D$ and the percentage distortion $D_{per}$ indicate the type of distortion, as shown in Figure 5.6.

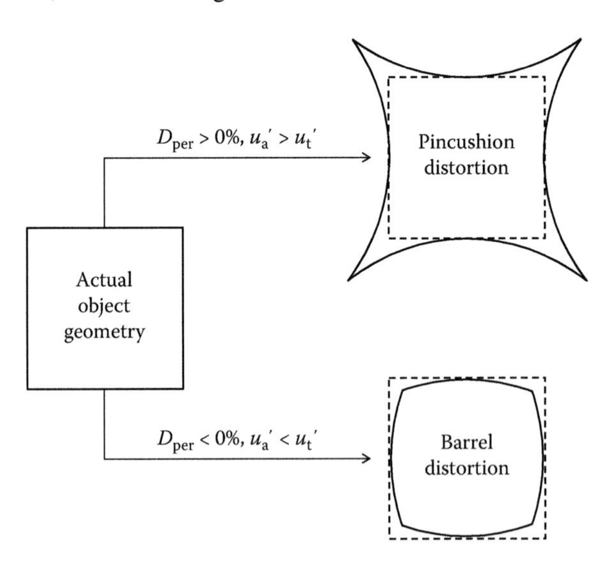


**FIGURE 5.6** Different types of distortion: a quadratic object can be transferred to either a pulvinated or a barrel-shaped image, depending on the amount of percentage distortion $D_{per}$ or the difference in actual image point coordinate $u_a'$ and target image point coordinate $u_t'$.

[6] Named after the German-Hungarian mathematician *Josef Maximilian Petzval* (1807–1891).

In a positive percentage distortion, $D_{per} > 0\%$, so-called *pincushion distortion* occurs, where a pulvinated image of a quadratic object is formed. Here, the actual image point coordinate is higher than the target image point coordinate, $u_a' > u_t'$. In contrast, a barrel-shaped image of a quadratic object is found for negative percentage distortion, $D_{per} < 0\%$, where $u_a' < u_t'$, which is referred to as *barrel distortion*.

Minimization of distortion can be realized by Steinheil aplanatic lenses and by the choice of an appropriate position of the aperture stop.

### 5.3.6 Ghost Images

Reflection of light at lens and detector surfaces and scattering at mounts or tubes can cause parasitic light, which propagates within the optical path of optical systems. Depending on the radii of curvature of the involved lens surfaces and the arrangement of the optical components, such parasitic light can get onto the detector and form so-called *ghost images*.[7] Ghost images are no classical aberration, but can be referred to as an optical defect of a system and needs to be considered in the course of optical system design and to be addressed by appropriate manufacturing steps. The description of ghost image formation is a complex task due to the high number of possible ghost image paths. As an example, an optical system with six lenses and 12 optical interfaces, respectively, features 132 possible ghost light paths for each light ray. Moreover, the image plane itself (i.e., the detector surface) can cause ghost light paths of high intensity due to the comparatively high reflection of some CCD-chips. For the evaluation of this effect, appropriate optical design software becomes crucial.

Ghost images can be classified into three different categories: first, ghost images where a final image is formed close to or even directly on the detector surface; second, flare where the final image is found at a sufficient distance from the detector surface. As a result of flare, light is spread and causes a general degradation of the image quality by the formation of blurred spots (a.k.a. orbs), secondary images or a reduction in contrast. Third, internal images are produced well away from the detector surface but are found within an optical element or at its surface. Such ghost images can even cause physical damage of optical components, for example, in the case of high-power laser optics.

Ghost images can be minimized by different actions such as dimming out incident light by the use of diaphragms, blackening of inner mount and tube surfaces, and applying antireflective coatings on lens surfaces as described in more detail in Chapter 11.

---

[7] A well-known example is the ghost image of the diaphragm of a camera, which can clearly be seen in some westerns where the camera lens is directly oriented toward the sun. This optical defect is even simulated in modern animated movies such as science-fiction productions since it has become very familiar. In the infancy of photography, the underlying mechanisms for the formation of ghost images were not well known or understood, and some ghost images were interpreted as images of real ghosts, which is the eponymous interpretation of this defect.

## 5.4 OPTIMIZATION

Once the required focal length and type of an optical component or system is determined, as described in Section 5.2, an optimization procedure is usually applied in order to minimize the optical aberrations just introduced. For this purpose, physical parameters are varied, and the impact of such variation on the resulting imaging quality is analyzed. Variable parameters are the radii of curvature, the center thickness, the material (i.e., the index of refraction and the Abbe number), and the apertures (i.e., the free diameters of optically active lens surfaces) for single components as well as air gaps between lenses in mounted optical systems. Moreover, defects of interest are defined and monitored where the choice of defects depends on the final functionality and application of an optical component or system. The most important defects are optical aberrations, which can be quantified by particular *Seidel sums* as follows: each optical interface (e.g., a lens surface) has a so-called *Seidel coefficient* $A$, named after the German mathematician and optician *Philipp Ludwig von Seidel* (1821–1896). This coefficient is calculated for a number of incident light rays on the basis of the object/aperture angle $w$ or image angle $w'$, the ray entrance height $h$, the involved index of refraction behind an optical interface $n$, and the radius of curvature $R$ of the interface. It is given by

$$A = n \cdot \left( h \cdot \frac{1}{R} + w \right). \tag{5.20}$$

The sum of all Seidel coefficients is referred to as Seidel sum, which allows the quantitative determination of aberrations. Low Seidel sums ($\ll 1$) represent low aberrations, where each optical aberration can be expressed by its particular Seidel sum. For spherical aberration, the Seidel sum $S_{\mathrm{I}}$ follows from

$$S_{\mathrm{I}} = -\sum_{i=1}^{k} A_i^2 \cdot h_i \cdot \left( \frac{w_{i+1}}{n_{i+1}} - \frac{w_i}{n_i} \right). \tag{5.21}$$

Coma is quantified by the Seidel sum $S_{\mathrm{II}}$ according to

$$S_{\mathrm{II}} = -\sum_{i=1}^{k} A_i \cdot \bar{A}_i \cdot h_i \cdot \left( \frac{w_{i+1}}{n_{i+1}} - \frac{w_i}{n_i} \right). \tag{5.22}$$

Here, the overbarred parameter $\bar{A}_i$ is considered in addition to $A_i$, consequently taking different construction rays into account.[8] $A_i$ refers to the chief ray, and $\bar{A}_i$ refers to the marginal ray; it is thus given by

---

[8] Construction rays are used in geometrical optics for the graphical determination of conjugated parameters. Starting at a specific object point, the chief ray passes through an optical component or system at its principal point (i.e., the intersection of the principal plane and the optical axis), whereas marginal rays represent the outer boundary of a bundle of light and are thus the parallel ray and the focal ray. The first propagates parallel to the optical axis in the object space and crosses the focal point in the image space. The second crosses the focal point in the object space and propagates parallel to the optical axis in the image space. Based on the construction of these rays, an object point can be transferred to a corresponding image point graphically, where the image point is given by the intersection of all construction rays, compare Figure 5.1.

$$\bar{A}_i = n_i \cdot \left( \bar{h}_i \cdot \frac{1}{R_i} + \bar{w}_i \right). \tag{5.23}$$

The third Seidel sum,

$$S_{\mathrm{III}} = -\sum_{i=1}^{k} \bar{A}_i^2 \cdot h_i \cdot \left( \frac{w_{i+1}}{n_{i+1}} - \frac{w_i}{n_i} \right), \tag{5.24}$$

represents and quantifies astigmatism and Petzval field curvature is covered by the fourth Seidel sum, given by

$$S_{\mathrm{IV}} = -\sum_{i=1}^{k} H_i^2 \cdot \frac{1}{R_i} \cdot \left( \frac{1}{n_{i+1}} - \frac{1}{n_i} \right). \tag{5.25}$$

Here, the parameter $H_i$ follows from

$$H_i = n_i \cdot \left( w_i \cdot \bar{h}_i - \bar{w}_i \cdot h_i \right), \tag{5.26}$$

where the overbarred parameters refer to the marginal ray (compare Seidel sum $S_{\mathrm{II}}$). Finally, distortion can be quantified on the basis of Seidel sums $S_{\mathrm{III}}$ and $S_{\mathrm{IV}}$ according to

$$S_{\mathrm{V}} = -\sum_{i=1}^{k} \frac{\bar{A}_i}{A_i} \cdot \left[ \left( S_{\mathrm{III}} \right)_i + \left( S_{\mathrm{IV}} \right)_i \right]. \tag{5.27}$$

It turns out that there is no actual Seidel sum for chromatic aberration, but this aberration can be quantified on the basis of Seidel Sum $S_{\mathrm{I}}$ for spherical aberration by taking the different wavelengths of the considered waveband into account.

Optical aberrations are not the only defects that can be considered for optimization. Other defects such as conjugated parameters, field parameters, aperture parameters, paraxial parameters, and ray data might be of interest for specific applications and thus be chosen in the course of an optimization process.

For the optimization of optical components or systems, a so-called *merit function* (*MF*) is defined on the basis of the defects $\Delta d$ of interest. A defect is generally given by the subtraction of an actual defect value $d_{\mathrm{a}}$ and the target defect value $d_{\mathrm{t}}$,

$$\Delta d = d_{\mathrm{a}} - d_{\mathrm{t}}. \tag{5.28}$$

Usually, the *MF* is set up applying weighted defects, which are defined by introducing a quantifier $Q$, resulting in

$$\Delta \bar{d} = Q \cdot (d_{\mathrm{a}} - d_{\mathrm{t}}). \tag{5.29}$$

However, due to the fact that defects of interest may have different units, relative defects $d_{\mathrm{rel}}$ are defined and used in practice according to

$$d_{\mathrm{rel}} = \left| \frac{d_{\mathrm{a}} - d_{\mathrm{t}}}{d_{\mathrm{tol}}} \right|. \tag{5.30}$$

Here, $d_{tol}$ is the acceptable fault tolerance, given by the predefined upper limit $d_{max}$ and the lower limit $d_{min}$ for a defect,

$$d_{tol} = \frac{d_{max} - d_{min}}{2}. \tag{5.31}$$

This value is specified for a particular defect and thus replaces the quantifier. As a result, $d_{rel}$ is a dimensionless quantity and is defined to be corrected if $d_{rel} \leq 1$. When using relative defects, a *MF* can finally be set up for a number (expressed by the additional index *i*) of different defects:

$$MF = \sum_i d_{i,rel}^2 = \sum_i \left( \frac{d_{i,a} - d_{i,t}}{d_{i,tol}} \right)^2. \tag{5.32}$$

The goal of an optimization process is then to minimize this *MF*. For this task, optical system design software is now used, where a vector matrix is defined for the calculation of the impact of variable parameters (e.g., radii of curvature, center thicknesses, etc.) on the considered defects (Schuhmann, 2005). There are three different types of *MF*s, the total *MF*, the working *MF*, and the concern *MF*. When applying a total *MF*, all possible relative defects are taken into account, whereas merely all activated relative defects are considered by a working *MF*. Finally, using a working *MF* and excluding all relative defects smaller than 1 gives the concern *MF*. This allows for defining the condition $MF = 0$ as stop criterion for calculation.

Figure 5.7 shows an example of the result of an optimization process, where the defect of interest was transverse ray aberration (TRA), and optimization was carried out employing a concern *MF*. After fulfilling the stop criterion $MF = 0$, the optimization simulation was stopped, and the result was displayed in the form of a TRA diagram.

As already mentioned, optical aberrations can be expressed by Seidel sums. Any optical component or system can thus be evaluated and optimized by defining the

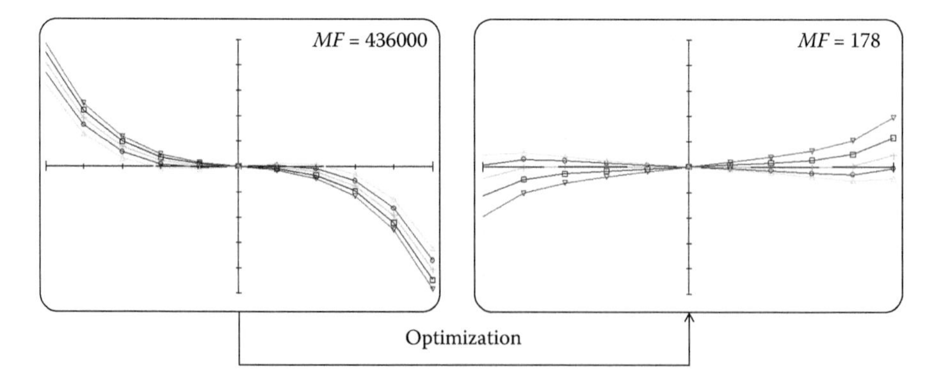


**FIGURE 5.7** Transverse ray aberration (TRA) diagram of an optical system for five wavelengths, including the particular Merit function (*MF*) value before (left) and after optimization (right) using optical design software (here: software WinLens3D from Qioptiq Photonics GmbH & Co. KG).

particularly relevant Seidel sums as target defect values, as shown by the following example:

An optical system with an object angle of $w = 31° = 0.54$ rad and an image angle of $w' = 20° = 0.35$ rad is applied for imaging an object on a CCD-chip. The pixels on this chip are quadratic and feature a lateral length of 14 µm; the pixel size is thus 196 µm². In order to achieve high resolution and imaging quality, each pixel should be illuminated separately by an image point. The spot diameter $D_{spot}$ (see Figure 5.8) of focused light arriving at the detector surface should thus correspond to the lateral length of one pixel. Generally, the spot diameter can be calculated on the basis of the Seidel sum for spherical aberration $S_I$ of the used optical system according to

$$D_{spot} = \frac{S_I}{n \cdot w'}. \tag{5.33}$$

Here, $n$ is index of refraction of the ambient medium in object space ($n = 1$ for air[9]), and $w'$ is the image angle.[10] The maximum Seidel sum for the present case then amounts to 0.0049 according to

$$S_I = D_{spot} \cdot n \cdot w'. \tag{5.34}$$

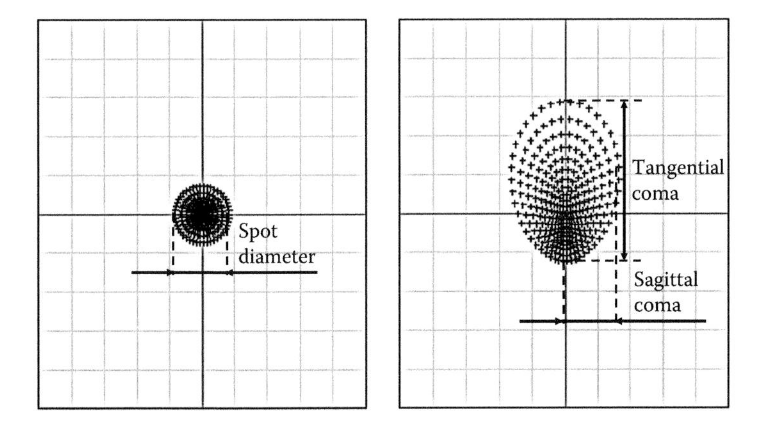


**FIGURE 5.8** Spot diagram of an optical system visualizing the impact of inclined light on the spot geometry and size. A circular image spot is found on the detector for an object angle of 0° (left) whereas coma occurs for inclined incidence of light (right, the object angle in this example is 5°), leading to a deformation and enlargement of the image spot. (Figure was generated using the software WinLens3D Basic from Qioptiq Photonics GmbH & Co. KG.)

[9] Usually, the value 1 can be used for the index of refraction of air for simple calculations or estimations. However, the temperature, air pressure, and humidity should be considered for precise calculation as suggested by the Swedish astrophysicist *Bengt Edlén* (1906–1993) (Edlén, 1966), see also Section 3.2.3.1.

[10] Note that for the calculation of Seidel sums, angles are entered in radians ($1° = \pi/180$ rad $\approx 0.018$ rad), and lateral dimensions such as the spot diameter are entered in millimeters. However, Seidel sums are unitless.

The optical system should consequently feature this maximum Seidel sum $S_I$ in order to ensure that incident focused light can be detected by one single pixel.

Another aberration that should be considered is coma, since due to the given object angle of 31° and the accompanying inclination of incident light, image spots are deformed, leading to an increase in lateral spot dimensions on the detector surface as shown in Figure 5.8.

It turns out that coma can be described by a horizontal and a vertical component, referred to as sagittal and tangential coma. Sagittal coma is given by

$$\delta_{sag} = \frac{1}{2} \cdot \frac{S_{II}}{n \cdot w}, \tag{5.35}$$

whereas tangential coma follows from

$$\delta_{tan} = \frac{3}{2} \cdot \frac{S_{II}}{n \cdot w}. \tag{5.36}$$

In the present case, tangential coma should be considered, since it represents the maximum lateral dimension of the aberrated image point. Defining this maximum dimension to correspond to the pixel size, the maximum acceptable Seidel sum for coma $S_{II}$ can be determined after solving Equation 5.36 for $S_{II}$,

$$S_{II} = \frac{2}{3} \cdot \delta_{tan} \cdot n \cdot w. \tag{5.37}$$

$S_{II}$ consequently amounts to 0.00504. Both Seidel sums, the one for spherical aberration ($S_I$=0.0049) and the one for tangential coma ($S_{II}$=0.00504), then represent the target defect values for setting up the *MF* applied for optimization as described above.

This approach can be applied to other aberrations in order to determine the particular Seidel sum as the target defect. For instance, the longitudinal shift of the image plane with respect to the Gaussian image plane,[11] commonly referred to as defocus $\varphi$, is also dependent on the Seidel sum $S_I$ for spherical aberration according to

$$\varphi = \frac{3}{8} \cdot \frac{S_I}{n \cdot w^2}. \tag{5.38}$$

For a predefined limit of defocus, the corresponding Seidel sum can thus be determined on the basis of Equation 5.38. Moreover, the maximum distance between the sagittal and the meridional focal plane $\varphi_{max}$, which occurs in the case of astigmatism, can be specified by defining the Seidel sum $S_{III}$ for astigmatism, where

$$\varphi_{max} = -\frac{S_{III}}{n \cdot w^2}. \tag{5.39}$$

[11] The Gaussian image plane is the image plane found for small aperture and field angles <5° (i.e., the paraxial imaging case).

Seidel sum $S_{IV}$ for Petzval field curvature $R_{Petzval}$ (i.e., the rotational-symmetric curvature of image planes) can be derived from

$$R_{Petzval} = \frac{H^2}{n \cdot S_{IV}}, \tag{5.40}$$

where the parameter $H$ is given by

$$H = n \cdot \left(w \cdot \bar{h} - \bar{u} \cdot h\right), \tag{5.41}$$

see above. Finally, the target defect for maximum distortion $D_{max}$ can be expressed by Seidel sum $S_V$, where

$$D_{max} = \frac{S_V}{2 \cdot n \cdot w}. \tag{5.42}$$

The result of any optimization process is the final definition of an optical component or system in terms of the radii of curvature, center thickness or thicknesses, material or materials, and—as the case may be—air gaps. However, the impact of slight variations of these parameters is again analyzed in a further step in order to determine the required manufacturing tolerances.

## 5.5 DETERMINATION OF MANUFACTURING TOLERANCES

For the determination of manufacturing tolerances via optical system design, the impact of material defects, manufacturing errors and assembly errors on one or more optical aberrations of interest is examined. The choice of the considered aberrations is based on the final application and use of the analyzed optical component or system. For example, spherical aberration may play an important role for a high-aperture laser objective, where the goal is to obtain small focus diameters, but chromatic aberration can be neglected due to the monochromatic nature of laser irradiation.

Generally, a large number of defects can influence the formation of aberrations and thus cause poor imaging quality of an optical component or system. The main defects that should be considered for the determination of manufacturing tolerances are classified into material defects, surface errors, and position errors. Material defects are for example variations in index of refraction and Abbe number (see Section 6.2), stress birefringence (see Section 6.2.1), bubbles and inclusions (see Section 6.2.2), and inhomogeneity in index of refraction as well as striae (see Section 6.2.3). Substantial surface errors are contour inaccuracies (see Section 6.3.1), surface damage such as scratches and digs (see Section 6.3.3), and high surface roughness (see Section 6.3.4). Finally, position errors of single optical components or cemented lens groups shall be considered. These errors are, for example, the centering error of a lens (see Section 6.3.2), tilt or lateral offset from the optical axis and distance errors between components of an optical system (see Section 6.5), cement wedges, and decenter of cemented components (see Section 9.3).

### 5.5.1 Sensitivity Analysis

For the determination of manufacturing tolerances, a so-called *sensitivity analysis* is performed as a first step in which critical surfaces are identified. This can be realized by the examination of so-called *Seidel bars*, which visualize the particular Seidel sums of the aberrations of interest for each surface as shown in Figure 5.9.

Per definition, critical surfaces feature high Seidel sums and high aberration, respectively. The manufacturing tolerances for such surfaces should thus be tighter than for less critical surfaces. In the example shown in Figure 5.9, where sensitivity analysis was performed relating to spherical aberration, the third lens surface turns out to be a critical surface, whereas the fifth lens surface does not significantly contribute to the formation of spherical aberration. Consequently, different manufacturing tolerances may be defined for each surface of an optical system.

### 5.5.2 Monte Carlo Simulation

The determination of manufacturing tolerances via computer-assisted optical system design is performed by so-called *Monte Carlo simulations*.[12] Here, the input parameters are the abovementioned manufacturing and position defects and errors, and the

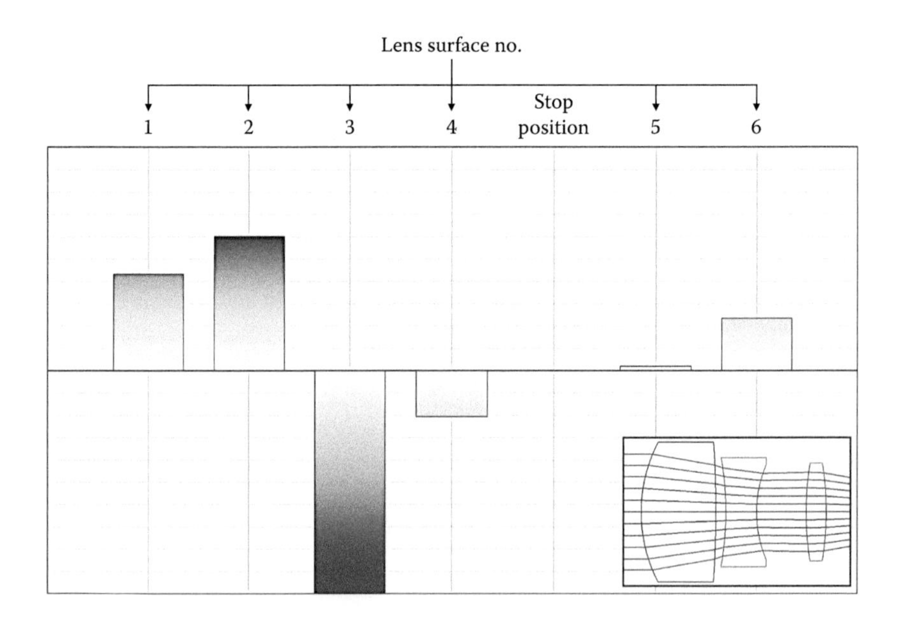


**FIGURE 5.9** Seidel bar chart of a lens triplet (inset) displaying the sensitivity of the involved surfaces for spherical aberration as expressed by the Seidel sums. (Figure was generated using the software WinLens3D Basic from Qioptiq Photonics GmbH & Co. KG.)

[12] Monte Carlo simulations are based on the repetition of random sampling via appropriate computational algorithms. This approach allows obtaining numerical results for mathematical tasks, where the result depends on a large number of input variables.

target value is low aberration, normally a minimum Seidel sum of the aberration of interest. For this purpose, the input parameters as well as lower and upper limits of these parameters are defined as listed in Table 5.2.

Once these lower and upper limits for the relevant defects are defined, the actual Monte Carlo simulation is carried out in order to analyze the impact of manufacturing tolerances on the imaging quality of the optical component or system as monitored by the chosen aberrations. This can be performed on the basis of the resulting Seidel sums or appropriate evaluation graphs as shown in Figure 5.10.

**TABLE 5.2**
**Defects of Material, Surfaces, and Position of Optical Components and Systems Considered for the Determination of Manufacturing Tolerances Including the Particular Tolerancing**

| Defect (Input Parameter) | Tolerancing (Lower and Upper Limit) |
|---|---|
| **Material** | |
| Variation in index of refraction and Abbe number | Definition of total tolerance values or percentage deviation from nominal value |
| Stress birefringence | Definition of maximum and minimum birefringence (by index number 0 in DIN ISO 10110) |
| Bubbles and inclusions | Definition of maximum and minimum number and size of bubbles and inclusions (by index number 1 in DIN ISO 10110) |
| Homogeneity in index of refraction and striae | Definition of maximum deviation in index of refraction and share of striae (by index number 2 in DIN ISO 10110) |
| **Surface** | |
| Contour accuracy | Definition of maximum deviation of radius of curvature from target value and its sphericity (by index number 3 in DIN ISO 10110) |
| Surface cleanliness | Definition of maximum number and size of surface damages (by index number 5 in DIN ISO 10110) |
| Surface roughness | Definition of maximum surface roughness by grade of polishing (P1 to P4) |
| **Position** | |
| Centering | Definition of maximum deviation of optical axis from mechanical axis (by index number 4 in DIN ISO 10110) |
| Tilt, lateral offset from optical axis, and distance error between components of an optical system | Definition of acceptable absolute deviation in angle or position |
| Cement wedge | Definition of acceptable absolute deviation in angle |
| Decenter of cemented components | Definition of acceptable absolute deviation in position (lateral offset) |

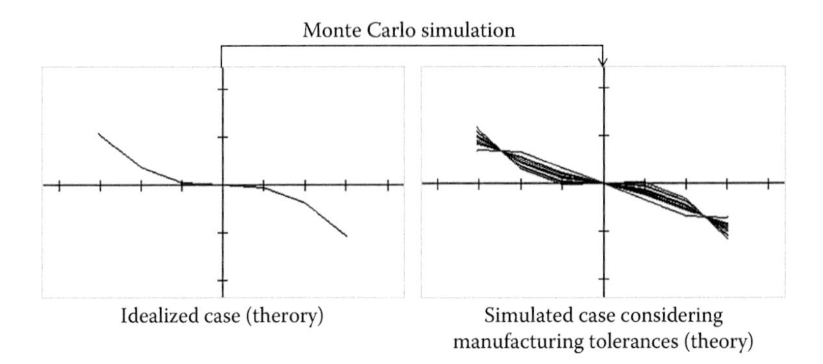


**FIGURE 5.10** TRA diagram of an optical system for one wavelength before (left) and after (right) Monte Carlo simulation of the impact of defect tolerances on the transverse ray aberration (TRA). It turns out that the predefined defect tolerances give rise to a broadening of the idealized curve. (Figure was generated using the software Tolerancer from Qioptiq Photonics GmbH & Co. KG.)

Here, the aberration of interest was transverse ray aberration (TRA); the result of the Monte Carlo simulation was thus displayed by the TRA graphs before and after simulation. It turns out that applying the predefined tolerances leads to a certain broadening of the theoretical TRA curve, which is displayed by the idealized case in Figure 5.10 (left). This case is calculated for fix lens parameters such as radii of curvature, center thickness, etc. In contrast, the predefined variations of these parameters were considered in the course of the Monte Carlo simulation, finally resulting in the broadening of the idealized curve as shown in Figure 5.10 (right).

In the example above, the broadening is quite moderate. In case of extensive broadening, the defined lower and upper tolerance limits may be readjusted in order to achieve low aberration and high image quality, respectively. This is shown by the comparison of the impact of different tolerance classes, listed in Table 5.3, on TRA in Figure 5.11 as determined by Monte Carlo simulation.

**TABLE 5.3**
**Tolerance Classes and Particular Tolerance Values for the Radii of Curvature, Center Thicknesses, Indices of Refraction, and Abbe Numbers as Applied for the Evaluation of the Impact of Manufacturing Tolerances on the TRA as Shown in Figure 5.11**

| | Tolerance Class and Particular Tolerance Value | | |
|---|---|---|---|
| **Parameter** | **Standard** | **Precision** | **Extra Precision** |
| Radius of curvature | ±10 fringes | ±5 fringes | ±1 fringe |
| Center thickness | ±250 μm | ±100 μm | ±50 μm |
| Index of refraction | ±0.001 | ±0.0005 | ±0.0002 |
| Abbe number | ±0.5 | ±0.2 | ±0.1 |

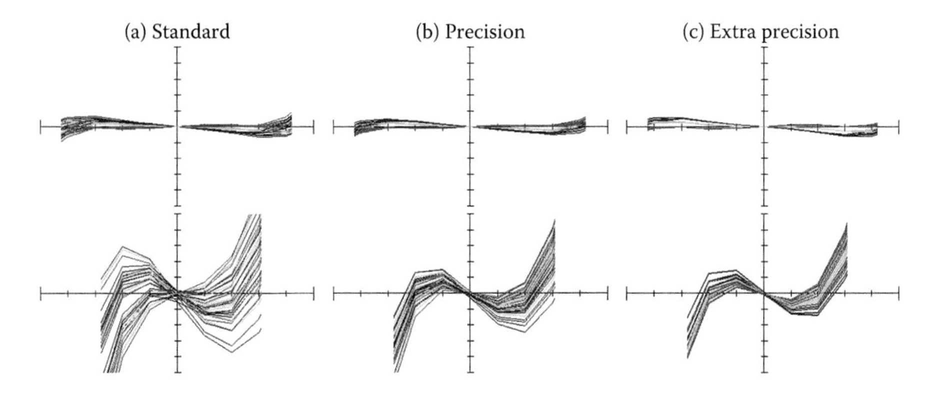


**FIGURE 5.11** Comparison of the impact of different manufacturing tolerance classes (standard, precision, and extra precision; see Table 5.3) on transverse ray aberration (TRA) for on-axis light rays (top) and the maximum field or aperture angle, respectively (bottom). (Figure was generated using the software Tolerancer from Qioptiq Photonics GmbH & Co. KG.)

Such Monte Carlo simulation is thus an iterative process, which is carried out until the residual aberrations of interest are within an acceptable range where the applied tolerance classes are successively refined. Finally, it provides the actually required tolerances and defines the production accuracy for the manufacture of an optical component and its glass. Hence, the result of the design of optical components is the basis of any manufacturing drawing, as shown by the example for a simple biconvex lens in Figure 5.12.

The specifications and tolerances shown here are compliant to DIN ISO 10110, entitled "Optics and photonics—Preparation of drawings for optical elements and systems," as explained in detail in Chapter 6.

## 5.6 SUMMARY

For the design of optical components or systems, unknown conjugated parameters are first determined on the basis of the given parameters of an imaging task. This basic consideration allows for identifying the required focal length and choosing the appropriate type of optical component or system. Second, the radii of curvature and center thickness are calculated, and the optical material is selected by means of its index of refraction and dispersion characteristics. As a result of these two steps, a so-called *start system* is defined. This system is subsequently evaluated and optimized regarding its optical aberrations via computer-assisted optical system design. Here, the impact of material defects, surface errors, and position errors on aberrations of interest, depending on the final use and application of the optical system, is investigated. This is realized by Monte Carlo simulations, where the lower and upper limits of defects are defined and readjusted iteratively until the residual aberrations are within specification. This procedure finally gives the manufacturing tolerances for glass production, optics manufacturing, and assembly of optomechanical systems as specified by manufacturing drawings.

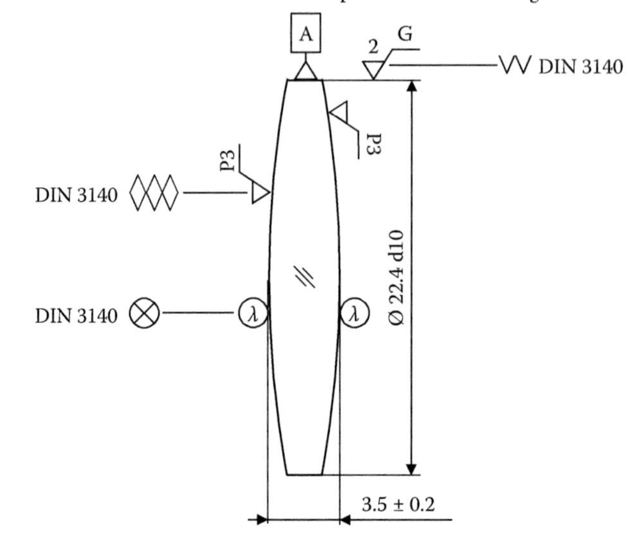


| fit size | max. size | min. size |
|---|---|---|
| Ø 22.4 d10 | −0.065 | −0.149 |

$f'$ = 101.695 mm

| Left surface | Material | Right surface |
|---|---|---|
| $R_c$: 104.41 CX ± 0.455%<br>Chamfers: 0.2 – 0.4<br>Coating: $R$ < 0.5% (VIS)<br>3/6(1)<br>4/10′<br>5/3 · 0.25 | Glass: N-BK7 (Schott)<br>$n_d$: 1.5168 ± 0.001<br>$V_d$: 64.17 ± 0.5%<br>0/20<br>1/3 · 0.25<br>2/0.3 | $R_c$: 104.41 CX ± 0.455%<br>Chamfers: 0.2 – 0.4<br>Coating: $R$ < 0.5% (VIS)<br>3/6(1)<br>4/10′<br>5/3 · 0.25 |

**FIGURE 5.12** Example of a manufacturing drawing for a biconvex lens according to DIN ISO 10110 including all relevant data and necessary specifications and tolerances as determined in the course of optical design.

## 5.7 FORMULARY AND MAIN SYMBOLS AND ABBREVIATIONS

**Imaging equation:**

$$\frac{1}{EFL} = \frac{1}{a} + \frac{1}{a'}$$

$a$ object distance
$a'$ image distance
$EFL$ effective focal length

**Magnification $\beta$ of an optical component or system:**

$$\beta = \frac{u'}{u} = \frac{a'}{a}$$

$u'$ image height
$u$ object height
$a'$ image distance
$a$ object distance

**Lensmaker's equation of a thin lens:**

$$\frac{1}{EFL} = (n-1) \cdot \left( \frac{1}{R_1} - \frac{1}{R_2} \right)$$

$EFL$ effective focal length
$n$ index of refraction of the lens material
$R_1$ radius of curvature of the first lens surface
$R_2$ radius of curvature of the second lens surface

**Lensmaker's equation of a thick lens:**

$$\frac{1}{EFL} = (n-1) \cdot \left( \frac{1}{R_1} - \frac{1}{R_2} \right) + \frac{(n-1)^2 \cdot t_c}{n \cdot R_1 \cdot R_2}$$

$EFL$ effective focal length
$n$ index of refraction of the lens material
$R_1$ radius of curvature of the first lens surface
$R_2$ radius of curvature of the second lens surface
$t_c$ lens center thickness

**Vergence equation:**

$$BFL = R \cdot \frac{n}{(n-1)}$$

$BFL$ back focal length
$R$ radius of curvature of the optical interface
$n$ index of refraction behind the optical interface

**Vergence equation for high ray entrance heights:**

$$BFL = R + \frac{h}{n \cdot \sin\left[ \arcsin\left( \frac{h}{R} \right) - \arcsin\left( \frac{h}{n \cdot R} \right) \right]}$$

$BFL$ back focal length
$h$ ray entrance height
$R$ radius of curvature of the optical interface
$n$ index of refraction behind the optical interface

**Longitudinal spherical aberration:**

$$\Delta BFL = BFL(h_{min}) - BFL(h_{max})$$

$\Delta BFL$ difference in BFL
$BFL(h_{max})$ back focal length for the maximum ray entrance height
$BFL(h_{min})$ back focal length for the minimum ray entrance height

**Total EFL $EFL_{tot}$ of a lens doublet without spacing:**

$$\frac{1}{EFL_{tot}} = \frac{1}{EFL_1} + \frac{1}{EFL_2}$$

$EFL_1$ effective focal length of the first lens
$EFL_2$ effective focal length of the second lens

**Longitudinal chromatic aberration:**

$$\Delta BFL = BFL(\lambda_{min}) - BFL(\lambda_{max})$$

$\Delta BFL$ difference in BFL
$BFL(\lambda_{max})$ back focal length for the maximum wavelength
$BFL(\lambda_{min})$ back focal length for the minimum wavelength

**Condition for achromatism:**

$$EFL_1 \cdot V_1 = -EFL_2 \cdot V_2$$

$EFL_1$ effective focal length of the converging lens
$V_1$ Abbe number of the converging lens
$EFL_2$ effective focal length of the diverging lens
$V_2$ Abbe number of the diverging lens

**Distortion $D$:**

$$D = u'_a - u'_t$$

$u'_a$ actual image point coordinate
$u'_t$ theoretical target image point coordinate

**Percentage distortion $D_{per}$:**

$$D_{per} = \frac{u'_a - u'_t}{u'_t} \cdot 100\%$$

$u'_a$ actual image point coordinate
$u'_t$ theoretical target image point coordinate

**Seidel coefficient $A$ of an optical interface:**

$$A = n \cdot \left( h \cdot \frac{1}{R} + w \right)$$

$n$ index of refraction before interface
$h$ ray entrance height
$R$ radius of curvature of interface
$w$ object/aperture angle

**Seidel sum $S_I$ (for spherical aberration):**

$$S_I = -\sum_{i=1}^{k} A_i^2 \cdot h_i \cdot \left( \frac{w_{i+1}}{n_{i+1}} - \frac{w_i}{n_i} \right)$$

$A$ Seidel coefficient
$h$ ray entrance height
$w$ object/aperture angle
$n$ index of refraction

**Seidel sum $S_{II}$ (for coma):**

$$S_{II} = -\sum_{i=1}^{k} A_i \cdot \overline{A}_i \cdot h_i \cdot \left( \frac{w_{i+1}}{n_{i+1}} - \frac{w_i}{n_i} \right)$$

with

$$\overline{A}_i = n_i \cdot \left( \overline{h}_i \cdot \frac{1}{R_i} + \overline{w}_i \right)$$

$A$ Seidel coefficient
$h$ ray entrance height
$w$ object/aperture angle
$n$ index of refraction
*Note:* Overbarred parameters refer to the marginal ray.

**Seidel sum $S_{III}$ (for astigmatism):**

$$S_{III} = -\sum_{i=1}^{k} \overline{A}_i^2 \cdot h_i \cdot \left( \frac{w_{i+1}}{n_{i+1}} - \frac{w_i}{n_i} \right)$$

$A$ Seidel coefficient
$h$ ray entrance height
$w$ object/aperture angle
$n$ index of refraction

**Seidel sum $S_{IV}$ (for Petzval field curvature):**

$$S_{IV} = -\sum_{i=1}^{k} H_i^2 \cdot \frac{1}{R_i} \cdot \left( \frac{1}{n_{i+1}} - \frac{1}{n_i} \right)$$

with

$$H_i = n_i \cdot \left( w_i \cdot \bar{h}_i - \bar{w}_i \cdot h_i \right)$$

$R$ radius of curvature
$n$ index of refraction
$w$ object/aperture angle
$h$ ray entrance height
*Note:* Overbarred parameters refer to the marginal ray.

**Seidel sum $S_V$ (for distortion):**

$$S_V = -\sum_{i=1}^{k} \frac{\bar{A}_i}{A_i} \cdot \left[ (S_{III})_i + (S_{IV})_i \right]$$

$A$ Seidel coefficient
$S_{III}$ Seidel sum for astigmatism
$S_{IV}$ Seidel sum for Petzval field curvature
*Note:* Overbarred parameters refer to the marginal ray.

**Defect $\Delta d$ (absolute):**

$$\Delta d = d_a - d_t$$

$d_a$ actual defect value
$d_t$ target defect value

**Defect $d_{rel}$ (relative):**

$$d_{rel} = \left| \frac{d_a - d_t}{d_{tol}} \right|$$

$d_a$ actual defect value
$d_t$ target defect value
$d_{tol}$ acceptable fault tolerance

**Acceptable fault tolerance $d_{tol}$:**

$$d_{tol} = \frac{d_{max} - d_{min}}{2}$$

$d_{max}$ upper defect limit
$d_{min}$ lower defect limit

**Merit function *MF*:**

$$MF = \sum_i d_{i,\mathrm{rel}}^2 = \sum_i \left( \frac{d_{i,\mathrm{a}} - d_{i,\mathrm{t}}}{d_{i,\mathrm{tol}}} \right)^2$$

$d_{\mathrm{rel}}$ relative defect
$d_{\mathrm{a}}$ actual defect value
$d_{\mathrm{t}}$ target defect value
$d_{\mathrm{tol}}$ acceptable fault tolerance
*Note:* The additional index $i$ quantifies the number of different considered defects.

**Spot diameter $D_{\mathrm{spot}}$:**

$$D_{\mathrm{spot}} = \frac{S_{\mathrm{I}}}{n \cdot w'}$$

$S_{\mathrm{I}}$ Seidel sum for spherical aberration
$n$ index of refraction
$w'$ image angle

**Sagittal coma $\delta_{\mathrm{sag}}$:**

$$\delta_{\mathrm{sag}} = \frac{1}{2} \cdot \frac{S_{\mathrm{II}}}{n \cdot w}$$

$S_{\mathrm{II}}$ Seidel sum for coma
$n$ index of refraction
$w$ object/aperture angle

**Tangential coma $\delta_{\mathrm{tan}}$:**

$$\delta_{\mathrm{tan}} = \frac{3}{2} \cdot \frac{S_{\mathrm{II}}}{n \cdot w}$$

$S_{\mathrm{II}}$ Seidel sum for coma
$n$ index of refraction
$w$ object/aperture angle

**Defocus $\varphi$:**

$$\varphi = \frac{3}{8} \cdot \frac{S_{\mathrm{I}}}{n \cdot w^2}$$

$S_{\mathrm{I}}$ Seidel sum for spherical aberration
$n$ index of refraction
$w$ object/aperture angle

**Maximum distance $\varphi_{\mathrm{max}}$ between focal planes in case of astigmatism:**

$$\varphi_{\mathrm{max}} = -\frac{S_{\mathrm{III}}}{n \cdot w^2}$$

$S_{\mathrm{III}}$ Seidel sum for astigmatism
$n$ index of refraction
$w$ object/aperture angle

**Petzval field curvature $R_{\text{Petzval}}$:**

$$R_{\text{Petzval}} = \frac{H^2}{n \cdot S_{\text{IV}}}$$

with

$$H = n \cdot \left(w \cdot \bar{h} - \bar{w} \cdot h\right)$$

$n$ index of refraction
$S_{\text{IV}}$ Seidel sum for Petzval field curvature
$w$ object/aperture angle
$h$ ray entrance height
*Note:* Overbarred parameters refer to the marginal ray.

**Maximum distortion $D_{\max}$:**

$$D_{\max} = \frac{S_{\text{V}}}{2 \cdot n \cdot w}$$

$S_{\text{V}}$ Seidel sum for distortion
$n$ index of refraction
$w$ object/aperture angle

## REFERENCES

Adams, G., Thöniß, T., and Gerhard, C. 2013. Designed to disperse: Easy modelling of prism and grating spectrometers and more. *Optik & Photonik* 8:50–53.

Edlén, B. 1966. The refractive index of air. *Metrologia* 2:71–80.

Gerhard, C., and Adams, G. 2010. Correction of chromatic aberration—From design to completed lens systems. *Imaging & Microscopy* 3:39–40.

Gerhard, C., and Adams, G. 2015. Easy-to-use software tools for teaching the basics, design and applications of optical components and systems. *Proceedings of SPIE* 9793:97930N.

Gerhard, C., Adams, G., and Wienecke, S. 2010. Design and manufacture of achromatic lenses. *LED Professional Review* 19:40–43.

Harendt, N., and Gerhard, C. 2008. Simulation and optimization of optical systems. *LED Professional Review* 10:40–42.

Kingslake, R. 1983. *Optical System Design*. New York: Academic Press.

Kingslake, R., and Johnson, R.B. 2009. *Lens Design Fundamentals*. Oxford: Elsevier Ltd.

Schuhmann, R. 2005. Entwicklung optischer Systeme. In *Technische Optik in der Praxis*, ed. G. Litfin, 95–125. Berlin, Heidelberg, and New York: Springer (in German).

Smith, W.J. 1990. *Modern Optical Engineering—The Design of Optical Elements*. New York: McGraw-Hill Professional.

Smith, W.J. 2004. *Modern Lens Design*. New York: McGraw-Hill Professional.

Thöniß, T., Adams, G., and Gerhard, C. 2009. Optical system design—Software tools cover envelope calculations to the final engineering drawings. *Optik & Photonik* 4:30–33.

van Albada, L.E.W. 1955. *Graphical Design of Optical Systems*. London: Sir Isaac Pitman & Sons, Ltd.

Welford, W. 1986. *Aberrations of Optical Systems*. Boca Raton, FL: CRC Press.

# 6 Tolerancing of Optical Components and Systems

## 6.1 INTRODUCTION

For the production of optical components, two quality aspects become crucial: first, the quality of the used raw material, usually glass, has to be specified in terms of purity and homogeneity of its optical properties. Second, the surface accuracy and geometric form deviation, as well as the surface cleanliness of the final product (lenses, prisms, etc.), need to be defined by appropriate tolerances. Both aspects are generally covered by the standard DIN ISO 10110 (which is based on the overridden and obsolete German standard DIN 3140). In addition, surface cleanliness can be characterized according to the new U.S.-standard ANSI/OEOSC OP1.002 and the older U.S.-standard MIL-PRF 13830B, respectively. Finally, the deviation in dimensions such as diameter or center thickness of lenses is specified by indicating absolute values or by appropriate tolerance classes as defined by DIN ISO 286. All these specifications are summarized and given in manufacturing drawings. In this chapter, the most important parameters indicated in such manufacturing drawings and specified by the abovementioned standards are introduced and explained in more detail.

## 6.2 TOLERANCING OF OPTICAL GLASSES

In the course of the production process of optical glasses (see Section 3.2.2.2) several glass defects can occur. For instance, inappropriate preparation of the batch (i.e., the mixing of the glass constituents in powder form) can result in a deviation from the target composition and thus a deviation from the target optical properties. Inadequate melting and plaining can further result in the formation of bubbles and inclusions and inhomogeneity in index of refraction (either large-scale inhomogeneity or short-range inhomogeneity, i.e., striae). Finally, uncontrolled cooling of the glass melt can give rise to mechanical tension within the glass bulk material, consequently resulting in stress birefringence and a reduction in mechanical strength. Moreover, partial crystallization and the formation of inclusions (crystallites) can occur.[1] Since the abovementioned defects (stress birefringence, bubbles, and inclusions and inhomogeneity in refractive index and striae) have a notable impact on the performance of a final optical component, appropriate tolerancing becomes necessary. These tolerances and specifications relating to the used material are exemplified by Figure 6.1 (i.e., a detail view of Figure 5.12).

[1] The effect of uncontrolled partial crystallization of a glass melt is referred to as devitrification.

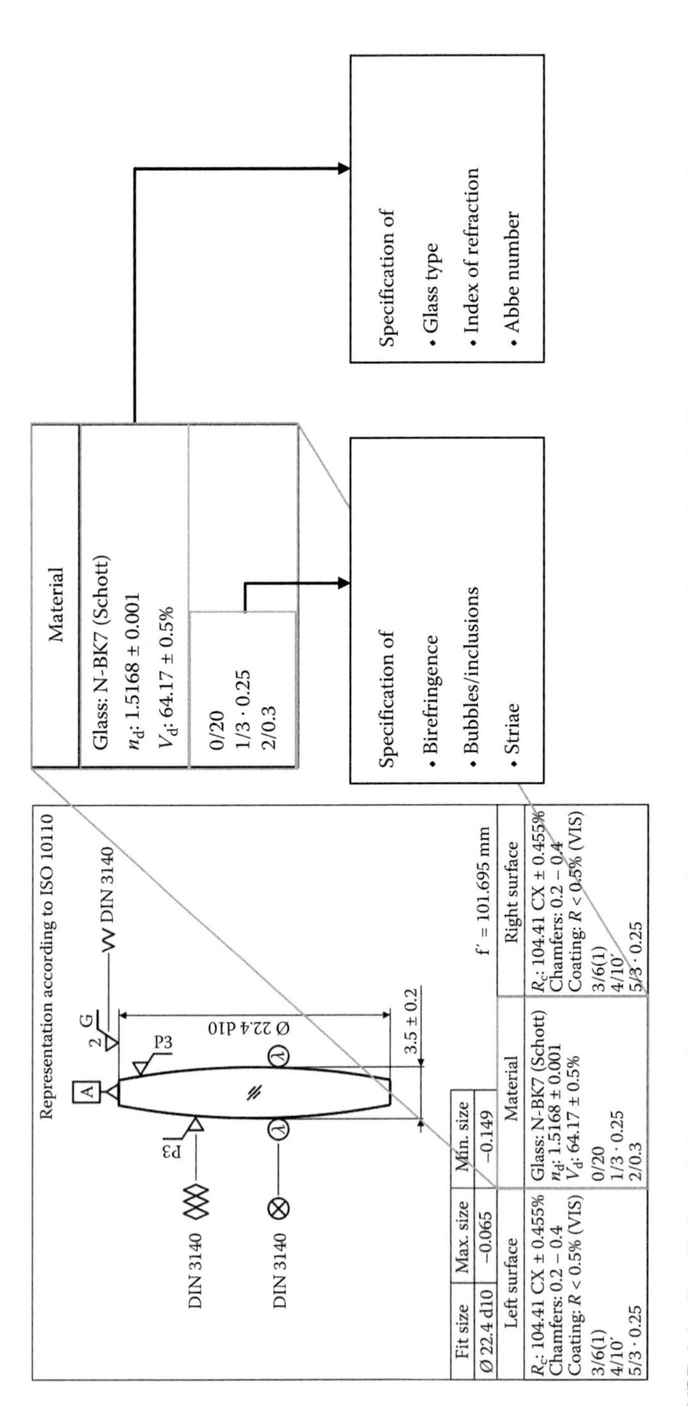


**FIGURE 6.1** Detail view of the manufacturing drawing for a biconvex lens shown in Figure 5.12, highlighting the box for specification of the used glass material and quality.

Here, the box for specification of the used glass material and quality is highlighted. First, the glass type and supplier, as well as its index of refraction and Abbe number including the maximum acceptable deviation from the nominal value, are specified. For the index of refraction, the tolerance is given by an absolute value (1.5168±0.001), and the Abbe number is specified by the percentage deviation from its nominal value (64.17%±0.5%). Second, birefringence, bubbles, and inclusions as well as striae are itemized by the particular index numbers according to DIN ISO 10110 as described in detail in the following sections.

### 6.2.1 Stress Birefringence

According to DIN ISO 10110, *stress birefringence* as induced by mechanical stress resulting from the cooling procedure is identified by the index number "0"[2] in manufacturing drawings. The full nomenclature is "0/A," where the identifier *A* indicates the acceptable maximum difference in optical path length *OPD* (and index of refraction, respectively) between the ordinary and extraordinary propagation direction of light.[3] This difference is directly related to the mechanical tension $\sigma_m$ within an optical component and can be determined by

$$OPD = 10 \cdot K \cdot t \cdot \sigma_m, \tag{6.1}$$

where *K* is the photo-elastic coefficient of the glass, and *t* is the thickness of the optical component.

The value *A* is given in nanometers per 10 mm optical path length (i.e., the reference length, e.g., the thickness of a glass sample). The example/specification given in Figure 6.1, "0/20," thus indicates that a maximum difference in optical path length of 20 nm per 10 mm reference optical path length is accepted. This difference finally amounts to $2 \cdot 10^{-5}$ mm/10 mm = $2 \cdot 10^{-6}$ = 0.0002%.

### 6.2.2 Bubbles and Inclusions

Due to insufficient plaining and uncontrolled cooling, bulk defects such as gas or air *bubbles* and *inclusions* (stones, crystallites, nonmolten agglomerates from the batch, etc.) can remain within the glass bulk material. Such residues are identified by the index number "1" (according to DIN ISO 10110) where the full nomenclature is "1/N · A." Here, *N* identifies the maximum number of bulk defects, and *A* represents the maximum acceptable cross-sectional area of each single bulk defect, given in $mm^2$, within a reference glass volume of 100 $cm^3$. The example given in Figure 6.1, "1/3 · 0.25," has thus the following meaning: a maximum of three bulk defects with a maximum cross-sectional area of 0.25 $mm^2$ per bulk defect is acceptable.

[2] Note that according to the overridden DIN 3140, the index number for stress birefringence was "6," whereas this index number is used for the specification of the LIDT according to DIN ISO 10110.

[3] Birefringence induces anisotropy in optically isotropic media. This means that such a medium features two different indices of refraction, one ordinary and one extraordinary, depending on the polarization and propagation direction of light. An incident nonpolarized light beam is thus split into two perpendicularly polarized light beams. This effect mainly occurs in optical crystals and is specifically employed for the realization of polarization prisms but undesirable in optical glasses.

In practice, the specified cross-sectional area can be divided: instead of one big bulk defect, several smaller defects are acceptable as long as the total area of these smaller inclusions does not exceed the specified cross-sectional area of one big bulk defect. For instance, a maximum of six bubbles or inclusions with a maximum cross-sectional area of 0.125 $mm^2$ also fulfills the abovementioned specification.

In contrast, it is not permissible to combine the specified number of bulk defects and the maximum cross-sectional area per defect to a lower number of defects with a higher cross-sectional area. Hence, one single bubble or inclusion with a cross-sectional area of 0.75 $mm^2$ ($3 \cdot 0.25$ $mm^2$) is not acceptable in the given example.

### 6.2.3 Inhomogeneity and Striae

*Inhomogeneity* and *striae* are addressed by the index number "2" according to DIN ISO 10110[4] and defined by particular classes as presented in Table 6.1 and Figure 6.3. The full nomenclature of inhomogeneity and striae is "2/A, B," where $A$ is the inhomogeneity class, and $B$ represents the striae class. Inhomogeneity can result from insufficient stirring of the glass melt; it is defined as large-scale deviations in index of refraction over the whole volume of an optical component. It is given by the peak-to-valley deviation as determined by interferometric measurements according to ISO 12123 and classified into six different so-called *inhomogeneity classes* as listed in Table 6.1.

The deviation in index of refraction $\Delta n$ can also be determined by measuring the deformation of a wave front $\Delta w$ after passing through a glass sample with a certain thickness $t$. It is then given by

$$\Delta n = \frac{\Delta w}{2 \cdot t}. \tag{6.2}$$

**TABLE 6.1**
**Inhomogeneity Classes and Acceptable Deviation in Index of Refraction**

| Inhomogeneity Class | Acceptable Deviation in Index of Refraction from Nominal Value | Acceptable Range, Nominal $n_e = 1.527685$ | |
|---|---|---|---|
| | | Lower Limit | Upper Limit |
| 0 | $\pm 50 \cdot 10^{-6}$ | 1.527635 | 1.527735 |
| 1 | $\pm 40 \cdot 10^{-6}$ | 1.527645 | 1.527725 |
| 2 | $\pm 10 \cdot 10^{-6}$ | 1.527675 | 1.527695 |
| 3 | $\pm 4 \cdot 10^{-6}$ | 1.527681 | 1.527689 |
| 4 | $\pm 2 \cdot 10^{-6}$ | 1.527683 | 1.527687 |
| 5 | $\pm 1 \cdot 10^{-6}$ | 1.527684 | 1.527686 |

*Source:* Schott, A.G., Optical Glass Data Sheets, 2015.
*Note:* Including the Acceptable Range of Index of Refraction at the Example of a Standard Crown Glass with a Nominal Index of Refraction of $n_e = 1.527685$.

[4] DIN ISO 10110 considers large-scale inhomogeneity in contrast to the obsolete DIN 3140, where merely striae were specified.

In contrast to inhomogeneity, striae are local short-range deviations in index of refraction with dimensions in the submillimeter range. As shown in Figure 6.2, they typically have the form of straight or curved lines and are embedded in the glass bulk material.

The orientation of such striae is directly related to the direction of streams that occur during the plaining process, due to local differences in temperature within the glass melt. The classification of striae is mainly based on the density (i.e., the share of striae in the total optically active surface area (i.e., the full aperture) of an optical component). Further, it has to be taken into account that by default, the wave front deformation of a plane reference wave front which passes such an optical component, is higher than 30 nm. An overview on the *striae classes* 1 to 4 including the particular percentaged share of striae in total optically active surface area of an optical component is shown in Figure 6.3.

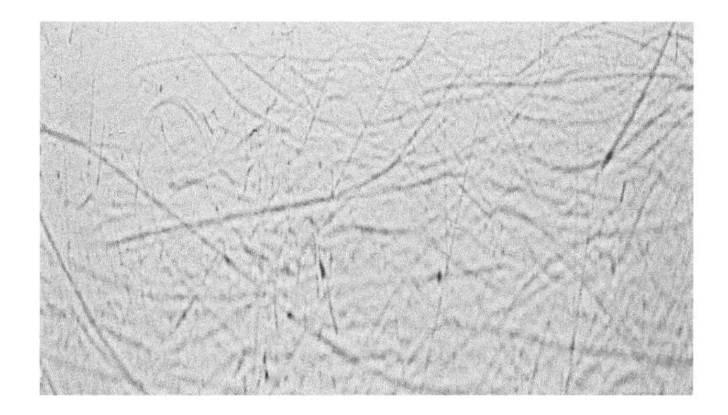

**FIGURE 6.2** Example for striae in optical glasses.

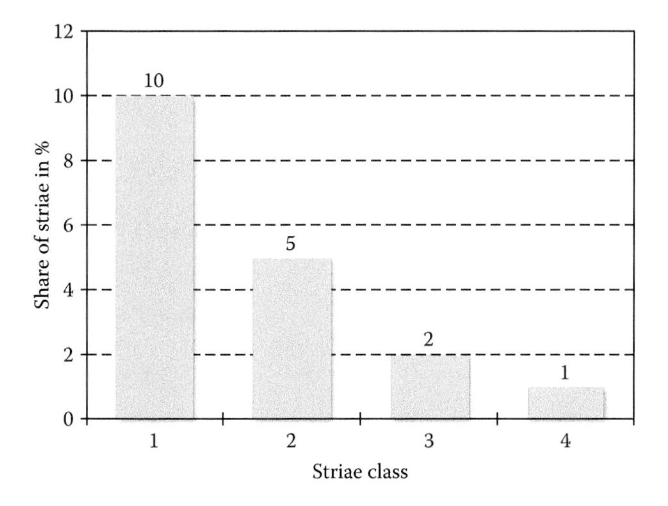


**FIGURE 6.3** Striae classes 1–4 including the particular percentaged share of striae in total optically active surface area of an optical component. (From Schott, A.G., Optical Glass Data Sheets, 2015.)

Moreover, a further striae class (striae class 5) is in hand. Here, the share of striae in the total optically active surface area of the optical component is lower than 1%, and the wave front deformation of a reference wave front passing through this optical component is lower than 30 nm.

The example shown in Figure 6.1, where inhomogeneity and striae are indicated by "2/0, 3," thus specifies the inhomogeneity class 0 and the striae class 3. The acceptable maximum deviation in index of refraction over the whole volume of the lens consequently amounts to $\pm 50 \cdot 10^{-6}$, and the maximum share of striae in the full lens aperture is 2%. Additionally, the wave front deformation induced by striae is higher than 30 nm.

### 6.2.4 Further Specifications

#### 6.2.4.1 Chemical Resistance and Stability

For some applications (e.g., in aggressive environments), the chemical resistance and stability of glass surfaces might be of interest. Against this background, glasses are classified in different resistance classes where climatic resistance, stain resistance, acid resistance, alkali/phosphate resistance, and visible surface changes due to glass corrosion are specified as defined in the following paragraphs.

##### *6.2.4.1.1 Climatic Resistance*

For the determination of *climatic resistance*, a polished glass sample is exposed to water vapor for 30 h. Subsequently, the change in haze $\Delta H$ is determined by transmission measurements. Finally, the glass is classified in one of four climatic resistance classes on the basis of the increase in haze as listed in Table 6.2.

##### *6.2.4.1.2 Stain Resistance*

*Stain resistance* is determined by exposing a polished glass surface to test solutions with different pH-values. As a result of chemical decomposition of the glass and the formation of stains, a colored thin layer is formed on the glass surface and becomes visible due to interferences. The particular stain resistance classes $FR$[5] (0–5) are

**TABLE 6.2**
**Climatic Resistance Classes and Acceptable Percentaged Increase in Haze**

| Climatic Resistance Class $CR$ | Increase in Haze $\Delta H$ in % |
|---|---|
| 1 | 0–0.3 |
| 2 | 0.3–1.0 |
| 3 | 1.0–2.0 |
| 4 | >2 |

*Source:* Schott, A.G., Optical Glass Data Sheets, 2015.

[5] The abbreviation "F" is based on the German term for stains, "Flecken."

then defined by the formation of stains and the grade of observed color change; they further depend on the exposure time and the type of test solution. For example, a glass of the stain resistance class $FR=0$ does not feature any stains or color change after being exposed to a slightly acid solution with a pH-value of 4.6 for 100h. In contrast, glasses that feature both stains and color change after an exposure time of merely 12min where the pH-value of the test solution is 5.6 are classified into stain resistance class $FR = 5$.

#### *6.2.4.1.3 Acid Resistance*

The *acid resistance* of glasses is determined and specified according to DIN 8424. For this purpose, the time $t_{AR}$ required for removal of a 100nm-thick surface layer from a glass sample by an acid is measured according to

$$t_{AR} = \frac{t_e \cdot \rho \cdot A}{(m_0 - m_e) \cdot 100}. \tag{6.3}$$

Here, $t_e$ is the total duration of the test given in hours, $\rho$ is the glass sample density given in g/cm$^3$, $A$ is the glass sample surface area in cm$^2$, $m_0$ is the initial mass of the sample, and $m_e$ is the mass after the experiment, both given in mg.

Acid resistance is then classified in two different main categories $SR$[6]; glasses of high acid resistance are classified in classes 1 to 4 as listed in Table 6.3. The pH-value of the acid used for testing is 0.3 in this case. Glasses with moderate or poor chemical stability and acid resistance, respectively, are indicated by classes 51 to 53 as shown in Table 6.4.

Here, the pH-value of the test acid amounts to 4.6 and is thus much higher (i.e., the acid is less acrid) than in the abovementioned case. The categories of acid resistance are separated by an intermediate point represented by the acid resistance class 5.

**TABLE 6.3**
**Acid Resistance Classes for Comparatively Resistant Glasses and Particular Time Required for the Removal of a Glass Surface Layer with a Thickness of 100 nm**

| Acid Resistance Class $SR$ | Time $t_{AR}$ Required for Removal of a 100 nm-thick Surface Layer in Hours |
|---|---|
| 1 | $t > 100$ |
| 2 | $10 < t < 100$ |
| 3 | $1 < t < 10$ |
| 4 | $0.1 < t < 1$ |
| 5 | $t < 0.1$ (for pH=0.3)<br>$t > 10$ (for pH=4.6) |

*Source:* Schott, A.G., Optical Glass Data Sheets, 2015.

[6] The abbreviation “S” is based on the German term for acid, “Säure.”

**TABLE 6.4**
**Acid Resistance Classes for Glasses with Comparatively Low Resistance and Particular Time Required for the Removal of a Glass Surface Layer with a Thickness of 100 nm**

| Acid Resistance Class *SR* | Time *t* Required for Removal of a 100 nm-Thick Surface Layer in Hours |
|---|---|
| 5 | $t < 0.1$ (for pH=0.3) |
| | $t > 10$ (for pH=4.6) |
| 51 | $1 < t < 10$ |
| 52 | $0.1 < t < 1$ |
| 53 | $t < 0.1$ |

*Source:* Schott, A.G., Optical Glass Data Sheets, 2015.

### *6.2.4.1.4 Alkali and Phosphate Resistance*

The resistance of glasses to aqueous alkaline solutions is specified by *alkali resistance (AR) classes* according to ISO 10629 and *phosphate resistance (PR) classes* according to ISO 9689. The first class becomes of essential importance during classical manufacturing processes, since the test liquids are quite comparable to cooling lubricants or polishing suspensions. Similar to the acid resistance, both alkali and phosphate resistance indicate the time required for removing a surface layer with a thickness of 100 nm from a glass sample. In the first case, sodium hydroxide (NaOH) is used as test liquid, whereas in the second case, pentasodium triphosphate ($Na_5P_3O_{10}$) is applied. The four different alkali and phosphate resistance classes determined in this way are listed in Table 6.5.

### *6.2.4.1.5 Visible Glass Surface Changes*

Acid, alkali, and phosphate resistance classes may be extended by digits in some cases. These digits indicate changes of a glass surface visible by the naked eye. The

**TABLE 6.5**
**AR Classes and Particular Time Required for the Removal of a Glass Surface Layer with a Thickness of 100 nm**

| AR Class and PR Class | Time *t* Required for Removal of a 100 nm-Thick Surface Layer in Hours |
|---|---|
| 1 | $t > 4$ |
| 2 | $1 < t < 4$ |
| 3 | $0.25 < t < 1$ |
| 4 | $t < 0.25$ |

*Source:* Schott, A.G., Optical Glass Data Sheets, 2015.

extension ".0" denotes no visible changes whereas ".4" describes deposits and thick layers on a glass surface. Moreover, a clear, but irregular surface is indicated by ".1," staining and visible interference color layers due to glass leaching are marked by ".2," and cloudy layers are indicated by the extension ".3."

#### 6.2.4.2 Hardness and Grindability

In addition to chemical resistance and stability, mechanical and thermodynamic glass properties are of importance for several manufacturing steps such as grinding or coating. During grinding, the *hardness* of a glass and its *grindability* are of special interest for the choice of tools and adjustment of appropriate process parameters.

The hardness of optical glass is usually specified by the Knoop hardness $HK$, which is determined on the basis of the measurement principle suggested by the American physicist *Frederick Knoop*. Here, a diamond tip is pressed on the test glass surface applying a constant and defined force $F$[7] according to DIN EN ISO 4545. Subsequently, the length $l$[8] of the resulting imprint of the diamond tip on the glass surface is measured. The Knoop hardness finally follows from

$$HK = 1.451 \cdot \frac{F}{l^2}. \tag{6.4}$$

As shown in Figure 6.4, optical glasses are classified into seven classes of hardness.

For the specification of grindability $G$ of any glass according to ISO 12844, the removed volume $\Delta V_{glass}$ during grinding for a duration of 30 s is measured and referred to the volume removed during grinding of a reference glass $\Delta V_{ref}$ (e.g., N-SK16 for SCHOTT glasses or S-BSL7 for OHARA glasses) according to

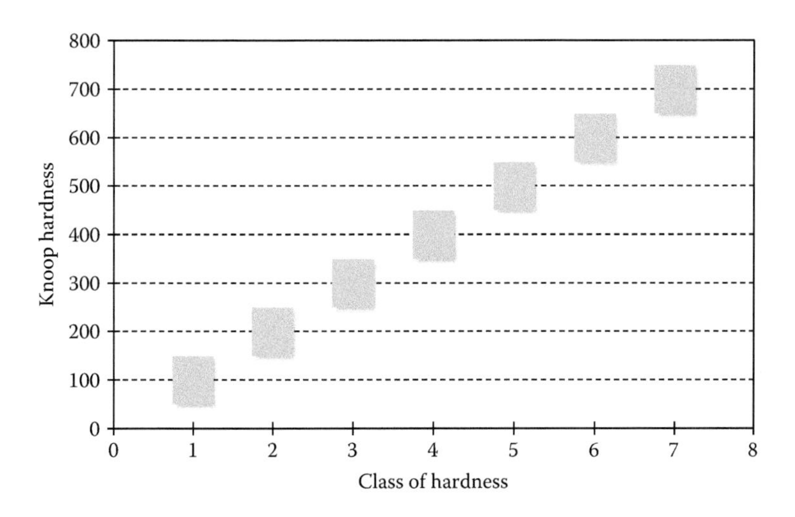


**FIGURE 6.4** Classes of hardness 1–7 including the particular Knoop hardness. (From Schott, A.G., Optical Glass Data Sheets, 2015.)

[7] The force is given in Newtons (N).

[8] The length is given in millimeters (mm).

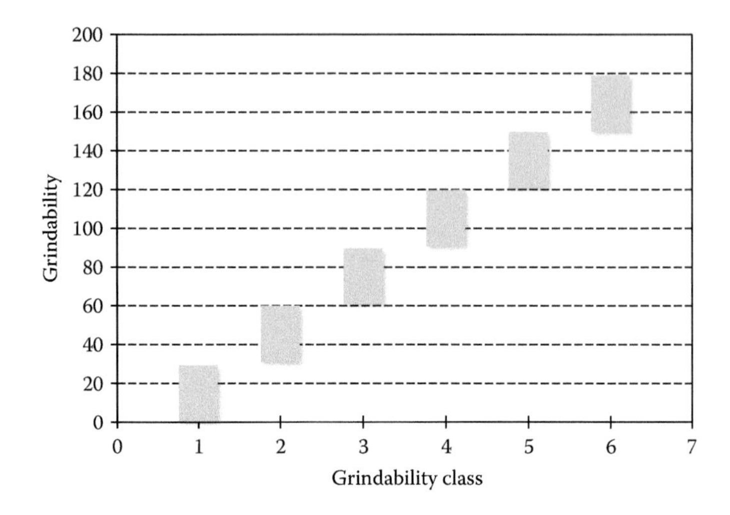


**FIGURE 6.5** Grindability classes 1–6 including the particular grindability. (From Schott, A.G., Optical Glass Data Sheets, 2015.)

$$G = \frac{\Delta V_{\text{glass}}}{\Delta V_{\text{ref}}}. \tag{6.5}$$

Based on this procedure, the glasses are classified into six grindability classes, where the grindability of the reference glass is arbitrary set to 100 per default. These grindability classes are visualized in Figure 6.5.

### 6.2.4.3 Thermodynamic Glass Properties

In the course of a coating process, an optical component may be placed in a heated coating chamber. Hence, thermodynamic glass properties such as the specific heat capacity $c_p$ or the *coefficient of thermal expansion* $\alpha$ are specified. The latter parameter is also of interest for the design of optomechanical systems that are exposed to extreme variations in temperature (as the case may be for telescope optics sent to space). It is given by

$$\alpha = \frac{1}{\Delta T} \cdot \frac{\Delta l}{l_0}, \tag{6.6}$$

where $\Delta T$ is the change in temperature, $\Delta l$ is the change in length, also referred to as elongation, and $l_0$ is the original length of the optical component at ambient temperature. A selection of the particular coefficients of thermal expansions of different optical media is listed in Table 6.6.

## 6.3 FABRICATION TOLERANCES OF OPTICS

Apart from plastics, crystals, and liquids, optical glass satisfying the requirements as introduced above and specified in manufacturing drawings represents the main raw material for the production of optical components.

**TABLE 6.6**
**Coefficient of Thermal Expansion of Selected Optical Media Including the Temperature Scope**

| Optical Medium | Coefficient of Thermal Expansion in $10^{-6}$/K | Temperature Scope in Centigrade |
|---|---|---|
| Fused silica | 0.57 | 0–200 |
| Aluminum borosilicate glass | 3.2 | 20–300 |
| Borosilicate glass (8250 from Schott) | 5.0 | 20–300 |
| Boron crown glass (BK7 from Schott) | 8.3 | 20–300 |
| Glass ceramics (ZERODUR from Schott) | 0±0.007–0±0.10 | 0–50 |

*Source:* Schaeffer, H.A., and Langfeld, R., *Werkstoff Glas*, Springer, Berlin and Heidelberg, 2014; Schott, A.G., ZERODUR® Zero Expansion Glass Ceramic Data Sheet, 2013.

The required specifications for manufacturing an optical component such as its dimensions and shape including the particular tolerances are itemized by appropriate elements and designations in manufacturing drawings. The most important specifications refer to contour accuracy, centering errors, and surface cleanliness. Figure 6.6 shows a detail view of Figure 5.12 where these parameters and further relevant information on the optically active surface are given.

First, the basic shape and radius of curvature including its tolerance range are specified. In the example shown in Figure 6.6, the basic shape is a convex (abbreviated by "CX") spherical segment with a radius of curvature of 104.41 mm. The tolerance for the radius of curvature is 0.455%; it can thus range from 104.36 to 104.45 mm. Second, the lens edge (i.e., the transient area between the optically active lens surface and its border cylinder) is specified as beveled or chamfered (see Section 7.6) where the bevel leg length is 0.2–0.4 mm. Third, the optically active lens surface is coated with an antireflective coating with a residual reflectance <0.5% within the visible (VIS) wavelength range.[9] Finally, the required contour accuracy, acceptable centering error, and desired surface cleanliness are indicated by appropriate index numbers (3, 4, and 5 in Figure 6.6) and extensively defined according to DIN ISO 10110 as described in detail hereafter.

### 6.3.1 Contour Accuracy

The *contour accuracy* specifies the acceptable deviation of the actual surface shape of an optical component's optically active surface with respect to the target surface shape. This target surface shape is a theoretical reference profile, for example, the

[9] The visible wavelength range is found between 380 and 780 nm. However, broadband VIS-antireflective coatings are usually specified for a smaller wavelength range from approximately 450–700 nm.

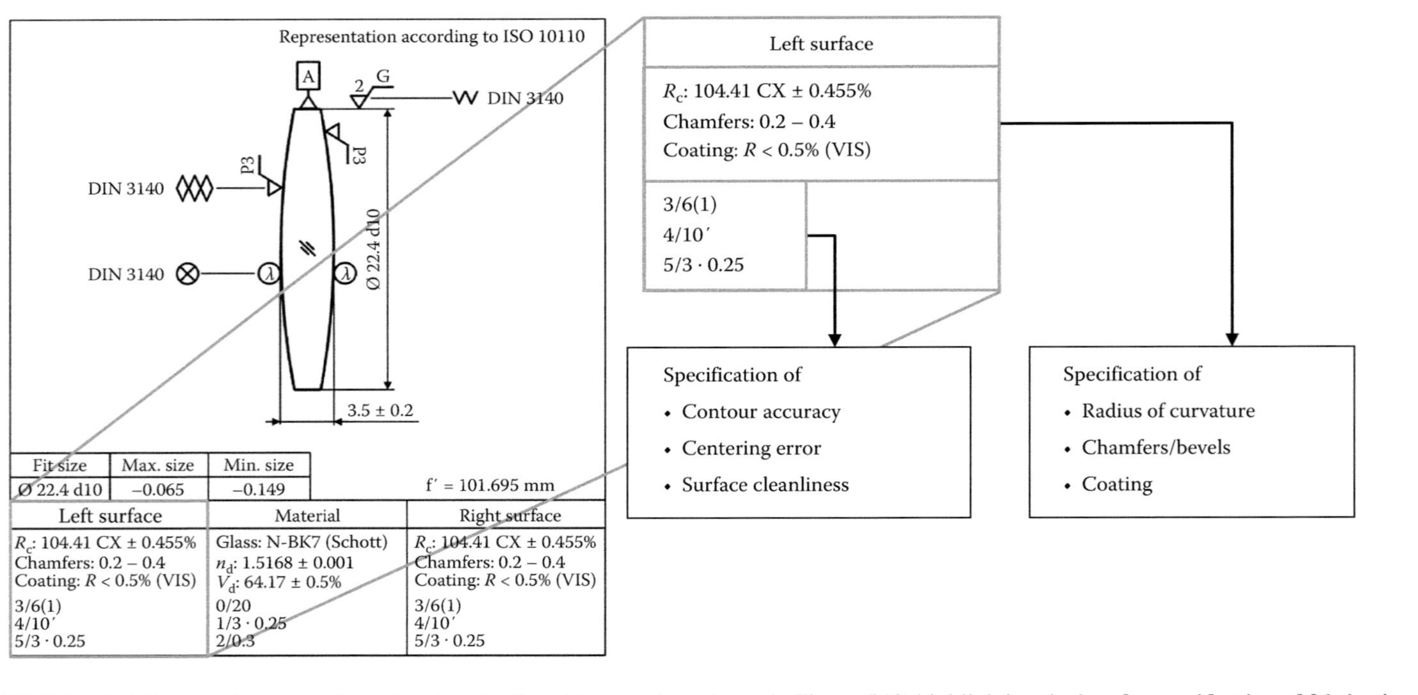


**FIGURE 6.6** Detail view of the manufacturing drawing for a biconvex lens shown in Figure 5.12, highlighting the box for specification of fabrication tolerances.

surface of a sphere in case of spherical lens surfaces or a plane for plane lens or prism surfaces. According to DIN ISO 10110, this deviation is indicated by the index number "3" and covers the maximum sagitta, irregularity, and fine contour error within a predefined test area. For rotational-symmetric optics such as lenses, the test area is usually further specified by the supplementary information "80% CA" (i.e., "80% centered area"). The test area is thus given by a concentric centered circular area that includes 80% of the total optically active surface area. The lens border, where it is finally beveled or mounted, is consequently not considered.

The contour accuracy can be measured in different ways where the basic underlying phenomenon is interference of light. In optical workshops, testing is performed by placing a *gauge glass* with known surface shape on the test object, for example, a lens. As shown in Figure 6.7, an interference pattern is then formed within the air gap between the gauge glass and the test object as long as the air gap is sufficiently thin.

This characteristic interference pattern is then evaluated regarding the maximum deviation between the gauge glass and the test object and irregularity. The maximum deviation corresponds to the sagitta (i.e., generally the height or depth of a circular arc). It is determined on the basis of the number of interference fringes as observed during testing where the distance of two fringes corresponds to half the measurement wavelength.[10] These fringes are formed by intensity maxima, where constructive interference occurs within the air gap between the gauge glass and the test object; they are also referred to as Newton's rings or Newton's fringes, named after the well-known English mathematician and physicist *Sir Isaac Newton* (1642–1726). Moreover, the regularity of the interference fringes is evaluated. For a perfect spherical surface tested by a spherical gauge glass, the resulting interference pattern features circular interference fringes as shown in Figure 6.8. The deviation in radius of curvature between the gauge glass and the test object, the sagitta, follows from the product of number of fringes and half the measurement wavelength.

In practice, the number of fringes is counted from the lens center to its border as shown in Figure 6.9. In a nonspherical test surface, the interference pattern additionally features certain irregularities. For example, a slightly toric lens surface gives rise to elliptical interference fringes when tested with a spherical gauge glass. As

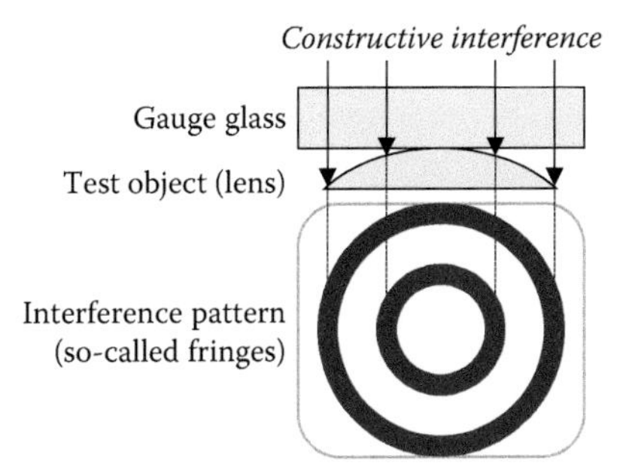


**FIGURE 6.7** Principle of interferometric surface inspection by gauge glasses.

[10] This approximation is valid for thin air gaps and a moderate number of interference fringes.

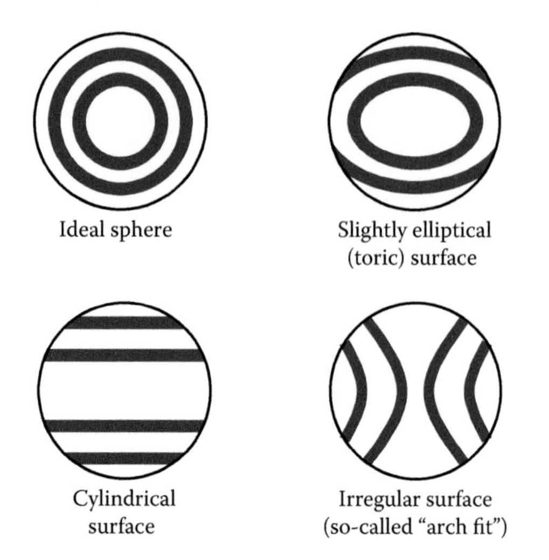


**FIGURE 6.8** Interference patterns of differently shaped lens surfaces: spherical (top left), elliptical, or toric (top right), cylindrical (bottom left), and irregular (bottom right).

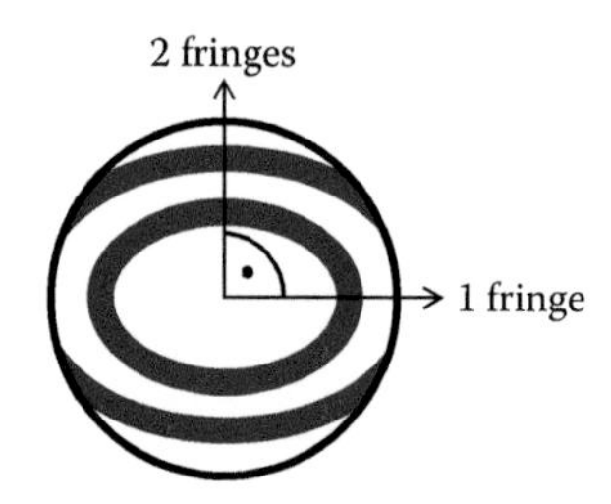


**FIGURE 6.9** Example for the determination of contour accuracy of a lens surface by the evaluation of an interference pattern; the observable fringes are counted in two directions perpendicular to each other.

shown in Figure 6.9, the number of fringes consequently differs when counting in two directions orthogonal to each other since such surfaces feature different radii of curvature in these directions of observation.

According to DIN ISO 10110, this effect is expressed and specified by the maximum acceptable irregularity of the interference pattern. It is given by the difference in number of fringes as determined by the procedure described in the previous paragraphs.

Finally, the *fine contour error* $C$ resulting from local peaks or pockets on the test surface is specified. It is given by two characteristic values: the distance between two interference fringes $a$ (which is approximately half the test wavelength) and the maximum deviation of a fringe from its basic shape $d$ according to

$$C = \frac{d}{a}. \tag{6.7}$$

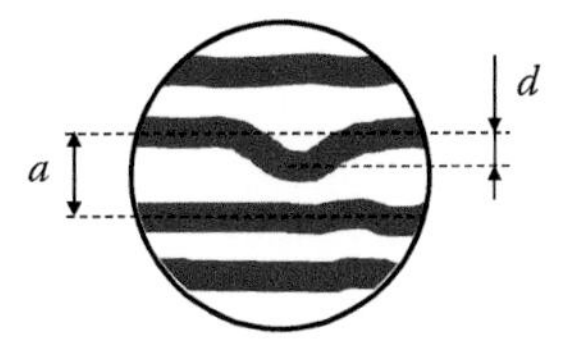


**FIGURE 6.10** Definition of the distance between two interference fringes $a$ and the maximum deviation of a fringe from its basic shape $d$ (here: a straight line) for the determination of the fine contour error $C$ at the example of a cylindrical lens surface with inaccuracies.

Both values are exemplified in Figure 6.10, where the interference pattern of a cylindrical lens surface is shown for better visualization. The basic shape of the interference pattern is thus a straight line, and the deviation $d$ is given by the maximum local bending, which deviates from this line.

The full nomenclature of surface accuracy according to DIN ISO 10110 is generally "3/A(B/C)." The parameter $A$ is the maximum acceptable sagitta given in interference fringes for a defined test wavelength, $B$ is the surface irregularity (the maximum acceptable deviation from the spherical shape with respect to $A$), and $C$ is the fine contour error. An example of an interference pattern featuring surface irregularity and a fine contour error is shown in Figure 6.11. Here, the value $d$ is given but could be measured by either computerized interferometers or by placing a mask on the test gauge in practice.

In this example, the maximum number of fringes, counted in $y$-direction in Figure 6.11, amounts to 2. The sagitta is thus $A=2 \cdot \lambda/2=600\,\mu m$ when measuring at a test wavelength of 600 nm (i.e., red interference maxima, which can be identified and evaluated quite well by the human eye). For the x-direction (perpendicularly to the y-direction), merely one fringe is found. The surface is thus not spherical, but slightly toric. The irregularity consequently amounts to $B=2-1=1$. Finally, the fine contour error $C=d/a$ follows from $d=30\,nm$ and $a=\lambda/2=300\,nm$. It accounts for $30\,nm/300\,nm=0.1$. The proper specification of surface accuracy of this example is thus 3/2(1/0.1) according to DIN ISO 10110.

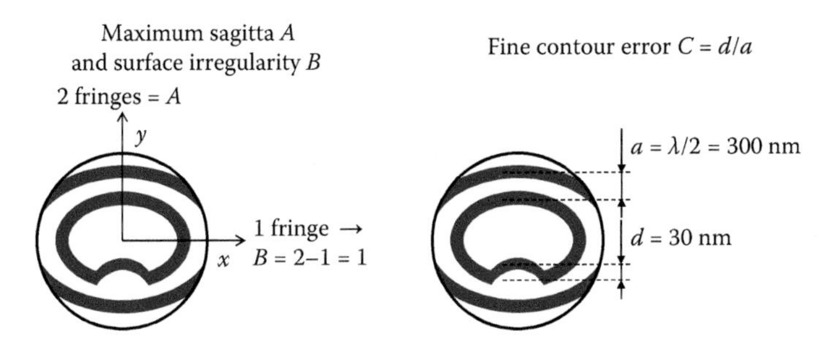


**FIGURE 6.11** Example of an interference pattern including fine contour error and principle of determination of the maximum sagitta $A$, the surface irregularity $B$ (left), and the fine contour error $C$ (right) according to DIN ISO 10110.

For the example shown in Figure 6.6, the contour accuracy is indicated by the specification "3/6(1)." The maximum sagitta is thus six interference fringes ($=6 \cdot \lambda/2 = 1.8\,\mu m$ at a test wavelength of 600 nm) with a maximum deviation of one fringe. A fine contour error is not specified. Figure 6.12 shows an example for the resulting interference pattern.

The method of placing gauge glasses on the test object (i.e., the polished lens surface) represents a contacting measurement method and may thus lead to damages on both the gauge glass surface and the test surface, for example, scratches. In practice, this is usually avoided but cannot be excluded. Further, the interpretation of the interferometric pattern formed by Newton's fringes requires a certain grade of experience of the craftsman. Finally, surfaces of high precision cannot be quantified properly by the more or less subjective evaluation of interference patterns. In some cases and especially in final inspection, the gauge glass method is thus replaced by the use of *interferometers* where the measurement procedure is based on the same principle; the comparison of the test surface to a reference surface is shown in Figure 6.13.

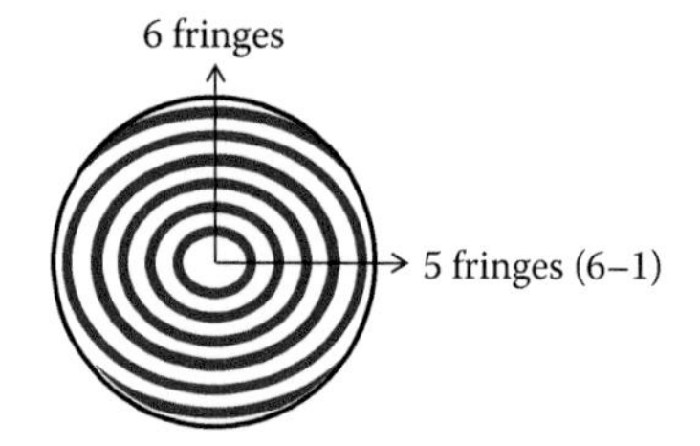


**FIGURE 6.12** Example of an interference pattern fulfilling the specification given in Figure 6.6.

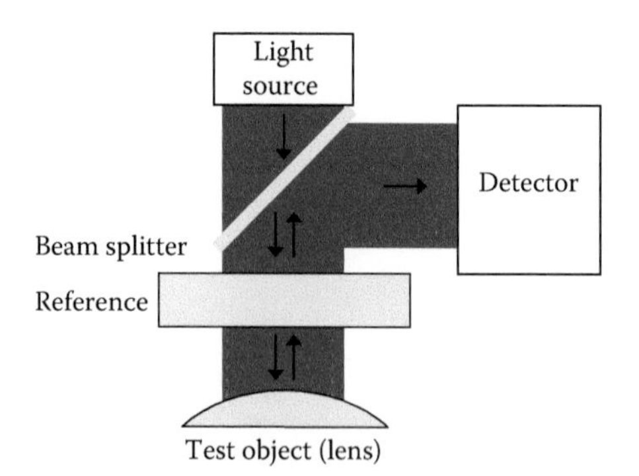


**FIGURE 6.13** Working principle of a Fizeau-interferometer[11]; a light beam passes through a reference device and is reflected by the test surface. The reflected light beam is detected and evaluated in terms of the deviation of the test surface from the reference.

[11] Named after the French physicist *Armand Hippolyte Louis Fizeau* (1819–1896).

Testing by interferometers is a contactless process; additionally, the evaluation of the measured interferograms by software allows a proper quantification of the surface shape of an actually tested lens surface.

### 6.3.2 Centering Error

The *centering error* of a lens is generally defined as the angular deviation between its optical and its mechanical axis (e.g., the lens border cylinder or the symmetry axis of a lens mount) as shown in Figure 6.14.

According to DIN ISO 10110, it is identified by the index number "4." Its total nomenclature is "4/X$^Y$," where $X$ is the value of maximum acceptable angular deviation and $Y$ its unit, either arc minutes (′) or arc seconds (″). The example "4/10′" as shown in Figure 6.6 thus specifies a maximum centering error of 10 arc min.

In practice, the centering error is determined as follows: as shown in Figure 6.15, the test object (i.e., a lens) is mounted at its mechanical axis, usually its border cylinder, and rotated. At the same time, a collimated laser light beam passes through the lens and is imaged on a detector by an optical setup. Since the optical axis is tilted with respect to the mechanical axis and the axis of rotation, respectively, the

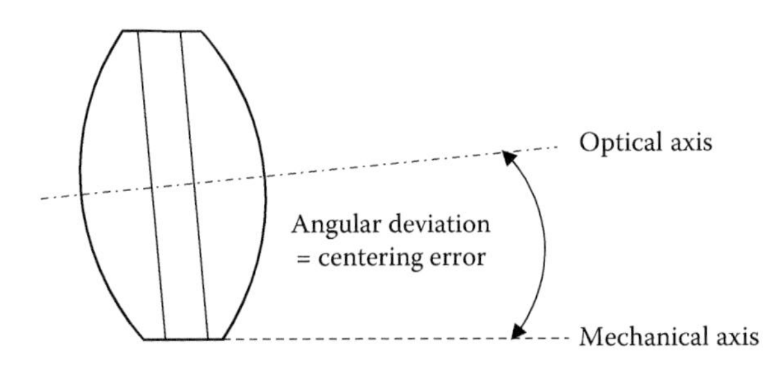


**FIGURE 6.14** Visualization of the lens centering error, given by the angular deviation between the optical axis and the mechanical axis (e.g., the border cylinder of a lens).

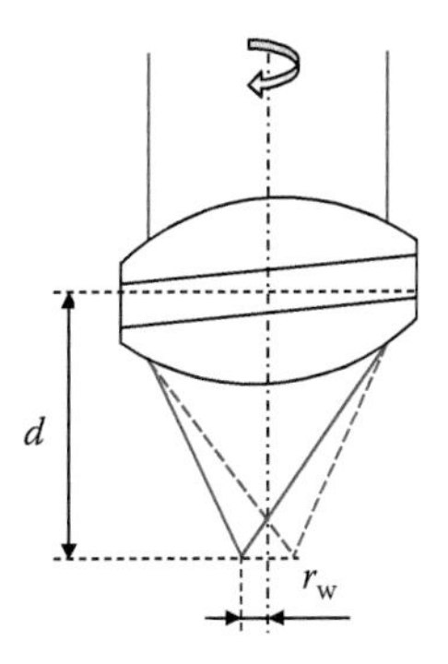


**FIGURE 6.15** Principle of determination of the lens centering error by rotating a lens and measuring the resulting wobble circle radius $r_w$ of incident collimated light focused by the lens.

focal point features a certain displacement from the axis of rotation, resulting in the formation of a *wobble circle* during rotation. The radius $r_w$ of this wobble circle is measured, and the centering error *CE*, given in arc minutes, is finally calculated according to

$$CE = \frac{1720 \cdot r_w}{m \cdot d \cdot (n-1)}. \tag{6.8}$$

Here, $m$ is the magnification of the used optical test setup, $d$ is the distance from the lens principal plane to the detector plane, and $n$ is the index of refraction of the lens material.

### 6.3.3 Surface Cleanliness

*Surface cleanliness* of finally polished optical components can be specified by two different main standards: the DIN ISO 10110 or the U.S.-standard ANSI/OEOSC OP1.002 and the older U.S.-standard MIL-PRF 13830B, respectively. Referring to DIN ISO 10110, surface cleanliness is indicated by the index number "5" in manufacturing drawings and quantifies the maximum number and size of surface defects such as scratches, digs, and stains. Such defects are usually measured visually (employing light microscopes in case of high required surface cleanliness) by the visual comparison of the surfaces in test to measurement standards. These standards are transparent masks that feature a number of structures of defined and standardized form and size and can be placed on the tested surface for direct comparison.

The full nomenclature of surface cleanliness is "5/A · B," where $A$ is the maximum acceptable number of defects, and $B$ indirectly quantifies the area of each single defect. By default, surface defects are assumed quadratic, where the parameter $B$ is given in millimeters and quantifies the maximum edge length of each square. In combination, the parameters $A$ and $B$ thus give the maximum acceptable total area of surface defects on a polished surface.

For the example given in Figure 6.6, where surface cleanliness is defined by the designation "5/3 · 0.25," a maximum of three scratches, digs, or stains is acceptable. Assuming those defects quadratic, the maximum edge length of each defect is 0.25 mm, resulting in an area per defect of 0.0625 mm$^2$. The maximum acceptable total surface defect area finally amounts to $3 \cdot 0.0625\,\text{m}^2 = 0.1875\,\text{mm}^2$.

In practice, the specified area per defect can be divided (compare the specification of bubbles and inclusions according to DIN ISO 10110): instead of one big surface defect, several smaller defects are acceptable as long as the total area does not exceed the area of one big defect as specified by parameter $B$. As an example, a maximum number of six defects with a maximum area per defect of 0.0313 m$^2$ are acceptable. In contrast, defects with an area bigger than 0.0625 mm$^2$ are not tolerable.

One has to consider that the defect area as specified by DIN ISO 10110 represents an absolute value and does not depend on the size of the optical component, for example the diameter of free aperture of a lens. In contrast, this has to be taken into account when specifying surface cleanliness according to MIL-PRF 13830B. This standard indicates the maximum acceptable length or dimensions but not the number of surface defects. The area of surface defects further depends on the test area and the area of the optically active surface, respectively. MIL-PRF 13830B thus represents a relative but not absolute description of surface cleanliness.

Here, scratches are distinguished from digs. The nomenclature of surface cleanliness according to MIL-PRF 13830B is "A-B scratches-digs" where the parameter $A$ is not definitively specified, but usually represents the width of scratches given in microns (Aikens, 2010), where the maximum length of scratches can amount to a quarter of the test area diameter. The parameter $B$ represents the maximum acceptable diameter of digs given in hundredths of a millimeter. In contrast to DIN ISO 10110, the number of scratches and digs is not clearly quantified.

As an example, surface defects specified by the designation "60-10 scratches-digs" can be determined as follows: the maximum width of scratches is $A = 60\,\mu\text{m}$ and has to be referred to the diameter of the tested optics surface. For a standard lens with a diameter of 1 in. = 25.4 mm, the effective test area diameter amounts to 20.32 mm when considering 80% of the total optically active surface area (as usually applied and indicated by the supplementary information "80% CA"). The maximum length of scratches is thus a quarter of the test area diameter, $20.32\,\text{mm}/4 = 5.08\,\text{mm}$. The maximum area of one scratch is finally given by the product of its length and width and amounts to $0.06 \cdot 5.08\,\text{mm} = 0.31\,\text{mm}^2$. Moreover, the maximum diameter of digs is 10 hundredths of a millimeter (since $B = 10$) and accounts for $10 \cdot 0.01\,\text{mm} = 0.1\,\text{mm}$. Assuming a circular dig, its area is thus approximately $0.008\,\text{mm}^2$. The total defect area finally amounts to $0.310 + 0.008\,\text{mm}^2 = 0.318\,\text{mm}^2$ when assuming merely one single scratch and one single dig.

It can be stated that in comparison to MIL-PRF-13830B, DIN ISO 10110 provides a more explicit and objective specification of acceptable surface defects by an absolute quantification of defect number and area and is therefore currently replacing MIL-PRF-13830B (Wang et al., 1999).

For coated optics surfaces, the designation of surface cleanliness is usually extended according to DIN ISO 10110. The full nomenclature is then "5/A · B;CM · N." The abbreviation $C$ indicates that the following identifiers $M$ and $N$ refer to the coating surface cleanliness similar to the identifiers $A$ and $B$ for the substrate's surface cleanliness. $M$ is thus the maximum acceptable number of defects, and $N$ quantifies the area of each single defect on a coating surface.

This designation may even be extended by further identifiers in order to take long scratches and chips at the lens edge[12] into account. Long scratches are indicated by the abbreviation $L$ and chips are abbreviated by the letter $E$. The specification "LQ · R" thus gives the maximum number ($Q$) and width ($R$) of scratches, and "ES" indicates the maximum excess length $S$ from the edge of an optical component toward the center of the optically active surface. Taking all these possible surface defects of a coated optical component into account, the full specification of surface cleanliness is then "5/A · B;CM · N;L Q · R, ES."

[12] In order to avoid the formation of chips, edges of optical components are usually beveled during each manufacturing step.

### 6.3.4 Surface Roughness

Apart from the surface accuracy and cleanliness, the residual roughness of polished optically active lens surfaces is an essential parameter that has to be specified since it directly impacts the transmission and reflection characteristics. As an example, the share of diffusively scattered light $R_{\text{diffuse}}$ in total reflection $R_{\text{total}}$ at an optics surface is given by

$$R_{\text{diffuse}} = TIS \cdot R_{\text{total}}. \tag{6.9}$$

The factor *TIS* is the so-called total integrated scatter, which mainly depends on the root mean squared surface roughness *Rq* according to

$$TIS = 1 - e^{-\left(\frac{4 \cdot \pi \cdot \cos AOI \cdot Rq}{\lambda}\right)^2}. \tag{6.10}$$

with $\lambda$ being the wavelength and *AOI* being the angle of incidence of the incident light (Bennett and Porteus, 1961).

Surface roughness is specified by appropriate symbols in manufacturing drawings as shown by the detail view of Figure 5.12 in Figure 6.16.[13]

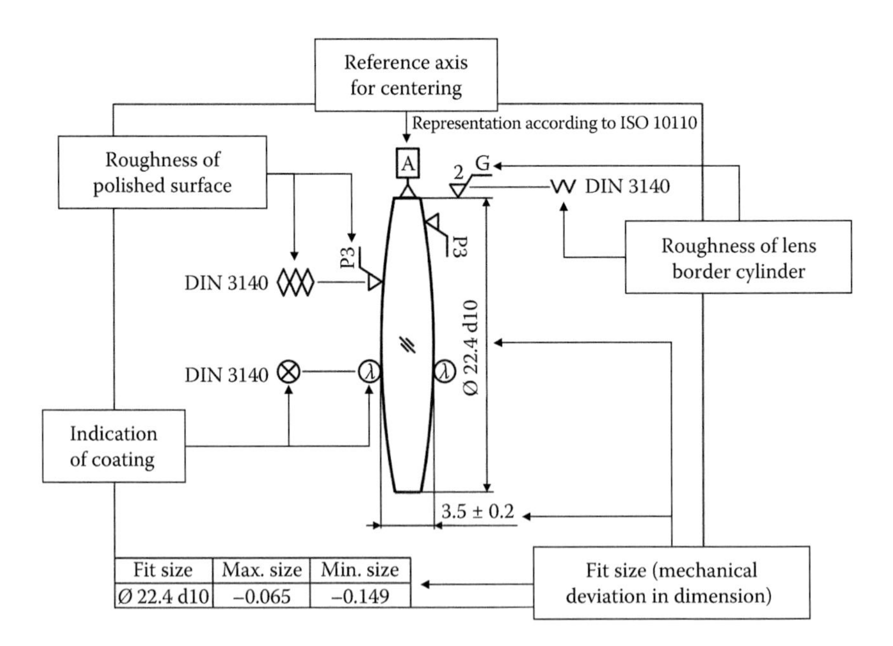


| Fit size | Max. size | Min. size |
|---|---|---|
| Ø 22.4 d10 | −0.065 | −0.149 |

**FIGURE 6.16** Detail view of the manufacturing drawing for a biconvex lens shown in Figure 5.12, showing supplementary information such as roughness of polished and ground surfaces, indication of coated surfaces, and mechanical fit sizes.

[13] Note that according to the overridden DIN 3140, the grade of polishing was indicated by rhombi. For example, three rhombi correspond to the notation “P3” according to DIN ISO 10110. Such rhombi are no longer used in DIN ISO 10110 but can still be found in some manufacturing drawings.

According to DIN ISO 10110, the nomenclature is "PX," where $P$ means "polishing" and $X$ is the so-called *grade of polishing* and residual surface roughness,[14] respectively, as listed in Table 6.7.

It should be noted that the grade of polishing indicated by P1–P4 specifies not only the surface roughness, but also the number of acceptable microdefects on an optics surface, such as microdigs, short scratches, and bubbles bared in the course of polishing as listed in Table 6.8. The indication of the grade of polishing thus represents supplemental information on surface cleanliness.

The impact of the surface roughness on the reflection characteristics becomes obvious when comparing the grades of polishing given in Table 6.7 and applying Equations 6.9 and 6.10. As shown in Figure 6.17, the share of diffusively scattered light in total reflection is significantly reduced when increasing the polishing quality (i.e., the grade of polishing).

Further, it can be seen that higher surface roughness leads to an increase in diffuse reflection and scattering, respectively, when reducing the wavelength of the incident

**TABLE 6.7**
**Grades of Polishing Including the Particular Residual Surface Roughness and Symbols Given in Manufacturing Drawings**

| Grade of Polishing | Residual Surface Roughness (nm) | Symbol Given in Manufacturing Drawing |
|---|---|---|
| Rough polished | 200–400 | P1 |
| Medium polished | 40–80 | P2 |
| Fine polished | 8–16 | P3 |
| Precision polished | 1.6–3.2 | P4 |

*Source:* Gerhard, C. et al., *Optik & Photonik*, 6, 35–38, 2011.

**TABLE 6.8**
**Grades of Polishing Including the Particular Acceptable Number of Microdefects and Symbols Given in Manufacturing Drawings**

| Grade of Polishing | Acceptable Number of Microdefects | Symbol Given in Manufacturing Drawing |
|---|---|---|
| Rough polished | 80–400 | P1 |
| Medium polished | 16–80 | P2 |
| Fine polished | 3–16 | P3 |
| Precision polished | <3 | P4 |

*Source:* Bliedtner, J., and Gräfe, G., *Optiktechnologie* (in German), Carl Hanser Verlag, München, Germany, 2008.

[14] The surface roughness indicated by the particular grade of polishing refers to a measuring length of 10 mm.

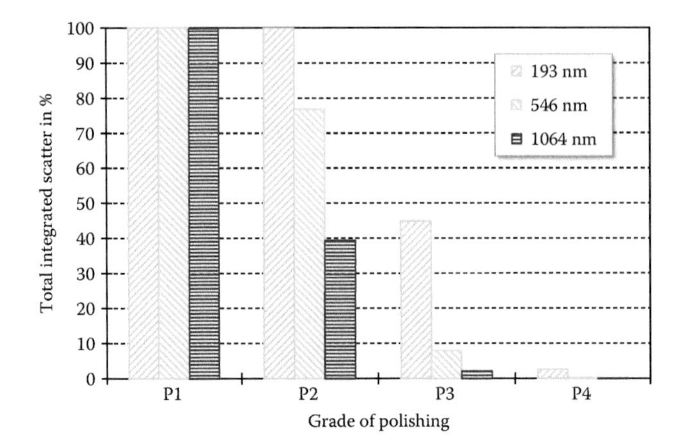


**FIGURE 6.17** Share of diffusively scattered light in total reflection expressed by the total integrated scatter *TIS* vs. grade of polishing for three selected wavelengths of light.

light. This fact is of essential importance for the production of UV optics, for example, excimer laser protection windows where usually even lower surfaces roughness as specified by the P4 grade of polishing are required and realized. Such surfaces are referred to as superpolished and feature roughness in the subnanometer range.

The reduction of scattering becomes also important when specifying the roughness of lens border cylinders, chamfers, or bevels. High surface roughness of such zones could give rise to increased scattering of vagabonding light and thus result in the formation of ghost images in optical systems. The roughness of nonoptically active surfaces of optical components is thus also specified by appropriate symbols in manufacturing drawings (compare Figure 6.16).[15] Here, the nomenclature is "GX" according to DIN ISO 10110, where *G* stands for "grinding," and the value *X* represents the *grade of grinding* and residual surface roughness, respectively. The particular values are listed in Table 6.9.

**TABLE 6.9**
**Grades of Grinding Including the Particular Residual Surface Roughness and Symbols Given in Manufacturing Drawings**

| Grade of Grinding | Residual Surface Roughness (µm) | Symbol Given in Manufacturing Drawing |
|---|---|---|
| Rough ground | 20–40 | G1 |
| Medium ground | 4–6 | G2 |
| Fine ground | 2–3 | G3 |
| Precision ground | <2 | G4 |

*Source:* Gerhard, C. et al., *Optik & Photonik*, 6, 35–38, 2011.

[15] Note that according to the overridden DIN 3140, the grade of grinding was indicated by chevrons. For example, three chevrons correspond to the notation "G3" according to DIN ISO 10110. Such chevrons are no longer used in DIN ISO 10110 but can still be found in some manufacturing drawings.

In the example shown in Figure 6.16, the grade of polishing of the optically active surfaces is P3. These surfaces should thus be fine polished and finally feature a residual roughness of 12±4 nm (8–16 nm, compare Table 6.7) and maximum 16 microdefects (compare Table 6.8). Further, the lens border cylinder roughness is marked G2 (where the number 2 is displayed separately). This means that its surface should be medium ground and consequently exhibit a roughness of 5±1 μm (4–6 μm, compare Table 6.9).

### 6.3.5 Geometrical Variations

For the geometry of an optical component, permissible geometrical variations are identified and specified where specific and relevant parameters of a particular component are addressed. For instance, appropriate tolerances for the wedge angle of prisms and wedges become of importance since geometrical variations may have a severe impact on the deviation (for deflection prism) or dispersion angle (for dispersion prisms). Appropriate tolerances, given in arc minutes (′) or arc seconds (″), are thus defined for the nominal wedge angle. For plane-parallel plates, the parallelism of the two optically active plane surfaces is specified according to DIN EN ISO 1101 and can be expressed as a tilt, given in arc minutes (′) or arc seconds (″), of both surfaces.

Relevant geometric parameters of a lens are its diameter and center thickness. In the first case, suitable tolerances are chosen in order to assure that the lens fits into a mount or tube during assembly without canting or excessive clearance (which might result in tilt or lateral decenter of the lens as described in Section 6.5). Appropriate tolerances are further specified to the lens center thickness since this parameter has a direct impact on the focal length of a lens (compare equations for thick lenses in Section 4.2.1).

Geometrical variations can be itemized by defining absolute lower and upper limits, by percentage deviation of the nominal value, or by applying appropriate tolerance classes. As an example, the same tolerance classes as for rotational-symmetric mechanic components such as shafts, spindles, and axles according to DIN ISO 286 can be applied for the specification of a lens diameter. Here, the corresponding maximum and minimum deviations depend on the nominal mechanical dimensions, which are classified into different ranges as exemplified by the selection in Table 6.10.

In the example shown in Figure 6.16, such a tolerance class is applied to the lens diameter, which is specified by the designation "22.4 d10." The corresponding maximum and minimum size for the nominal diameter of 22.4 mm and the tolerance class d10 are additionally shown and amount to −0.065 mm and −0.149 mm, respectively. The lower limit of the lens diameter is thus 22.251 mm, whereas the upper limit is 22.335 mm. With respect to the nominal diameter, the actual lens diameter should thus be undersized in any case.

The center thickness indicated in Figure 6.16 is specified by its nominal value of 3.5 mm and an allowance of ±0.2 mm. It can thus range from 3.3 to 3.7 mm. Taking this tolerance into account, the tolerance for the lens focal length can be specified indirectly. At the lower limit, the focal length is 101.665 mm, whereas a focal length of 101.725 mm results for the upper limit. The given tolerance for the center thickness of ±0.2 mm thus corresponds to a variation in focal length by ±60 μm.

**TABLE 6.10**
**Examples for Tolerance Classes and the Corresponding Maximum and Minimum Deviation from the Nominal Mechanical Dimension (Outer Diameter) according to DIN ISO 286**

| Nominal Dimension Range | Tolerance Class | | | |
|---|---|---|---|---|
| | f6 | h6 | k6 | r6 |
| | Maximum/minimum deviation (values given in μm) | | | |
| 10–18 mm | −16/−27 | 0/−11 | +12/+1 | +34/+23 |
| 18–30 mm | −20/−33 | 0/−13 | +15/+2 | +41/+28 |
| 30–50 mm | −25/−41 | 0/−16 | +18/+2 | +50/+34 |

*Source:* Data taken from DIN EN ISO 286.

### 6.3.6 Coatings

As already introduced above, the coatings applied to optically active surfaces are mentioned and specified in the box for specification of fabrication tolerances. The surfaces to be coated are identified by the symbol $\lambda$ (or ⊗ according to the obsolete DIN 3140, see Figure 6.16). In the present case, both lens surfaces are to be coated with an antireflective coating with a residual reflectance smaller than 0.5% in the visible wavelength range.

## 6.4 SPECIFICATION OF LASER-INDUCED DAMAGE THRESHOLD

One important field of application for optical components is laser technology, where collimating, defocusing, and—of course—focusing systems are set up on the basis of single lenses or where prisms are used for laser beam guidance. Here, optics surfaces can be damaged due to the high power or energy density of laser beams.[16] Once a glass surface is irreversibly modified by an incident laser beam, for example, in the form of rippling or wrinkling, the coupling of laser power or energy into the surface is supported by multiple reflections (i.e., scattering) and absorption. Further laser irradiation then leads to material removal by ablation and surface damage, respectively, as shown in Figure 6.18.

Against this background, the *laser-induced damage threshold* (LIDT) is defined by the index number "6" according to DIN ISO 10110.[17] This value depends on a number of influencing factors. First, surface cleanliness has a considerable impact since incident laser irradiation can be absorbed at organic surface contaminations

[16] As an example, the LIDT of fused silica exposed to pulsed laser irradiation is approximately 6–10 J/cm$^2$, depending on the laser parameters.

[17] Note that according to DIN 3140, the index number "6" indicated stress birefringence. The use of this index number for two different parameters may cause confusion; the underlying standard used for the generation of a manufacturing drawing should thus be checked since older and nonmodified drawings are still in use.

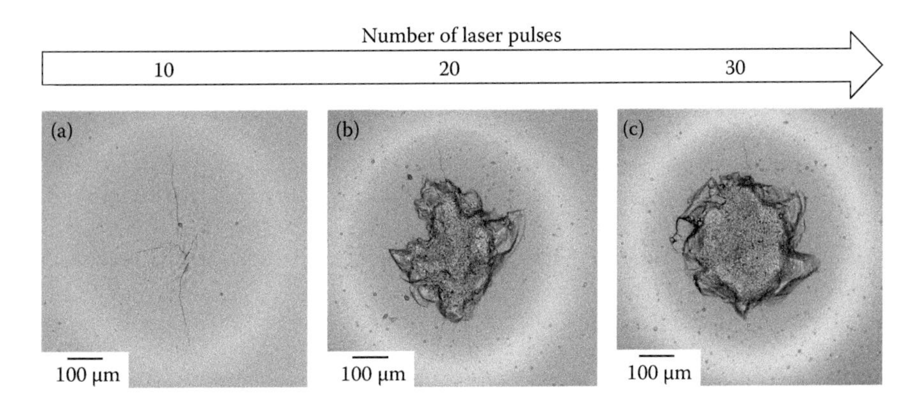


**FIGURE 6.18** Evolution of laser-induced surface damage at the example of an optical crown glass surface when increasing the number of applied laser pulses. After 10 laser pulses, the surface is rippled and cracks are formed due the thermal impact of the laser beam (a). For higher number of pulses (b, c), ablation occurs.

(Bien-Aimé et al., 2009; Cheng et al., 2014; Gerhard et al., 2017) such as hydrocarbons, or residues from grinding or polishing agents (Boling and Dubé, 1973; Neauport et al., 2005). In addition, subsurface damage, for example, microcracks from grinding, lapping, and polishing, represent precursors for laser damage (Neauport et al., 2009a) since such microcracks give rise to scattering of laser irradiation and can be described as cavities for accumulation of contaminants (Camp et al., 1998), for example, coolants and lubricants used for grinding (Neauport et al., 2009b). Even trace contaminations, which can occur during laser operation in industrial production lines (e.g., impurities from process gases, dust or metallic microparticles), can cause laser damage (Hovis et al., 1994; Genin et al., 1997) as shown in Figure 6.19.

Second, surface roughness plays an important role where high roughness supports the initiation of laser damage (Ihlemann et al., 2007; Uteza et al., 2007) due

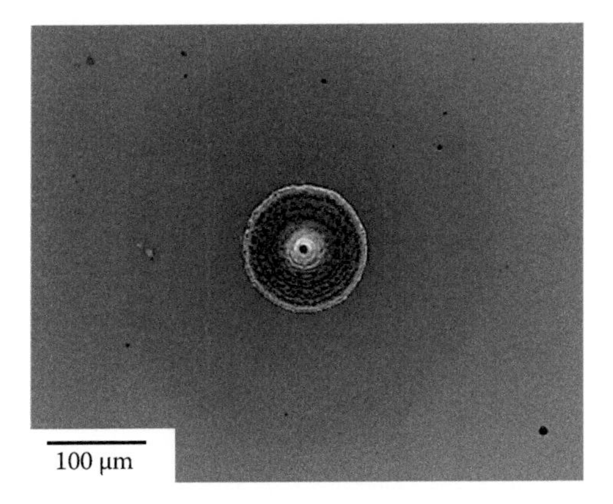


**FIGURE 6.19** Initial stage of laser damage due to the absorption of laser irradiation at dust particles (black/dark spots), leading to circularly shaped damage sites.

to scattering and multiple reflection of laser irradiation at roughness peaks. Laser-induced damage is thus directly dependent on surface cleanliness and the grade of polishing. Laser optics components are thus usually superpolished where the surface roughness is in the range of some angstrom.[18] However, the surface cleanliness and roughness of coatings applied to optics surfaces may notably differ from the values specified for the substrate (see Section 6.3.3). It is thus obvious that a proper definition of the LIDT is required for laser optics. Here, it is important to distinguish between two types of laser irradiation, continuous and pulsed, since they differ in terms of the particular damage mechanism.

For continuous laser irradiation, the LIDT is specified by the full nomenclature "6/P;λ;N." Here, $P$ is the maximum power density given in W/cm², $\lambda$ is the laser wavelength, and $N$ is the number of test sites on the optical component's surface. For pulsed laser irradiation, the full nomenclature is "6/E;λ;Δτ;f;N · n," where $E$ is the maximum energy density given in J/cm², $\Delta\tau$ indicates the pulse duration group as specified by ISO 11254, $f$ is the laser pulse repetition rate given in Hz, $N$ is the number of test sites, and $n$ is the number of laser pulses applied to each test site.

One has to consider that the LIDT differs significantly for glasses and materials commonly used in optics manufacturing. A selection of such media including the particular LIDT (valid at a laser wavelength of 1064 nm and a pulse duration in the nanosecond range) is listed in Table 6.11.

Further, the LIDT decreases with a decreasing laser wavelength due to the higher photon energy,[19] as well as the higher absorption coefficient and the accompanying improved absorption of incoming laser light in the particular glass or material in this case.

**TABLE 6.11**
**LIDT of Selected Media Used in Optics Manufacturing (Data Valid for a Laser Wavelength of $\lambda = 1064$ nm and a Laser Pulse Duration of Some Nanoseconds)**

| Material | LIDT (J/cm²) |
|---|---|
| Optical fine cement | ≈2 |
| Metallic mirror coatings | ≈3 |
| Fused silica | ≈6 |
| Dielectric antireflective coatings | ≈4–50 (depending on type of coating and substrate) |

[18] The angstrom or ångström (symbol: Å) is a unit of length and amounts to 100 picometers. It is named after the Swedish physicist *Anders Jonas Ångström* (1814–1874) and represents the current unit for atomic radii. For example, the atomic radius of silicon amounts to 1.1 Å = 110 pm (Slater, 1964).

[19] The photon energy is given by $E_{photon} = h \cdot f$. Here, $h$ is the Planck constant of $h \approx 6.626 \cdot 10^{-34}$ Js, and $f$ is the frequency of light. The lower the wavelength $\lambda$, the higher the frequency of light according to $f = c/\lambda$ and the higher the photon energy, respectively.

## 6.5 TOLERANCING OF OPTOMECHANICAL ASSEMBLIES

Generally, *position tolerances* in optomechanical systems and assemblies have a significant impact on the performance (i.e., the imaging quality of such devices). As shown in Figure 6.20, several defects due to inaccuracies of mechanical elements such as mounts or tubes can occur in the course of mounting optical components: (1) tilts $\Delta\varepsilon$ of the optical axes of optical components with respect to mechanical axes of the mount due to tilted bearing surfaces,[20] (2) lateral offsets $\Delta y$ (a.k.a. decenters) due to decentered bearing surfaces, and (3) distance errors $\Delta z$ (a.k.a. longitudinal decenters) between lenses and lens groups, for example, due to inaccurately structured mounts or insufficient tolerancing of spacers (in stacked systems).

Such defects have to be considered and specified by appropriate position tolerances. Mechanical dimensions such as inner diameters of optical mounts including the tolerances are specified by DIN ISO 286. This standard defines tolerance classes for rotational-symmetric mechanic hollow components and the inner diameter of such elements. As shown by the examples listed in Table 6.12, each tolerance class

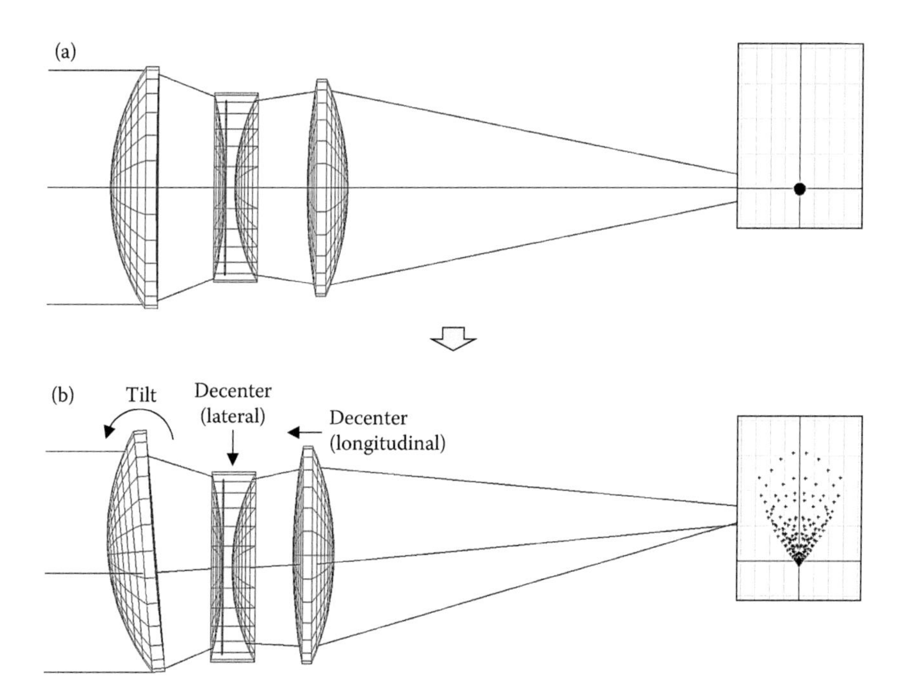


**FIGURE 6.20** Comparison of a Cooke-triplet without (a) and with positioning errors (b) (i.e., tilt, as well as lateral and longitudinal decenter, a.k.a. lateral offset and distance error), including impact of these errors on the resulting imaging quality visualized by the particular spot diagram (inset). (Figure was generated using the software WinLens3D Basic from Qioptiq Photonics GmbH & Co. KG.)

[20] Tilt is due not only to mechanical elements such as mounts; it can also result from poor lens centering (i.e., a high deviation of the optical axis from the mechanical lens border cylinder as specified by index number "4" according to DIN ISO 10110).

**TABLE 6.12**
**Examples for Tolerance Classes and the Corresponding Maximum and Minimum Deviation from the Nominal Mechanical Dimension (Inner Diameter) according to DIN ISO 286**

| Nominal Dimension Range (mm) | Tolerance Class | | | |
|---|---|---|---|---|
| | E6 | H6 | K6 | R6 |
| | Maximum/minimum deviation (values given in μm) | | | |
| 10–18 | +43/+32 | +11/0 | +2/–9 | –20/–31 |
| 18–30 | +53/+40 | +13/0 | +2/–11 | –24/–37 |
| 30–50 | +66/+50 | +16/0 | +3/–13 | –29/–45 |

*Source:* Data taken from DIN EN ISO 286.

indicates the fit size (i.e., the maximum and minimum deviation from the nominal mechanical dimension). These deviations are related to dimension ranges; each tolerance class thus indicates different absolute maximum and minimum deviations dependent on the particular inner diameter.

In addition to the tolerancing of inner diameters of lens mounts and tubes, tilts, lateral offsets and distance errors are specified by the Geometric Dimensioning and Tolerancing standard as defined by either the US-Geometric Product Specification-Standard ASME Y 14.5 or the ISO 1101 by the International Organization for Standardization.

These standards cover the form (i.e., straightness, flatness, circularity, etc.), orientation (i.e., perpendicularity, angularity, parallelism), and location (i.e., symmetry, position, concentricity) of mechanical components.

## 6.6 SUMMARY

All relevant information required for the manufacture of optical components is given in manufacturing drawings where the representation is usually according to DIN ISO 10110. Here, the glass quality and production accuracy are indicated by appropriate symbols. Glass quality is specified in terms of stress birefringence, bulk defects such as bubbles and inclusions, and inhomogeneity of the index of refraction and striae. An overview on these parameters and the particular tolerance indications is shown in Table 6.13.

The production accuracy is further determined regarding surface contour accuracy, acceptable centering error, and surface cleanliness, as listed in Table 6.14. The contour accuracy describes the deviation between the actual shape of a surface and the target geometry; it is tested interferometrically by evaluating interference patterns. The angular deviation between the optical axis and the mechanical axis of a lens is referred to as centering error. Surface cleanliness is described by the maximum number and size of surface effects. In the last case, the cleanliness of either the

**TABLE 6.13**
**Tolerance Indications for Glass Quality according to DIN ISO 10110**

| Parameter | Index Number | Full Nomenclature | Identifier | Meaning |
|---|---|---|---|---|
| Stress birefringence | 0 | 0/A | A | Maximum optical path difference |
| Bubbles and inclusions | 1 | 1/N · A | N | Maximum number of bulk defects |
| | | | A | Maximum cross-sectional area of one bulk defect |
| Inhomogeneity and striae | 2 | 2/A, B | A | Inhomogeneity class |
| | | | B | Striae class |

**TABLE 6.14**
**Tolerance Indications for Production Accuracy according to DIN ISO 10110**

| Parameter | Index Number | Full Nomenclature | Identifier | Meaning |
|---|---|---|---|---|
| Contour accuracy | 3 | 3/A(B/C) | A | Maximum sagitta |
| | | | B | Surface irregularity |
| | | | C | Fine contour error |
| Centering error | 4 | $4/X^Y$ | X | Maximum angular deviation between optical and mechanical axis |
| | | | Y | Unit of angular deviation |
| Surface cleanliness (uncoated component) | 5 | 5/A · B | A | Maximum number of glass surface defects |
| | | | B | Maximum area of one glass surface defect |
| Surface cleanliness (coated component) | 5 | 5/...; CM · N | C | Indicator for coating |
| | | | M | Maximum number of coating surface defects |
| | | | N | Maximum area of one coating surface defect |

final polished glass surface or the surface of coatings applied to optical components can be specified.

In addition, the roughness of polished and ground surfaces, geometrical variations and the type, and residual reflectivity of coatings are particularized. For laser optics, the LIDT is specified as listed in Table 6.15.

Finally, dimensions and position errors of optomechanical assemblies such as tilts, lateral offsets, or distance errors between mounted optical components are enumerated on the basis of DIN ISO 286, ASME Y 14.5, and ISO 1101.

**TABLE 6.15**
**Tolerance Indications for LIDT for Continuous-wave (CW) or Pulsed Laser Irradiation (P) according to DIN ISO 10110**

| Parameter | Index Number | Full Nomenclature | Identifier | Meaning |
|---|---|---|---|---|
| LIDT (CW) | 6 | 6/P;λ;N | P | Maximum power density |
| | | | λ | Laser wavelength |
| | | | N | Number of test sites |
| LIDT (P) | 6 | 6/E;λ;Δτ;f;N · n | E | Maximum energy density |
| | | | λ | Laser wavelength |
| | | | Δτ | Pulse duration group |
| | | | f | Laser pulse repetition rate |
| | | | N | Number of test sites |
| | | | n | Number of laser pulses applied to each test site |

## 6.7 FORMULARY AND MAIN SYMBOLS AND ABBREVIATIONS

**Difference in optical path length *OPD* (following from stress birefringence):**

$$OPD = 10 \cdot K \cdot t \cdot \sigma_{\mathrm{m}}$$

$K$ photo-elastic coefficient (of optical component)
$t$ thickness (of optical component)
$\sigma_{\mathrm{m}}$ mechanical tension

**Deviation in index of refraction $\Delta n$ (determination via wave front measurement):**

$$\Delta n = \frac{\Delta w}{2 \cdot t}$$

$\Delta w$ wave front deformation
$t$ sample thickness

**Etching time $t_{\mathrm{AR}}$ for determination of acid resistance class:**

$$t_{\mathrm{AR}} = \frac{t_{\mathrm{e}} \cdot \rho \cdot A}{(m_0 - m_e) \cdot 100}$$

$t_{\mathrm{AR}}$ time required for removal of 100 nm-thick surface layer
$t_{\mathrm{e}}$ total test duration
$\rho$ glass density
$A$ sample surface area
$m_0$ initial mass of sample
$m_{\mathrm{e}}$ mass of sample after experiment

**Knoop hardness *HK*:**

$$HK = 1.451 \cdot \frac{F}{l^2}$$

$F$ force
$l$ length

**Glass grindability *G*:**

$$G = \frac{\Delta V_{\text{glass}}}{\Delta V_{\text{ref}}}$$

$\Delta V_{\text{glass}}$ removed volume of tested glass
$\Delta V_{\text{ref}}$ removed volume of reference glass (with $G = 100$ per default)

**Coefficient of thermal expansion *α*:**

$$\alpha = \frac{1}{\Delta T} \cdot \frac{\Delta l}{l_0}$$

$\Delta T$ change in temperature
$\Delta l$ change in length
$l_0$ original length

**Centering error *CE*:**

$$CE = \frac{1720 \cdot r_{\text{w}}}{m \cdot d \cdot (n-1)}$$

$r_{\text{w}}$ measured wobble circle radius
$m$ magnification of the used optical test setup
$d$ distance principal plane to detector plane
$n$ index of refraction of lens material

**Total integrated scatter *TIS*:**

$$TIS = 1 - e^{-\left(\frac{4 \cdot \pi \cdot \cos AOI \cdot Rq}{\lambda}\right)^2}$$

$AOI$ angle of incidence
$Rq$ root mean squared surface roughness
$\lambda$ wavelength

## REFERENCES

Aikens, D.M. 2010. The truth about scratch and dig. *International Optical Design Conference and Optical Fabrication and Testing, OSA Technical Digest*, paper No. OTuA2.

Bennett, H.E., and Porteus, J.O. 1961. Relation between surface roughness and specular reflectance at normal incidence. *Journal of the Optical Society of America* 51:123–129.

Bien-Aimé, K., Néauport, J., Tovena-Pecault, I., Fargin, E., Labrugère, C., Belin, C., and Couzi, M. 2009. Laser induced damage of fused silica polished optics due to a droplet forming organic contaminant. *Applied Optics* 48:2228–2235.

Bliedtner, J., and Gräfe, G. 2008. *Optiktechnologie*. München, Germany: Carl Hanser Verlag (in German).

Boling, N.L., and Dubé, G. 1973. Laser-induced inclusion damage at surfaces of transparent dielectrics. *Applied Physics Letters* 23:658–660.

Camp, D.W., Kozlowski, M.R., Sheehan, L.M., Nichols, M.A., Dovik, M., Raether, R.G., and Thomas, I.M. 1998. Subsurface damage and polishing compound affect the 355-nm laser damage threshold of fused silica surfaces. *Proceedings of SPIE* 3244:356.

Cheng, X., Miao, X., Wang, H., Qin, L., Ye, Y., He, Q., Ma, Z., Zhao, L., and He, S. 2014. Surface contaminant control technologies to improve laser damage resistance of optics. *Advances in Condensed Matter Physics* 2014:974245.

Genin, F.Y., Michlitsch, K., Furr, J., Kozlowski, M.R., and Krulevitch, P.A. 1997. Laser-induced damage of fused silica at 355 and 1064 nm initiated at aluminum contamination particles on the surface. *Proceedings of SPIE* 2966:126.

Gerhard, C., Tasche, D., Munser, N., and Dyck, H. 2017. Increase in nanosecond laser-induced damage threshold of sapphire windows by means of direct dielectric barrier discharge plasma treatment. *Optics Letters* 42:49–52.

Gerhard, C., Wienecke, S., and Lotz, S. 2011. Was ist genau? Fertigungstoleranzen optischer Komponenten und Systeme. *Optik & Photonik* 6:35–38.

Hovis, F.E., Shepherd, B.A., Radcliffe, C.T., Bailey, A.L., and Boswell, W.T. 1994. Optical damage at the part per million level: The role of trace contamination in laser-induced optical damage. *Proceedings of SPIE* 2114:145.

Ihlemann, J., Schulz-Ruhtenberg, M., and Fricke-Begemann, T. 2007. Micro patterning of fused silica by ArF- and F2-laser ablation. *Journal of Physics: Conference Series* 59:206–209.

Neauport, J., Ambard, C., Cormont, P., Darbois, N., Destribats, J., Luitot, C., and Rondeau, O. 2009a. Subsurface damage measurement of ground fused silica parts by HF etching techniques. *Optics Express* 17:20448–20456.

Neauport, J., Cormont, P., Legros, P., Ambard, C., and Destribats, J. 2009b. Imaging subsurface damage of grinded fused silica optics by confocal fluorescence microscopy. *Optics Express* 17:3543–3554.

Neauport, J., Lamaignere, L., and Bercegol, H. 2005. Polishing-induced contamination of fused silica optics and laser induced damage density at 351 nm. *Optics Express* 13:10163–10171.

Schaeffer, H.A., and Langfeld, R. 2014. *Werkstoff Glas*. Berlin and Heidelberg: Springer Vieweg.

Schott, A.G. 2013. ZERODUR® Zero Expansion Glass Ceramic Data Sheet.

Schott, A.G. 2015. Optical Glass Data Sheets.

Slater, J.C. 1964. Atomic radii in crystals. *Journal of Chemical Physics* 41:3199–3205.

Standard DIN EN ISO 286-2:2010-11. 2010. Geometrical product specifications (GPS)—ISO code system for tolerances on linear sizes—Part 2: Tables of standard tolerance classes and limit deviations for holes and shafts (ISO 286-2:2010).

Uteza, O., Bussière, B., Canova, F., Chambaret, J.-P., Delaporte, P., Itina, T., and Sentis, M. 2007. Laser-induced damage threshold of sapphire in nanosecond, picosecond and femtosecond regimes. *Applied Surface Science* 254:799–803.

Wang, D.Y., English, R.E., and D.M. Aikens. 1999. Implementation of ISO 10110 optics drawing standards for the National Ignition Facility. *Proceedings of SPIE 3782, Optical Manufacturing and Testing III*, 502–508.

# 7 Shape Forming

## 7.1 INTRODUCTION

It is said that the Italian sculptor *Michelangelo di Lodovico Buonarroti Simoni* (1475–1564), commonly simply called *Michelangelo*, once was asked how he managed to do his famous masterpiece sculpture *David*. He replied, "David was always there in the marble. I just took away everything that was not David." In principle, this legendary reply describes the classical shape forming process of any optical component. Starting with a simple block of glass, the target device, for example, a high-precision lens, is successively excavated in classical manufacturing. This is performed by different subsequent machining steps where both the surface shape and smoothness are successively approximated to the target values. First, lens blanks or preforms are produced in the course of preshaping. Second, the actual surface shape, for example a spherical segment, is realized. The last process consists of different significant machining steps that differ in terms of the tools used and size of abrasives.

In this chapter, the particular manufacturing steps for shape forming (i.e., preshaping, rough grinding, and finish grinding or lapping) are primarily shown for lenses. However, the presented manufacturing methods and procedures, as well as the tools and operating materials used, are also applied for other components such as windows and prisms.

## 7.2 PRESHAPING

*Preshaping* describes the process of making blanks or preforms from the raw material, generally glass. Here, two main approaches can be identified: molding and classical shape forming. Such classical forming can be realized in different ways, as summarized in Figure 7.1 and explained in detail in the following sections.

### 7.2.1 Compression Molding of Pressed Blanks

For the production of lens blanks via *compression molding*, a certain volume of glass raw material is heated up to its processing temperature (see Section 3.2.2.2 and Table 3.8), i.e., approximately 450°C–950°C for multicomponent glasses or taken directly from a glass melt for the melt cooling procedure. Subsequently, a drop of heated glass, a so-called *gob*, is brought into a mold as shown in Figure 7.2.

The mold represents the negative of the target surface shape of the final preform. The glass gob is then compression molded by a plunger, cooled to ambient temperature, and finally removed from the mold. As a result, a pressed blank with nearly the target geometry of the optical component, but adequate allowances for subsequent machining steps, is obtained.

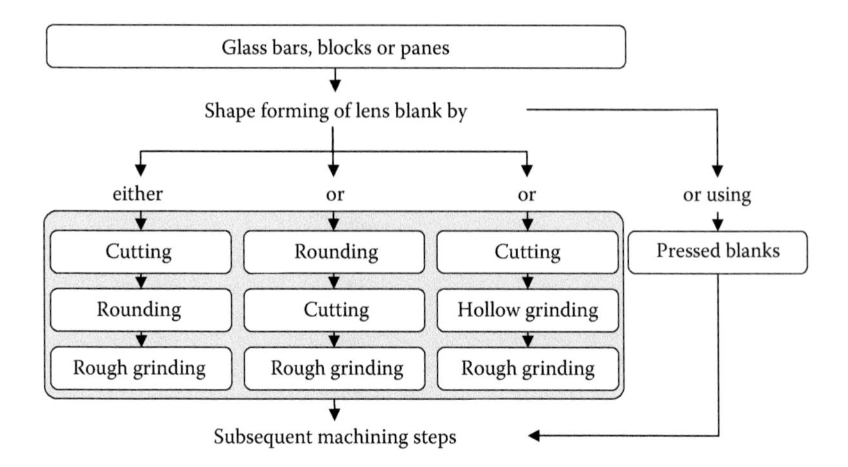


**FIGURE 7.1** Flow diagram of approaches for preshaping lens blanks.

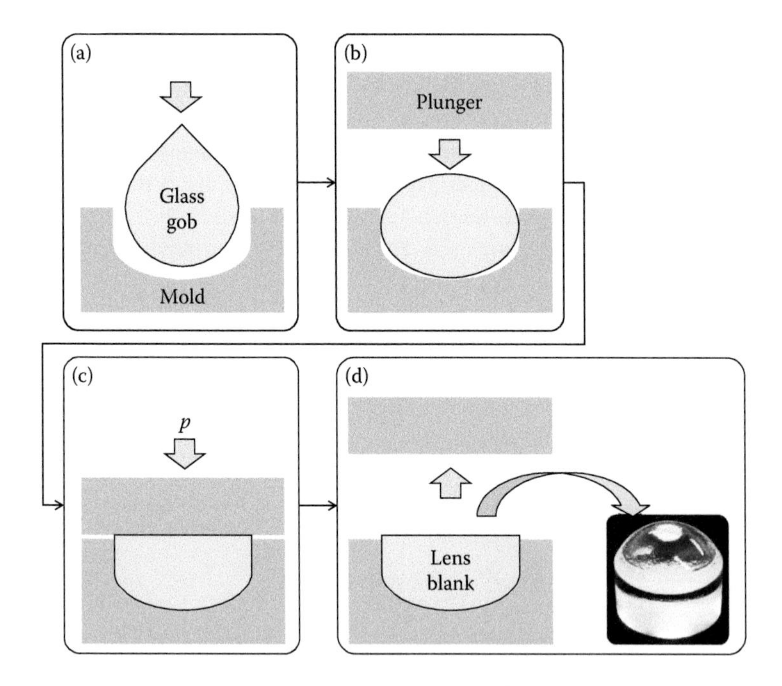


**FIGURE 7.2** Preshaping of a lens blank by compression molding; a glass gob is filled into a mold (a) and then molded by a plunger (b, c), resulting in the generation of a preshaped blank of a plano-convex lens after cooling (d).

Due to the high process velocity, compression molding is a powerful method for large-scale production of blanks. It further allows easy realization of comparatively complex shapes such as aspherical or free-form surfaces (compare Section 13.4.4). However, there are some challenges in the process. First, shrinking of the glass gob during cooling has to be taken into account and compensated for by appropriate design of the molds, dies, and plungers. This task might become complicated for complex surface geometries. Second, the cooling process needs to be well controlled in order to avoid devitrification or the formation of internal stress birefringence within the final pressed preform. Third, the thermal impact on the glass material may induce defects due to hydrolytic scission or cause deviations and inhomogeneities in the index of refraction. Actually, not every optical glass can be preshaped via compression molding. For such glasses, the classical approach of cutting and rounding as explained in the following section is employed.

### 7.2.2 Cutting and Rounding

#### 7.2.2.1 Cutting

*Cutting* is the first step of classical optics manufacturing; blanks of optical components are cut out of glass blocks or bars supplied by glass manufacturers. For this purpose, classic *circular saw benches* as shown in Figure 7.3 are used.

The cutting edge of the saw blades features a special coating consisting of a matrix made of copper, brass, plastics, or other suitable materials and grains of the actual cutting medium that is embedded within this matrix. Well-established cutting media are diamond, corundum, and silicon carbide. As listed in Table 7.1, these media feature a significantly higher hardness than optical glasses, where the Knoop hardness ranges from approximately 50–750 (compare Section 6.2.4.2).

Diamond and silicon carbide are among the hardest available materials, and corundum can be found in the group of second hardest substances known. This becomes obvious when classifying hardness according to the Mohs scale of mineral hardness as suggested by the German-Austrian mineralogist *Carl Friedrich Christian Mohs* (1773–1839) and listed in Table 7.2. For comparison, glasses have an intermediate Mohs hardness of approximately 5.5.

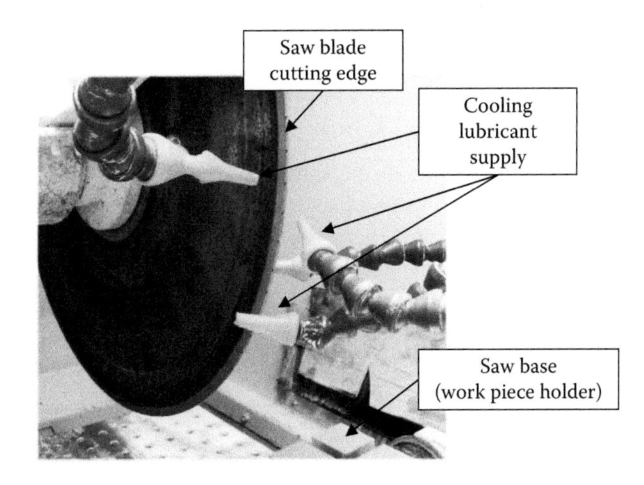


**FIGURE 7.3** Circular saw bench as used for preproduction in optics manufacturing consisting of a saw base that acts as work piece holder, a saw blade with coated cutting edge, and adjustable cooling lubricant supply tubes.

**TABLE 7.1**
**Selection of Cutting Media Used for Saw Blades for Glass Cutting Including the Particular Knoop Hardness**

| Material | Symbol/Total Formula | Knoop Hardness |
|---|---|---|
| Diamond | C | 7000 |
| Corundum | $\alpha$-$Al_2O_3$ | 2000 |
| Silicon carbide | SiC | 2500 |

*Source:* Bliedtner, J., and Gräfe, G., *Optiktechnologie*, Carl Hanser Verlag, München, Germany, 2008 (in German).

**TABLE 7.2**
**Classification of Hardness of Different Materials according to Mohs Scale of Mineral Hardness**

| Material | Mohs Hardness |
|---|---|
| Talc | 1 |
| Gypsum | 2 |
| Calcite | 3 |
| Fluorite | 4 |
| Apatite | 5 |
| Orthoclase | 6 |
| Quartz | 7 |
| Topaz | 8 |
| Corundum ($\alpha$-$Al_2O_3$) | 9 |
| Diamond (C) and silicon carbide (SiC) | 10 |

*Note:* For the materials used for cutting in optics manufacturing as listed in Table 7.1, the symbol or total formula is additionally given in brackets.

During cutting, the glass block or bar is attached to the saw base, which consequently acts as work piece holder. Such fixing can be realized by (1) pressing the work piece on a stop as applicable for glass blocks and bars with high net weight and inertia, respectively, (2) mechanical clamping, or (3) cementing the work piece on a carrier glass and clamping the carrier glass. The latter approach is mainly used for small work pieces and has the advantageous effect of less chipping by the saw blade on the rear side.

The cement used in this case is referred to as *raw cement*.[1] This working material is a mixture of different temperature-sensitive natural materials, for example, wax, colophony, wood resins, or pitch, as shown in Table 7.3.

[1] A distinction is drawn between raw cement and fine cement. The first type of cement is usually non-transparent and comparatively viscous. It is used for fixing optical components during manufacturing, for example, in the course of cutting or grinding. In contrast, fine cement is transparent and highly fluid and features a well-defined index of refraction; it is applied to optically active surfaces in order to cement lenses and prisms to groups.

**TABLE 7.3**
**Example for the Composition of Raw Cement**

| Medium | Content in % |
|---|---|
| Colophony | 60 ± 10 |
| Natural resin | 7.5 ± 2.5 |
| Synthetic paint resin | ≈2 |
| Shellac | 25 ± 5 |
| Colophony oil | 10 ± 5 |

*Source:* Pforte, H., *Der Optiker*, Verlag Gehlen, Homburg, Germany, 1995 (in German).
*Note:* Here, the composition of shellac-containing raw cement is shown.

**TABLE 7.4**
**Overview on Raw Cement Materials Including Selected Mechanical and Thermal Properties**

| Cement Material | Shear Strength in g/cm² | Tensile Strength in g/cm² | Flow Temperature (°C) | Condensing Temperature (°C) |
|---|---|---|---|---|
| Dental wax | 7,200 | 9,000 | 55 | 56 |
| Beeswax and resin | 20,000 | 12,000 | 56 | 62 |
| Pitch | 54,000 | 4,500 | 60 | 130 |
| Tan wax | >54,000 | >54,000 | 60 | 125 |

*Source:* Fynn, G.W., and Powell, W.J.A., *The Cutting and Polishing of Electro-optic Materials*, John Wiley & Sons, New York, 1979.

Raw cement features a low melting point and resolidifies at ambient temperature. It is thus easy to apply to and to remove[2] from glass surfaces by heating, where heating is carried out with the aid of Bunsen burners or conventional hotplates. Raw cement is thus the most important adhesive in classical optics manufacturing and is generally applied where temporal fixing is required. It is suitable not only for cutting processes, but also for other subsequent manufacturing steps such as rounding, grinding, or centering, where machine vibrations are compensated for by the cement due to its elasticity. At some extent, it thus contributes to process stability. Different cement materials including selected relevant mechanical[3] and thermal properties are listed in Table 7.4.

Another working material used for cutting but also for rounding, grinding, and centering is the *cooling lubricant*, an emulsion of water and synthetic or semisynthetic

[2] Raw cement can be solved by different chemical solvents, see Section 14.2.1.
[3] The mechanic properties listed here refer to a cement layer with a thickness of 20 μm and an area of 1 mm² applied to polished glass and ground stainless steel.

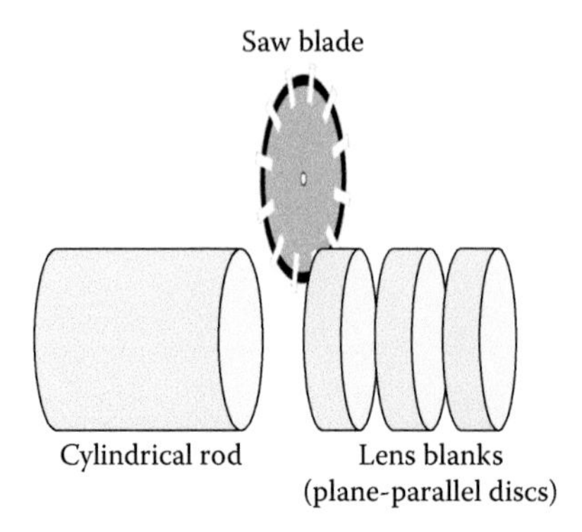


**FIGURE 7.4** Preproduction of lens blanks by cutting plane-parallel discs from a cylindrical glass rod.

mineral oils. It is supplied to the cut area (i.e., the contact zone of the saw blade cutting edge and the glass) where it fulfills different functions. First, it cools the glass surface and the saw blade, which heat up due to friction. Second, it effects a certain lubrication and thus allows increasing the cutting quality. Third, it removes glass debris from the cutting point.

#### 7.2.2.2 Rounding

The process of *rounding* is applied during the preproduction of lens blanks in order to produce plane-parallel discs. This can be achieved by three different approaches. The easiest way is to slice cylindrical rods supplied by the glass manufacturer or produced by the rounding of rectangular rods prior to cutting. The rod diameter corresponds to the diameter of the lens to be produced with adequate allowance for later centering. As shown in Figure 7.4, plane-parallel cylindrical discs are cut off from the rod (adequate allowances have to be considered for the disc thickness since the comparatively high surface roughness and deep microcracks resulting from cutting have to be removed or minimized in the course of subsequent rough grinding, finish grinding, and polishing).

Another method for rounding is to cut quadratic glass plates from blocks, bars, or rectangular rods. This approach is applied, since usually the raw material glass is provided in this form, which also allows the realization of any optical component, not only lenses, but also prisms or wedges.[4] For the production of lens blanks, quadratic plates are cut so that the plate thickness corresponds to the final lens center thickness with adequate allowance. As shown in Figure 7.5, the cut plates are subsequently agglutinated to a rod using raw cement.

Rounding of this rod is then performed with a *turning lathe* (i.e., a turning machine featuring two opposite spindles with coaxially arranged mechanical axes). The cemented rod is clamped between both spindles, which are then driven at the same rotation velocity $\omega$ as shown in Figure 7.6a.

[4] More complex blanks of optical components such as roof prisms or other 3D-geometries are cut or realized by rough grinding from a solid glass block applying multiple-axes grinding machines.

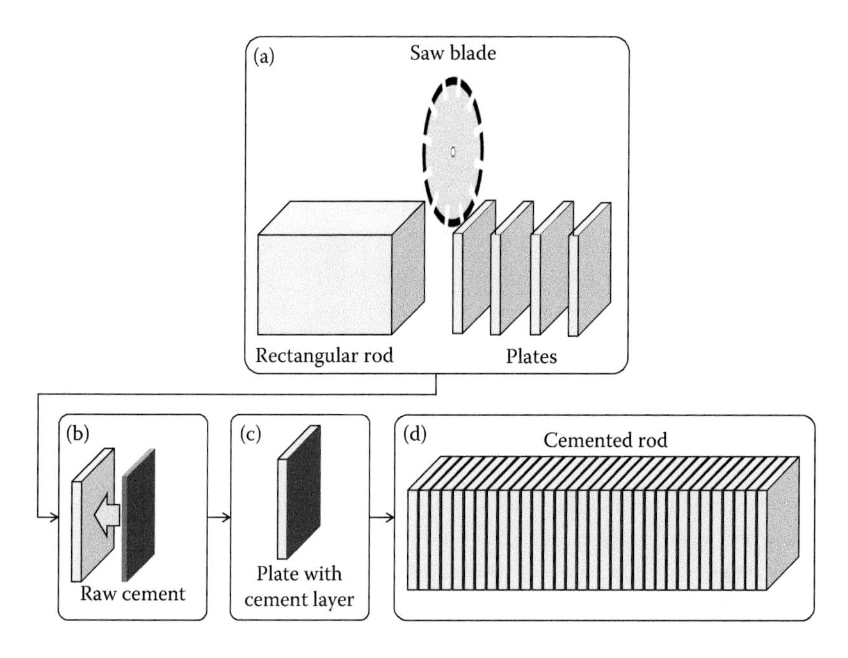


**FIGURE 7.5** First step of preproduction of lens blanks by cutting plates from a rectangular glass rod (a). The cut plates are subsequently cemented to a rod using raw cement (b–d).

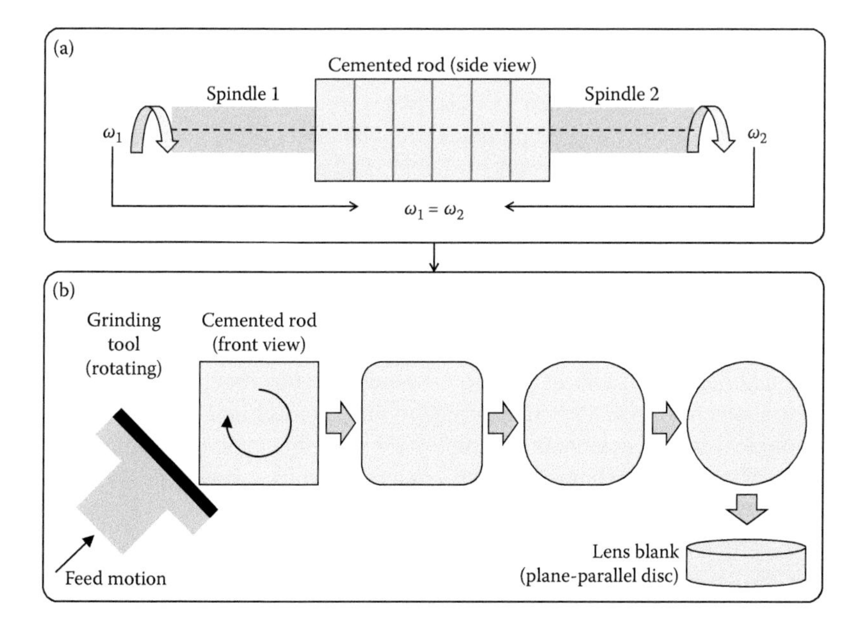


**FIGURE 7.6** Second step of preproduction of lens blanks by cutting plates from a rectangular glass rod. After cementing cut plates to a rod as shown in Figure 7.5, the cemented rod is clamped in a turning lathe (a), rotated at a certain rotation velocity $\omega$, and successively rounded by a grinding tool (b). Finally, the cement is solved, and the actual lens blank is obtained.

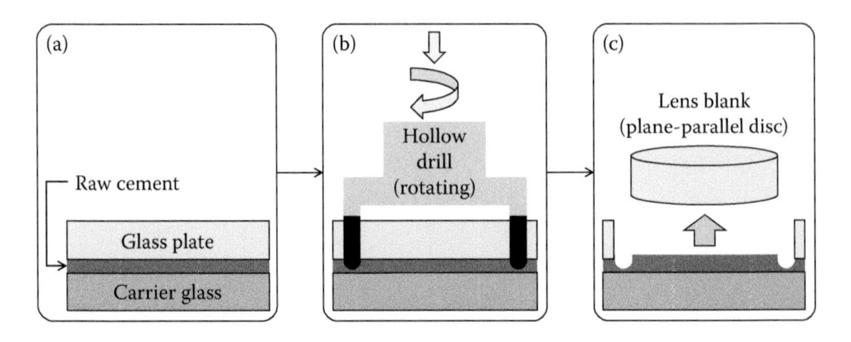


**FIGURE 7.7** Preproduction of lens blanks by hollow drilling; a glass plate is cemented on a carrier glass (a) and subsequently drilled by a hollow drill (b). Finally, the lens blank is obtained after solving the raw cement (c).

During rotation, a grinding tool is approached at moderate feed motion toward the rod, where cooling lubricant is applied in the course of the process. The initially quadratic rod is thus successively rounded as shown in Figure 7.6b. After dissolving the cement, plane-parallel discs (the actual lens blanks) are obtained.

Lens blanks can also be generated by *hollow drilling* as follows: as a first step, plates are cut from a glass block or bar. The glass plates are cemented on a carrier glass, and plane-parallel discs are generated by a rotating hollow drill, applying cooling lubricant, as shown in Figure 7.7.

The choice of the particular approach for preproducing lens blanks mainly depends on the available supply forms. Even though most glass manufacturers offer semimanufactured products such as rods, plates, round plates, cut prisms, or even pressed preforms, blanks may be produced from blocks or bars for economic reasons or because the particular glass is not available as a half-finished product.

## 7.3 ROUGH GRINDING OF LENSES

### 7.3.1 Machines and Tools for Rough Grinding

After preproduction as described above, the actual lens shape (i.e., a spherical segment of spherical lens surfaces) has to be generated. For this purpose, cylindrical plane-parallel lens blanks are rough ground in milling machines. Usually, computerized numerical control machines are employed for this manufacturing step, but it can also be performed on mechanical profile grinding machines. Both types of machines consist of two (or more) axes and at least two motorized spindles: the tool spindle and the work piece spindle, as shown in Figure 7.8.

Different parameters can be adjusted in order to control the grinding process and surface shape generation. The adjustable machine parameters and the resulting impact on lens parameters are listed in Table 7.5.

In order to obtain gentle machining and to reduce mechanical stress on the glass, an appropriate feed motion of the tool spindle should be chosen, and cooling lubricant should be added to the process. Moreover, the rotation velocity of the tool spindle should be much higher than the rotation velocity of the work piece spindle. As an

**FIGURE 7.8** Grinding machine for rough and fine grinding of lenses consisting of a mechanical chuck (placed on the work piece spindle), a grinding tool (mounted on the tool spindle), and adjustable cooling lubricant supply tubes.

**TABLE 7.5**
**Adjustable Machine Parameters for Controlling the Resulting Lens Parameters during Rough Grinding**

| Machine Parameter | Resulting Lens Parameter |
|---|---|
| Setting angle between work piece spindle and tool spindle | Radius of curvature |
| Distance from work piece spindle to tool spindle | Center thickness |
| Rotation velocities of spindles | Surface waviness and roughness |

example, a drive of approximately 20,000 rpm for the tool spindle and approximately 180 rpm for the work piece spindle have turned out to be suitable parameters for standard lens surfaces. The work piece spindle feed motion and the spindle's rotation velocities define the process and machining time, as well as the machining quality of the ground lens surface. These parameters depend on a number of influencing factors such as the type of glass, the work piece and tool sizes, the lens surface geometry, and the grain size of the tool.

Figure 7.9 shows the principal configuration of the essential elements of a *rough grinding* process. The work piece (i.e., the lens blank) is held by a work piece carrier and ground by a cylindrical cup wheel. Fastening of the work piece in the milling machine is realized by different types of chucks, either by vacuum exhaust or by mechanical clamping. Alternatively, it can also be cemented on the work piece carrier using raw cement.

The virtual material removal during rough grinding is achieved by *cylindrical cup wheels* (i.e., hollow cylinders made of metal with coated cutting edges as shown in Figure 7.10). Here, the abrasive coatings are similar to the ones used for saw blades.

In terms of geometry, a cylindrical cup wheel is specified by its diameter $D_{cw}$ and the radius of its cutting edge $r$. Moreover, cylindrical cup wheels are classified on the basis of the grain size of the abrasive grains embedded in the coating matrix

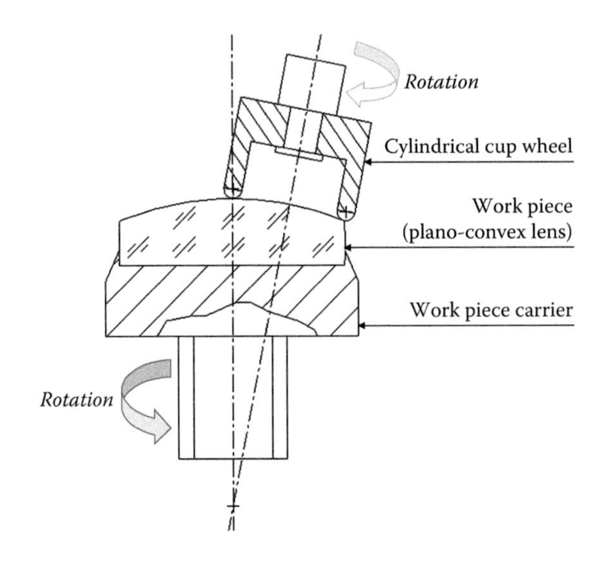


**FIGURE 7.9** Principal configuration of a rough grinding process; the work piece is placed on a work piece carrier and ground by a rotating cylindrical cup wheel for shape forming.

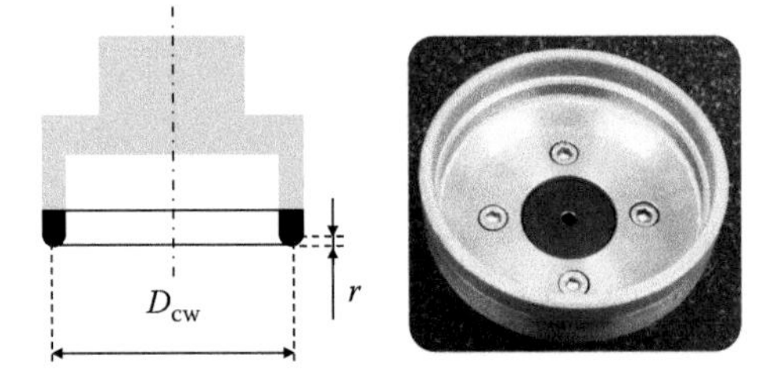


**FIGURE 7.10** Schematic (left) and photo (right) of a cylindrical cup wheel used for rough grinding including the definition of its diameter $D_{cw}$ and the radius of its cutting edge $r$.

where two different official classification systems are applied. The American standard ASTM-E-11-70 refers to the mesh (i.e., the unit of the mesh width of sieves used for filtering grains where the nominal mesh width is specified by ISO 6106). For example, a sieve with a value of 400 meshes features 400 meshes per inch. Considering the wire thickness of the mesh, the grain size amounts to 37 μm. The European equivalent to ASTM-E-11-70 is the FEPA[5] standard. A selection of classification and denomination of cylindrical cup wheels (and grinding tools in general) according to ASTM-E-11-70 and FEPA and the corresponding nominal mesh width is given in Table 7.6.

As mentioned above, the grain size given in mesh can be directly converted into the grain size in microns, where the interrelationship is not linear, but can be described by a potential function as shown in Figure 7.11.

[5] FEPA is the abbreviation of "Fédération Européenne des Fabricants de Produits Abrasifs" = Federation of European Producers of Abrasives.

**TABLE 7.6**
**Classification/Denomination of Grinding Tools according to ASTM-E-11-70 and FEPA and the Corresponding Nominal Mesh Width**

| Classification/Denomination according to | | |
|---|---|---|
| **US Standard ASTM-E-11-70** | **FEPA Standard** | **Nominal Mesh Width (μm)** |
| 60/70 | D251 | 250/212 |
| 70/80 | D213 | 212/180 |
| 80/100 | D181 | 180/150 |
| 100/120 | D151 | 150/125 |
| 120/140 | D126 | 125/106 |
| 140/170 | D107 | 106/90 |
| 170/200 | D91 | 90/75 |
| 200/230 | D76 | 75/63 |
| 230/270 | D64 | 63/53 |
| 270/235 | D54 | 53/45 |
| 325/400 | D46 | 45/38 |

*Source:* Winter Diamantwerkzeuge/Saint-Gobain Abrasives GmbH & Co. KG, Diamantwerkzeuge zur Bearbeitung feinoptischer, brillenoptischer und technischer Bauelemente, *Technical Information Sheet*, 2006 (in German).

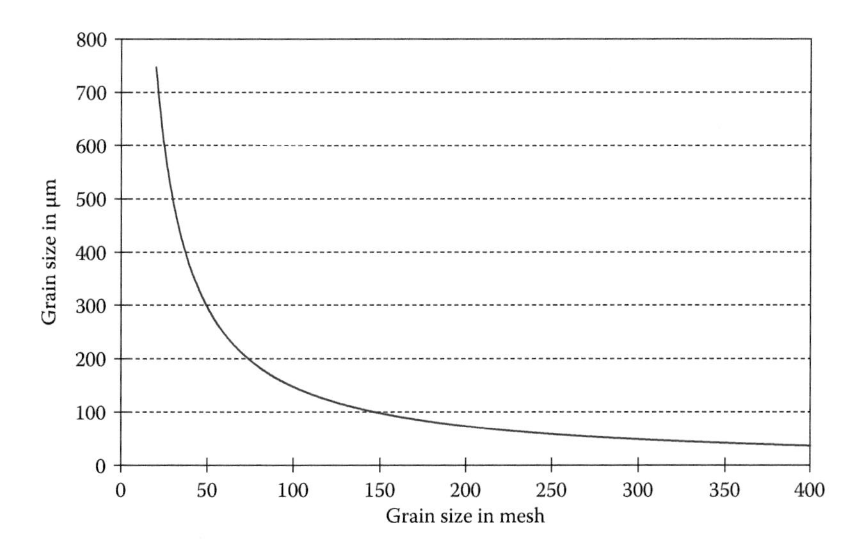


**FIGURE 7.11** Dependency of the abrasive grain size given in microns on the abrasive grain size given in mesh. The minimum size of abrasive grains sorted out via sieving is approximately 37 μm. Smaller grains as used for precision grinding or lapping are generated by sedimentation in liquids.

As shown here and listed in Table 7.7, it turns out that the higher the mesh value, the lower the grain size in microns.

In practice, the denomination of cylindrical cup wheels and grinding tools in general is based on the actual grain size according to the FEPA standard. This denomination is usually used by manufacturers and suppliers of grinding tools, especially for small grain sizes as listed in Table 7.8.

**TABLE 7.7**
**Overview on Abrasive Grain Sizes Expressed in Mesh and Microns**

| Abrasive Grain Size in | |
|---|---|
| **Mesh** | **Microns** |
| 60 | 250 |
| 70 | 210 |
| 80 | 177 |
| 100 | 149 |
| 120 | 125 |
| 140 | 105 |
| 170 | 88 |
| 200 | 74 |
| 230 | 63 |
| 270 | 53 |
| 325 | 44 |
| 400 | 37 |

**TABLE 7.8**
**Example for Grinding Tool Denominations and Corresponding Abrasive Grain Sizes**

| Grinding Tool Denomination | Abrasive Grain Size (μm) |
|---|---|
| D25 | 42 ± 10 |
| D20B | 35 ± 5 |
| D20A | 27.5 ± 2.5 |
| D15 | 22.5 ± 7.5 |
| D15C | 22.5 ± 2.5 |
| D15B | 17.5 ± 2.5 |
| D15A | 12.5 ± 2.5 |
| D7 | 7.5 ± 2.5 |
| D3 | 3.5 ± 1.5 |

*Source:* Winter Diamantwerkzeuge/Saint-Gobain Abrasives GmbH & Co. KG, Diamantwerkzeuge zur Bearbeitung feinoptischer, brillenoptischer und technischer Bauelemente, *Technical Information Sheet*, 2006 (in German).

**TABLE 7.9**
**Example for Grinding Tools and Resulting Surface Roughness on Fused Silica as Reported in Literature**

| Denomination of Used Grinding Tool | Arithmetic Mean Roughness *Ra* (μm) |
|---|---|
| D181 | 1.220 ± 0.140 |
| D64 | 0.286 ± 0.179 |
| D20 | 0.030 ± 0.015 |

*Source:* Neauport, J. et al., *Optics Express* 22, 20448–20456, 2009.

Relating to the surface of an optical component, its degree of grinding (i.e., rough ground, medium ground, fine ground, or precision ground) and the resulting surface roughness depends on the abrasive grain size of the used tool as shown by the example in Table 7.9.

However, the degree of grinding follows from the grain size and from a number of process parameters such as the feed motion velocity and the resulting pressure of the tool on the work piece, the rotation velocities of the tool and work piece spindle, and the hardness and grindability of the particular glass.

### 7.3.2 Rough Grinding Process

For actual rough grinding, the plane-parallel lens blank is placed on a work piece carrier, which is mounted on the work piece spindle of a grinding machine. Then, both the work piece spindle and the tool spindle are rotated at different velocities as described in the previous section. As shown in Figure 7.12, the work piece is then

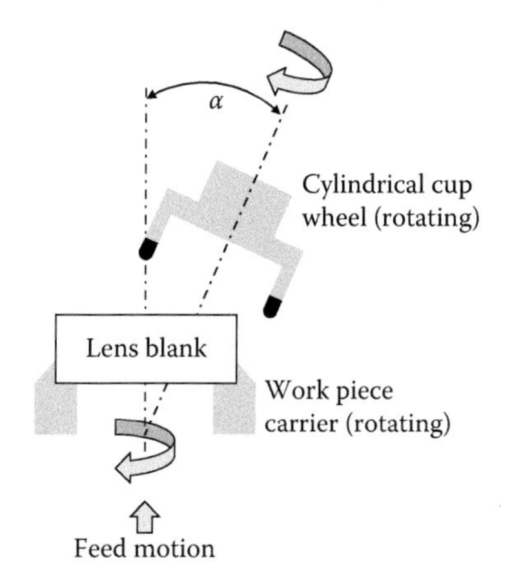


**FIGURE 7.12** Definition and visualization of the setting angle $\alpha$ between the work piece and the tool axes, adjusted during rough grinding where both the tool spindle and the work piece spindle are rotated. The latter is additionally moved toward the grinding tool by moderate feed motion.

slowly moved toward the rotating grinding tool by moderate feed motion (or vice versa, depending on the setup of the grinding machine), where both spindle axes feature a certain inclination to each other, the so-called setting angle $\alpha$. In the course of this process, cooling lubricant is applied to the contact zone of the grinding tool and the work piece.

As a result of the movement of the work piece toward the tool, the surface is successively ground, where grinding starts at its edge when generating convex lens surfaces and at its center in the case of concave ones as shown in Figure 7.13.

The radius of curvature of a lens surface generated in this way can be adjusted by the *setting angle* $\alpha$ between the axes of the work piece spindle and the tool spindle. For a given target radius of curvature $R_c$, the required setting angle for generating a convex (index: CX) lens surface can thus be calculated according to

$$\alpha_{CX} = \arcsin\frac{D_{cw}}{2\cdot(R_c + r)}, \tag{7.1}$$

where $D_{cw}$ is the diameter of the used cylindrical cup wheel and $r$ is the radius of its cutting edge. For a concave lens surface (index: CC), Equation 7.1 becomes

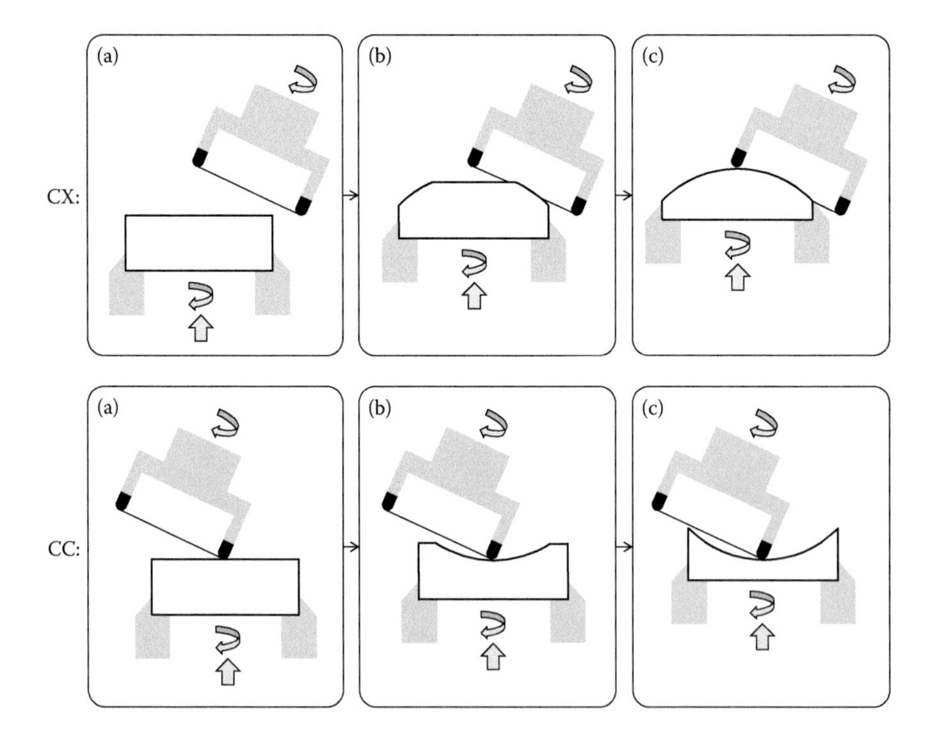


**FIGURE 7.13** Successive shape forming of convex (top, abbreviated CX) and concave (bottom, abbreviated CC) lens surfaces by rough grinding, starting with a plane-parallel disc as lens blank (a). Grinding then starts at the blank edge or center respectively (b), and the entire surface is ground as a result of the work piece spindle feed motion until the target lens center thickness including appropriate allowance is realized (c).

$$\alpha_{CC} = \arcsin\frac{D_{cw}}{2\cdot(R_c - r)}, \tag{7.2}$$

since in this case $r$ is defined to be a negative value.

In order to ensure that the whole work piece surface is covered and ground by the cylindrical cup wheel, its diameter should be adapted to the diameter of the work piece. This is realized if

$$D_{cw} > (R_c \pm r)\cdot\sin 2\cdot\alpha. \tag{7.3}$$

The bracket term in Equation 7.3 is $R+r$ for convex surfaces and $R-r$ for concave ones. Moreover, the cutting edge of the cylindrical cup wheel has to be properly aligned in order to avoid the formation of noses on the top of the ground lens surface as shown in Figure 7.14. This is achieved if the cutting edge's contact point is found right at the center of the lens. The cylindrical cup wheel should thus not feature any positive or negative lateral offset with respect to the lens center axis.

The approach for surface shaping as described above applies for spherical lens surfaces. In contrast, *zonal machining* has to be applied in order to generate aspherical surfaces. This can be performed by the use of either cylindrical cup wheels or grinding tools with spherical (ball-shaped) heads. In the first case, not the entire cutting edge ring of a cylindrical grinding tool, but merely one point of the cutting edge is in contact with the work piece surface. In the second case, grinding is performed by the contact point of the ball-shaped grinding head, where its diameter depends on the curvatures of the target geometry. In both cases, the work piece is rotated for the

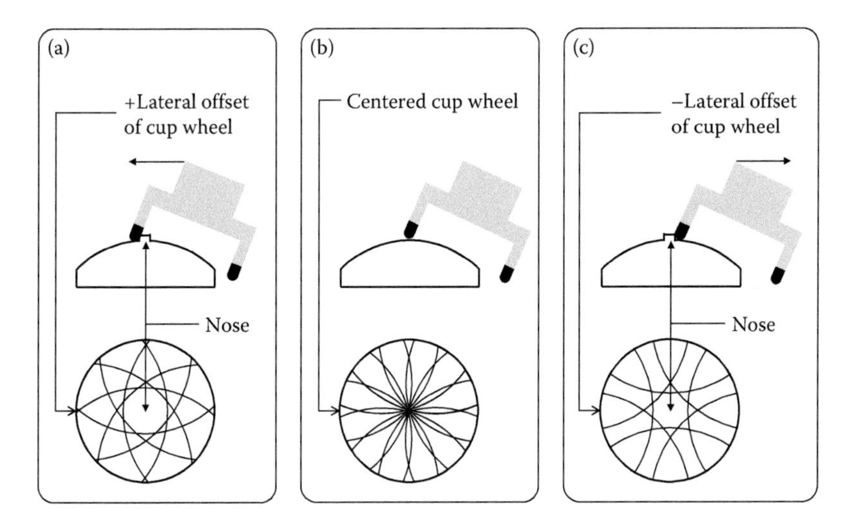


**FIGURE 7.14** Impact of positive (+) or negative (–) lateral offset of cylindrical cup wheels on the surface grinding pattern. For proper alignment, the circular cutting edge marks form a crosshatch pattern (b, bottom), whereas the lens center is not affected by the cutting edge in case of lateral offsets, leading to the formation of noses on the ground surface (a, c).

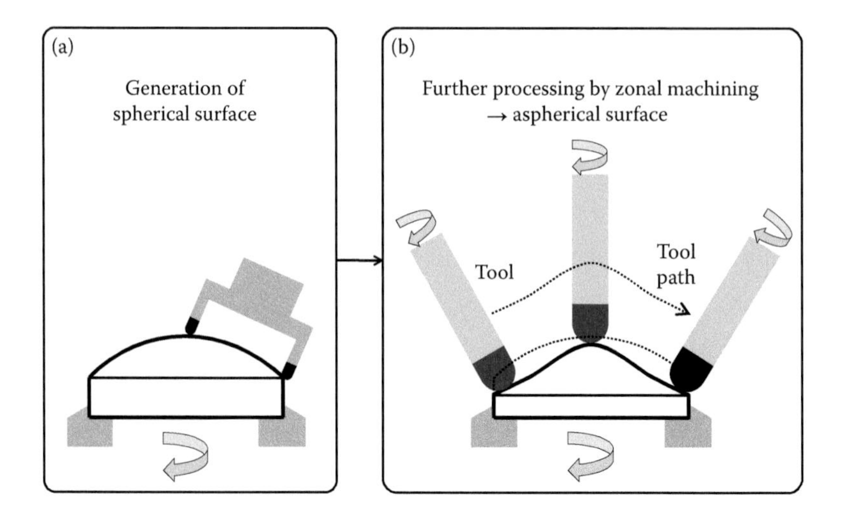


**FIGURE 7.15** Zonal machining for the realization of aspherical lens surfaces. First, a spherical start surface is generated (a), and the aspherical surface is subsequently ground by a rotating grinding tool with ball-shaped head, where the tool path corresponds to the target surface geometry (b).

generation of rotation-symmetric surface shapes such as hyperbolas, parabolas, or ellipses (see Section 4.2.2). It becomes obvious that precise guidance of the particular grinding tool by the grinding machine is required where the tool path principally corresponds to the target geometry of the lens surface. As shown in Figure 7.15, the generation of an aspheric lens surface usually starts with the production of a spherical surface as described above. This surface is subsequently further shaped by zonal machining. The generation of aspherical surfaces may also be performed on plane-parallel lens blanks without previous grinding of a spherical start surface.

### 7.3.3 Roughing

Another approach for shape generation is *roughing* in spherical cup wheels. This manufacturing step can be performed with the aid of bound abrasive grains or loose abrasive grains. In the first case, a spherical cup wheel (for definition see Section 7.4.1.1) with an abrasive coating is used where the grains embedded in the coating matrix are comparatively large (i.e., in the range of a few hundred microns). This process is referred to as roughing by grinding. In the second case, spherical cup wheels made of cast iron without any coating are employed and material removal is achieved by adding a mixture of water and large loose abrasive grains. This approach is called lapping as explained in detail in Sections 7.4.1.2 and 7.4.2.2. The spherical cup wheel used in both cases features a radius of curvature, which corresponds to the target value with adequate allowance (compare Table 7.10). As shown in Figure 7.16, a lens blank is pressed into such a spherical cup wheel, and the radius of curvature is successively ground.

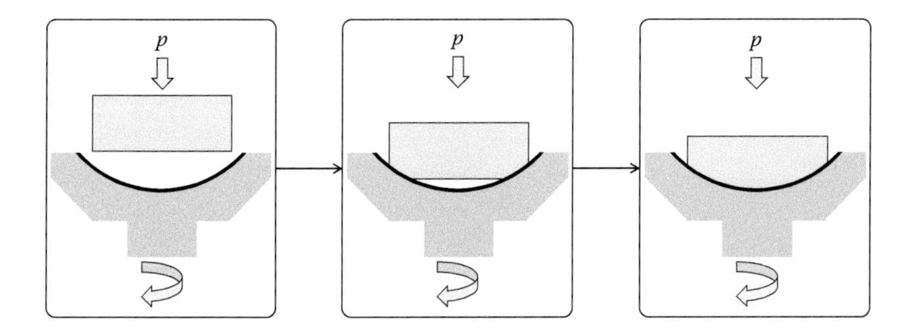


**FIGURE 7.16** Principle of roughing in spherical cup wheels at the example of a convex lens surface; a lens blank is successively ground and the radius of curvature is generated from the lens border to its center.

This method can principally be applied in order to generate convex and concave lens surfaces. However, it is quite unstable in the latter case and thus requires solid guidance of the work piece. In contrast, guidance of the work piece occurs automatically for convex lens surfaces since the surface shape is ground from the border of the blank to its center, resulting in edge support of the blank on the tool surface.

Such self-guidance as found during rough grinding of convex surface is of essential importance for the process stability of subsequent machining steps. Against this background, the radius of curvature after surface shape generation is ground slightly higher (approximately 10–12 µm for standard optics with a diameter of 21.5 mm (Bliedtner and Gräfe, 2008)) than the final target value, independent of the method. As a result, the tools used for subsequent finish grinding and lapping rest on the lens border, leading to stable self-guidance of the particular tool due to edge support as shown in Figure 7.17.

This approach is also applied for polishing, so the work piece radius of curvature after finish grinding is slightly higher (approximately 2–3 µm) than the final target value. Moreover, sufficient allowance is realized for the center thickness in order to ensure that roughness valleys, digs, and subsurface damage such as microcracks are minimized or even removed by the subsequent manufacturing steps.

Subsurface damage occurs in the course of rough grinding (and during subsequent finish grinding and even polishing) due to the mechanical shock waves induced by

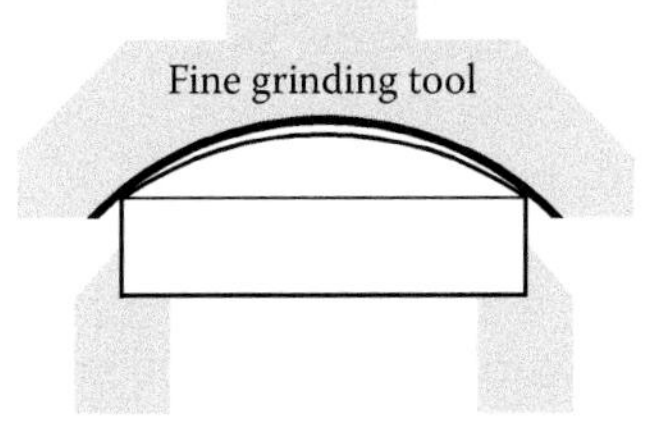


**FIGURE 7.17** Visualization of edge support of a fine grinding tool resulting from the strategy of generating a lens surface with higher radius of curvature than the final target radius of curvature in the course of rough grinding.

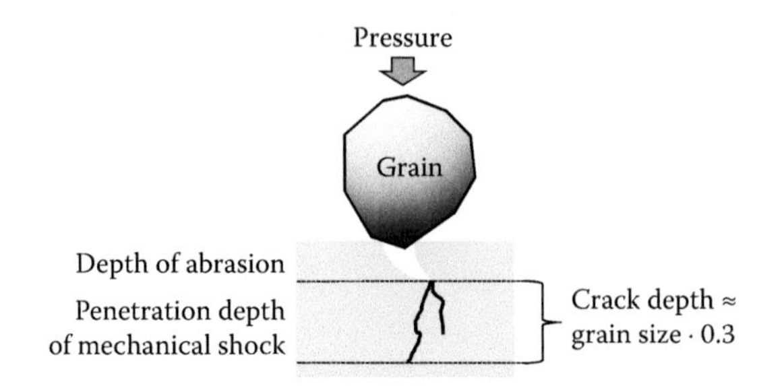


**FIGURE 7.18** Formation of micro cracks due to mechanical shock induced by an abrasive grain.

**TABLE 7.10**
**Allowances for the Radius of Curvature and Center Thickness for Rough Grinding and Finish Grinding, Referring to a Lens with a Diameter of 21.5 mm**

| Process Step | Allowance for Radius of Curvature (μm) | Allowance for Center Thickness (μm) |
|---|---|---|
| Rough grinding | 10–12 | 700 |
| Finish grinding | 1–2 | 100 |

*Source:* Bliedtner, J., and Gräfe, G., *Optiktechnologie*, Carl Hanser Verlag, München, Germany, 2008 (in German).

the abrasive grains. As shown in Figure 7.18, such shock waves propagate into deeper regions and lead to the formation of microcracks with a certain depth $d_{mc}$, approximately given by

$$d_{mc} \approx 0.3 \cdot D_g. \tag{7.4}$$

The depth of microcracks[6] thus exclusively depends on the grain grit size (i.e., its equivalent diameter $D_g$).

This layer of microcracks is successively minimized or even removed by further grinding or lapping using tools or grains with reduced grit size diameter. Depending on the grit size diameter (see Table 7.13), this further grinding can be classified into three different categories with different grades of grinding (see Section 6.3.4): medium ground, fine ground, and precision ground.

As listed in Table 7.10, allowances for the center thickness mentioned above are thus defined not only for rough grinding, but also for subsequent finish grinding, where the subsequent manufacturing step is polishing.

[6] The depth of microcracks on glass surfaces produced by classical grinding, lapping, and polishing typically ranges from 1 to 20 μm (Suratwala et al., 2011) but may even reach 100 μm in some cases (Papernov and Schmid, 2008).

It turns out that the geometry and dimensions of a lens or an optical component in general nearly correspond to the target values after finish grinding. This is due to the fact that during final polishing, moderate material removal is achieved by the comparatively small grit size of the used polishing grains.

## 7.4 FINISH GRINDING

*Finish grinding* is applied in order to remove or minimize the subsurface damage layer and microcracks induced by precedent rough grinding and to reduce the surface roughness to achieve a polishable surface. Moreover, the surface shape further approaches the target shape in the course of this process.

There are two types of finish grinding: grinding with bound abrasive grains or grinding with loose abrasive grains. In the first case, the abrasive grains are embedded in a basic matrix, thus forming a coating that is applied to the tool surface in saw blades or cylindrical cup wheels. In contrast, the abrasive grains are not embedded within any matrix, but are rather provided in the form of a lapping suspension during loose abrasive grinding. As shown in Figure 7.19, the abrasive grains can thus perform a rolling motion between the tool and the work piece surface.

By definition, the term “grinding” describes grinding with bound abrasive grains, whereas grinding with loose abrasive grains is referred to as lapping.

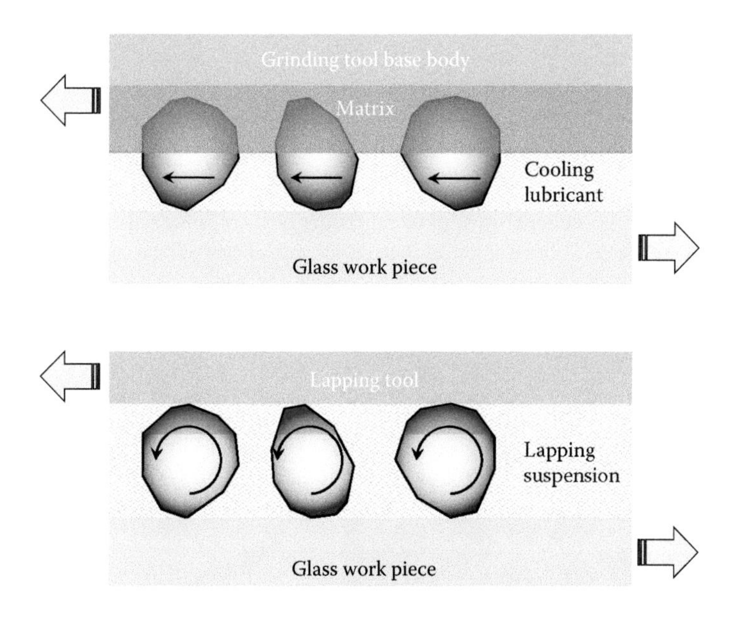


**FIGURE 7.19** Comparison of grinding with bound abrasive grains (top) and with loose abrasive grains, (a.k.a. lapping (bottom)). In the latter case, the abrasive grains perform a rolling motion between the glass and the lapping tool surface.

### 7.4.1 Machines and Tools for Finish Grinding

#### 7.4.1.1 Bound Abrasive Grinding

For *bound abrasive grinding*, the machines and tools used for rough grinding can be applied. It can thus be performed by employing cylindrical cup wheels with appropriate grit size of the abrasive grains embedded in the cutting edge matrix. However, *spherical cup wheels*, also referred to as "bowls," are usually used. Such spherical cup wheels consist of a spherical base body made of aluminum, brass, or other metal. As shown in Figure 7.20, the spherical surface is either coated with a closed abrasive layer (matrix and embedded grains) or provided with abrasive pellets.

When using spherical cup wheels with abrasive pellets as shown in Figure 7.21, the cooling lubricant applied in the course of the process can get between the pellets, leading to enhanced cooling and a lower process temperature, respectively. This advantage with respect to spherical cup wheels with closed abrasive coatings is of notable interest when machining temperature-sensitive glasses.

The process efficiency (i.e., the material removal rate) can be influenced by the pellet density or degree of coverage on the tool surface, resulting in different effective contact areas of the tool and the work piece. The degree of coverage $C_\mathrm{p}$ of pellets on a cup wheel surface is given by

$$C_\mathrm{p} = \frac{4 \cdot N \cdot \left(\frac{D_\mathrm{p}}{2}\right)^2}{D_\mathrm{cw}^2}, \tag{7.5}$$

**FIGURE 7.20** Cross-sectional view of two different types of spherical cup wheels used for fine grinding with bound abrasive grains; cup wheel with closed abrasive coating (black line, left), and cup wheel with abrasive pellets (black structures, right).

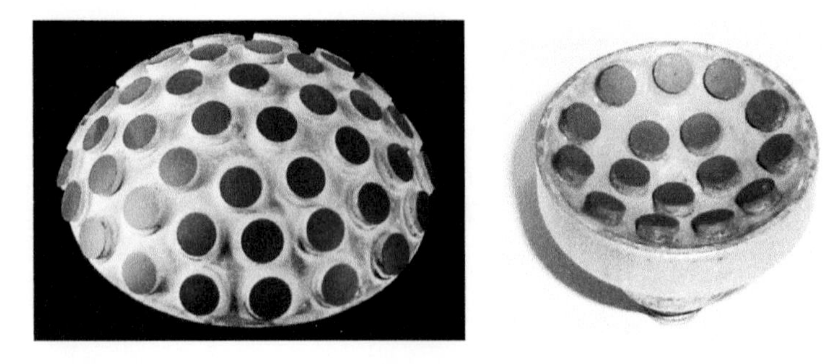

**FIGURE 7.21** Spherical cup wheels with abrasive pellets for the generation of concave lens surfaces (left) and convex lens surfaces (right).

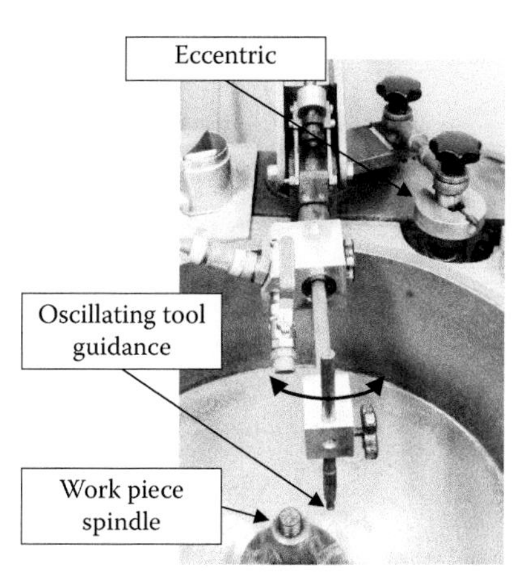


**FIGURE 7.22** Lever arm machine used for loose abrasive grinding (and polishing) with eccentric spindle drive.

where $N$ is the number of pellets, $D_p$ is the pellet diameter, and $D_{cw}$ is the diameter of the cup wheel.

#### 7.4.1.2 Loose Abrasive Grinding

*Loose abrasive grinding* or *lapping* is usually performed on lever arm machines with eccentric spindle drives, as shown in Figure 7.22.

The work piece is mounted on a separate work piece spindle, where the rotation velocity of both spindles, the eccentric tool spindle, and the work piece spindle can be adjusted independently of each other. The tool spindle transfers its eccentricity on the lever arm, finally resulting in an oscillation of the tool guidance, which is placed at the end of the lever arm. The lapping tool thus performs an oscillating motion over the work piece, where the eccentric position allows adjusting the amplitude of this oscillation motion. In the course of this process, the tool is not actively rotated, but indirectly driven by the work piece rotation. This setup can principally be inversed (i.e., the work piece can be oscillated and driven by the eccentric spindle drive). This approach allows improving lapping results in some cases, for example, for increasing the radius of curvature of a lens surface.[7]

For lapping, plane or spherical solid bodies made of brass, aluminum, or gray cast iron as shown in Figure 7.23 are employed where gray cast iron is preferentially used as material due to its relatively high porosity. Consequently, lapping suspension and

[7] Usually, optical surfaces are machined "from the edge to the center" when being produced on lever arm machines in order to obtain self-guidance of the used grinding, lapping, and polishing tools in subsequent machining steps. The radius of curvature after each step is thus slightly higher (for convex surfaces) or lower (for concave surfaces) than the target radius. If, for example, the radius of curvature of a convex surface becomes lower than the target radius, material needs to be removed at the lens center in order to increase the radius again (referred to as "lap back" or "polish back" in opticians slang). Depending on the lens geometry, this can become a quite challenging task, which can be solved by the above-mentioned inverted use of the spindles.

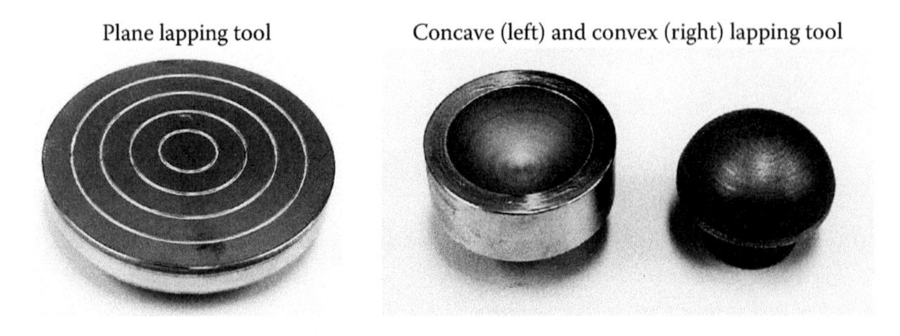


**FIGURE 7.23** Examples for lapping tools made of gray cast iron used for loose abrasive grinding.

grains can accumulate in the pores of the tool surface, which act as reservoirs for the lapping agent. This behavior contributes to the lapping process stability.

As already mentioned, the lapping agent is provided in the form of an aqueous suspension. It is continuously brought between the tool and the work piece surface in the course of the lapping process. This can be realized either manually or by automatic lapping suspension supply via tubes. The lapping suspension (a.k.a. lapping slurry) is a mixture of abrasive grains made of corundum ($Al_2O_3$), silicon carbide (SiC), diamond (C), boron carbide ($B_4C$), or other media, water, and—as the case may be—further additives. The choice of the particular lapping grain medium mainly depends on the hardness and grindability (see Section 6.2.4.2) of the work piece material; it can be carried out on the basis of the so-called *lapping coefficient*, a well-established effectiveness indicator for lapping media, as listed in Table 7.11.

The lapping coefficient depends on the hardness of the grain medium and the shape of the lapping grains, where the highest lapping coefficient is found for hard and sharp-edged lapping grains. It is referred to the lapping coefficient of pure sand ($SiO_2$), which is set to 1 per default.

As listed in Table 7.12, lapping grains are available in different sizes, where similar to grinding with bound abrasive grains, the resulting grade of lapping (rough, medium, fine, or precision), and the lens surface roughness, respectively, follow from the grain size as listed in Table 7.13.

**TABLE 7.11**
**Lapping Coefficient of Different Lapping Media**

| Lapping Medium | Lapping Coefficient |
|---|---|
| Corundum ($Al_2O_3$) | 4.8–5.7 |
| Boron carbide ($B_4C$) | 8.7 |
| Diamond (C) | 12.5 |

*Source:* Bliedtner, J., and Gräfe, G., *Optiktechnologie*, Carl Hanser Verlag, München, Germany, 2008 (in German).

**TABLE 7.12**
**Mean Grain Size and Type Designation of the Lapping Grain Materials Silicon Carbide and Aluminum Oxide**

| | Lapping Grain Material and Type Designation | |
|---|---|---|
| Mean Grain Size (μm) | Silicon Carbide | Aluminum Oxide |
| 250 | F60 | |
| 177 | F80 | |
| 125 | F100 | |
| 105 | F120 | |
| 88 | F150 | |
| 74 | F180 | |
| 63 | F220 | |
| 53 | F230 | |
| 45 | F240 | |
| 40 | | WCA40 |
| 37 | F280 | |
| 35 | | WCA35 |
| 30 | | WCA30 |
| 29 | F320 | |
| 25 | | WCA25 |
| 23 | F360 | |
| 20 | | WCA20 |
| 18 | | WCA18 |
| 17 | F400 | |
| 15 | | WCA15 |
| 13 | F500 | |
| 12 | | WCA12 |
| 9 | F600 | WCA9 |
| 7 | F800 | |
| 5 | F1000 | WCA5 |
| 3 | F1200 | WCA3 |
| 2 | | WCA2 |
| 1 | | WCA1 |

*Note:* According to the standard designation, silicon carbide is indicated by the prefix F and aluminum oxide is marked by the prefix WCA (where its trade name is Microgrit WCA).

Lapping abrasives do not exclusively consist of the nominal grain material (e.g., silicon carbide or aluminum oxide) but contain a number of other elements as listed in Table 7.14. This fact might be considered when choosing an abrasive, since some of the elements of a lapping grain batch, for example, carbon, may lead to severe surface contamination of selected glasses or other media used in optics manufacturing such as crystals.

**TABLE 7.13**
**Different Grades of Lapping and the Corresponding Particular Lapping Grain Sizes**

| Grade of Lapping | Grain Size (μm) | |
|---|---|---|
| | Lower Limit | Upper Limit |
| Rough | 115 | 230 |
| Medium | 45 | 75 |
| Fine | 17 | 29 |
| Precision | 3 | 13 |

*Source:* Karow, H.K., *Fabrication Methods for Precision Optics*, John Wiley & Sons, Hoboken, NJ, 2004.

**TABLE 7.14**
**Example for the Typical Composition of Lapping Grain Batches**

| | Chemical Composition/Content of Lapping Grain Material in Mass % | |
|---|---|---|
| Element | Silicon Carbide | Aluminum Oxide |
| Silicon carbide | 98.60–99.00 | — |
| Aluminum oxide | — | 99.60 |
| Silicon dioxide | 0.25–0.34 | 0.02 |
| Iron oxide | 0–0.05 | 0.03 |
| Sodium oxide | — | 0.40 |
| Calcium oxide | — | 0.05 |
| Silicon | 0.15–0.27 | |
| Iron | 0–0.05 | |
| Aluminum | 0–0.04 | |
| Carbon | 0.15–0.30 | |

### 7.4.2 Finish Grinding Process

Finish grinding is the last manufacturing step before polishing. The aim of this step is thus to produce a polishable surface in terms of its roughness and the deviation of the radius of curvature of the finish ground surface from the target radius after polishing. Against this background, finish grinding is performed employing spherical cup wheels in most cases. Here, full cup wheels with closed abrasive coating or with abrasive pellets can be applied for bound abrasive finish grinding, whereas in loose abrasive grinding, lapping media, and so-called bowls made of gray cast iron are used as introduced above.

#### 7.4.2.1 Bound Abrasive Grinding

For bound abrasive finish grinding, the rough-ground work piece is placed on a work piece carrier such as a mechanical chuck mounted on a rotating work piece spindle. The grinding tool is rotated by the tool spindle, and both spindles feature a certain angle of inclination. The tool may also perform an oscillation motion in order to achieve high surface regularity and uniformity. In this grinding process, cooling lubricant is applied. This approach is well established due to its stability and reliability.

#### 7.4.2.2 Loose Abrasive Grinding

For loose abrasive grinding, the work piece is mounted on the work piece spindle of a lever arm machine and rotated. Then, lapping suspension is brought onto the glass surface and continuously applied in the course of the entire lapping process. The gray cast iron tool—the “bowl”—is placed directly on the work piece, performing an oscillating motion over its surface as schematically shown in Figure 7.24, where the eccentric position allows for the adjustment of the amplitude of the oscillation motion.

Material removal by loose abrasives occurs because lapping grains can roll between the tool and the work piece where relative motion is realized by the oscillation motion. During rolling, the abrasive grains cut into glass, cant, and quarry out shell-shaped glass particles mechanically as shown in Figure 7.25.

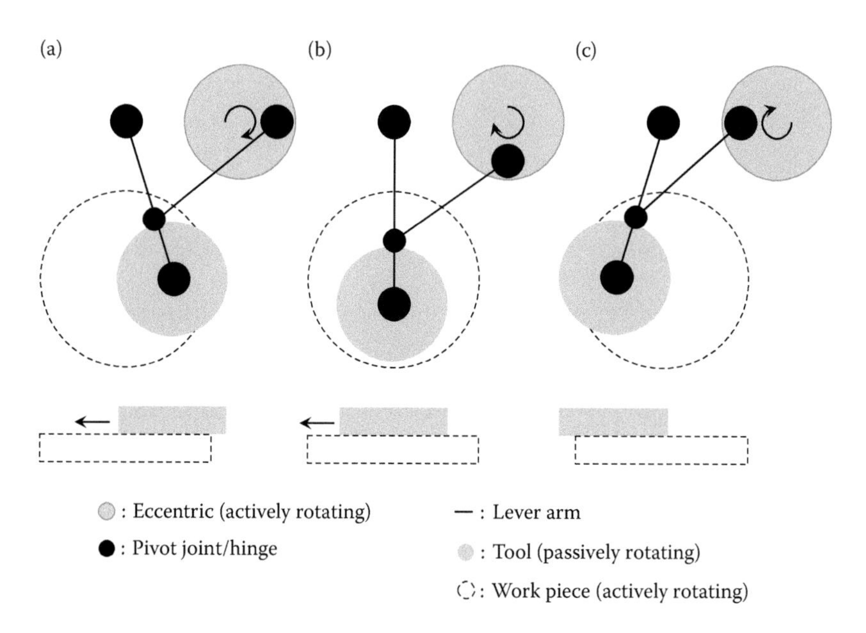


**FIGURE 7.24** Principle of tool drive by eccentric spindles on a lever arm machine; because of the rotation of the eccentric, the lever arm, or the tool, respectively, is guided over the work piece surface (a–c) (top: top view, bottom: side view).

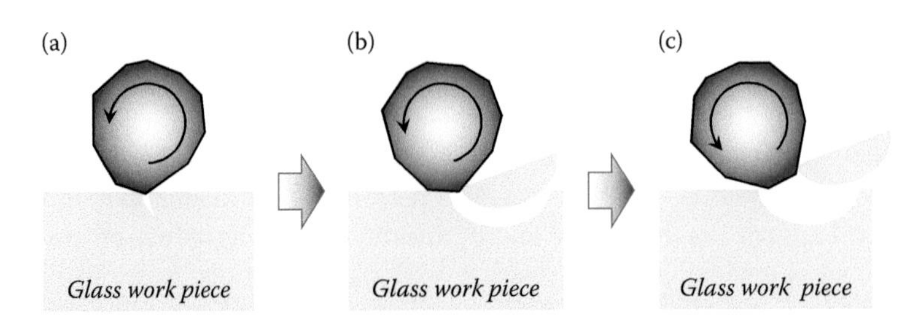


**FIGURE 7.25** Material removal during loose abrasive grinding; abrasive grains cut into the surface of the glass work piece (a) and cant and quarry out shell-shaped glass particles (b, c).

**TABLE 7.15**
**Allowances for the Center Thickness for Different Steps of Finish Grinding**

| Finish Grinding Step | Mean Lapping Grain Size (μm) | Corresponding Lapping Type Designation | Allowance for Center Thickness (μm) |
|---|---|---|---|
| Medium | 53 | F230 | 700 |
| Fine | 17–37 | F400–F280 | 300 |
| Precision | 9–13 | F500–F600 | 100 |

*Source:* Bliedtner, J., and Gräfe, G., *Optiktechnologie*, Carl Hanser Verlag, München, Germany, 2008 (in German).

The size of glass particles and amount of material removal and resulting surface roughness, respectively, depend on the abrasive grain size; it may thus be successively reduced in the course of a loose abrasive finish grinding process in order to reduce surface roughness of the work piece surface. Consequently, different allowances for the center thickness have to be applied depending on the particular finish grinding step as listed in Table 7.15.

Such allowances ensure that the center thickness will not underrun its target value at the end of the manufacturing process, since the lapping grain size impacts not only the surface roughness, but also the material removal height and the center thickness, respectively, as well as the material removal volume. These dependencies are shown in Figures 7.26, 7.27 and 7.28.

In addition to the center thickness, the radius of curvature has to be controlled and tested at regular intervals during finish grinding as described in more detail in Section 7.7. On the basis of such testing, the process parameters might be corrected or readjusted in order to meet the required target radius of curvature. For manipulating the lens radius and approaching the target value during loose abrasive grinding, the amplitude and speed of oscillation motion of the lapping tool and the tool pressure can be varied as shown in Figure 7.29.

For decreasing the radius of curvature of a convex lens surface, high amplitude of oscillation motion at low rotation velocity of the eccentric spindle, resulting in slow

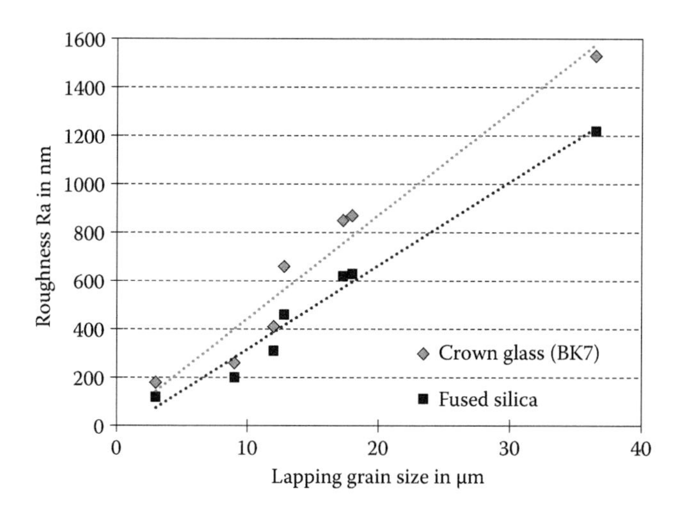


**FIGURE 7.26** Dependency of arithmetic mean surface roughness Ra on lapping grain size. (From Bliedtner, J., Gräfe, G., *Optiktechnologie*, Carl Hanser Verlag, München, Germany, 2008, in German.)

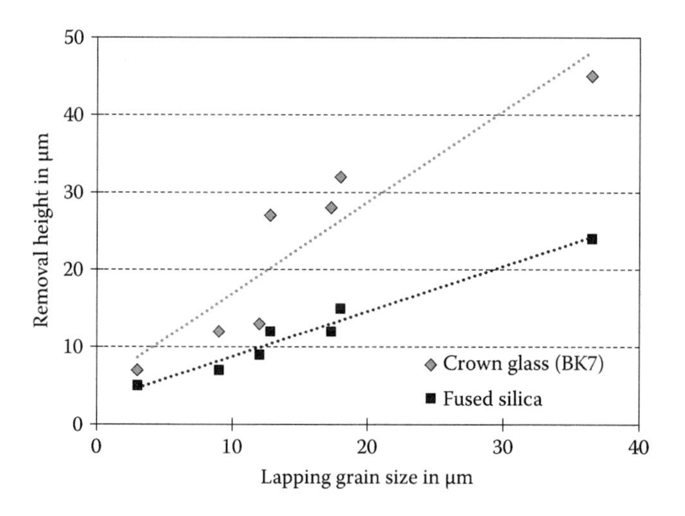


**FIGURE 7.27** Dependency of removal height on lapping grain. (From Bliedtner, J., Gräfe, G., *Optiktechnologie*, München, Germany: Carl Hanser Verlag, 2008, in German).

oscillation motion of the tool, and high pressure of the tool on the work piece surface should be applied. In doing so, the lens surface becomes more convex in order to approach the target radius as shown in Figure 7.29. In contrast, the radius of curvature is increased if a low amplitude of oscillation motion at a high rotation velocity of the eccentric spindle, leading to fast oscillation motion of the tool and low pressure of the tool on the work piece, are chosen. As shown in Figure 7.29, the lens surface then becomes more concave in order to approach the target radius.

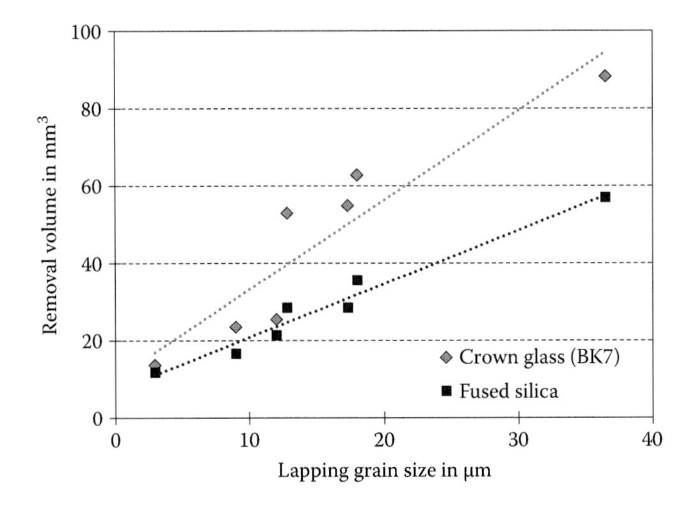


**FIGURE 7.28** Dependency of removal volume on lapping grain size. (From Bliedtner, J., and Gräfe, G., *Optiktechnologie*, Carl Hanser Verlag, München, Germany, 2008, in German).

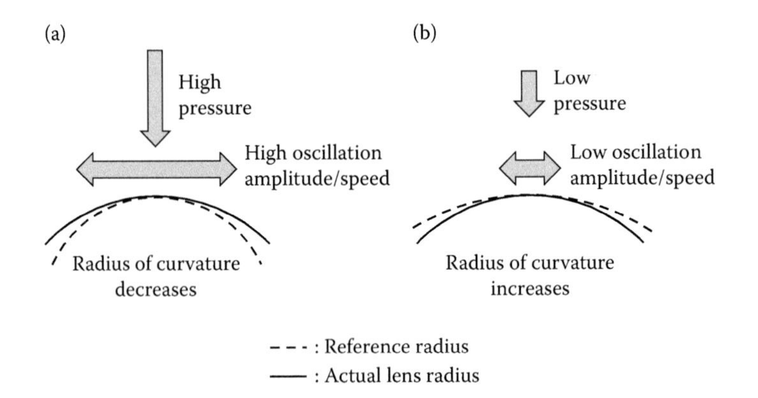


**FIGURE 7.29** Impact of tool pressure and speed as well as oscillation amplitude on the surface shape of a convex spherical lens surface; high pressure, speed, and amplitude lead to a decrease in radius of curvature (a), whereas an increase in radius of curvature results from low pressure, speed, and amplitude (b). This behavior is used for approximating the actual lens radius of curvature to the target reference radius of curvature.

This principally applies for both single lens surfaces and lens carriers as shown in Figure 7.30. Here, several[8] lenses are cemented on a carrier body using raw cement and can be machined simultaneously as shown in Figure 7.30.

Such lens carriers are preferably employed for loose abrasive grinding (and for subsequent polishing) since the used cement would not resist the forces arising

[8] Hundreds of single lenses may be cemented on a carrier, depending on the radius of curvature and the diameter of the lenses, where the degree of coverage of lenses on a carrier can be determined on the basis of Equation 7.5.

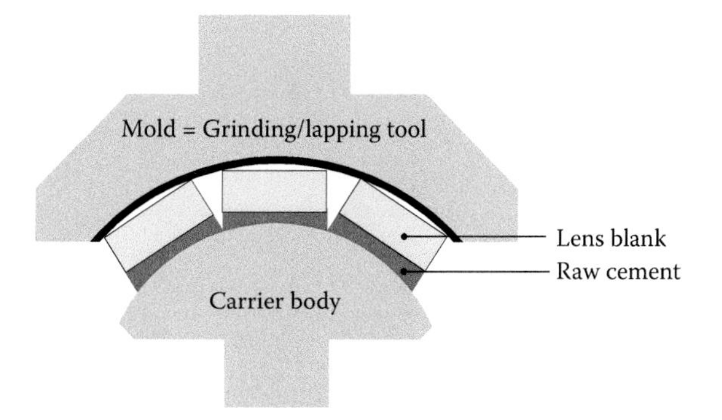


**FIGURE 7.30** Carrier body for simultaneous machining of several lenses; the lenses are aligned in a mold (i.e., a grinding or lapping tool) and then cemented onto the carrier.

during bound abrasive grinding. The use of carriers thus makes loose abrasive grinding more economic than bound abrasive grinding in some cases.

## 7.5 FLAT GRINDING

Up to now, rough and fine grinding of spherical or aspherical lens surfaces was presented. The manufacturing steps introduced and explained in this context can also be applied for grinding flat surfaces, such as plane lens or prism surfaces, where either plane grinding tools with abrasive coating or pellets or plane lapping tools made of gray cast iron are employed. For *flat grinding*, the tool diameter $D_t$ is usually twice the diameter or lateral dimension (e.g., the cathetus of a prism) of the work piece surface $D_{wp}$. In this case, the cutting velocity $v_c$ is given by

$$v_c = D_t \cdot \pi \cdot n_t, \tag{7.6}$$

with $n_t$ being the drive of the tool. The removal volume $V$ can then be determined according to

$$V = \frac{\pi \cdot D_t^2}{4} \cdot \Delta z, \tag{7.7}$$

where $\Delta z$ is the feed motion of the grinding tool. The change in work piece volume $\Delta V$ per machining time $t$ finally gives the material removal rate

$$MRR = \frac{\Delta V}{t}. \tag{7.8}$$

The process of flat grinding is shown with a dispersion prism in Figure 7.31. This principal procedure as described hereafter is applied not only for rough or finish grinding but also for polishing.

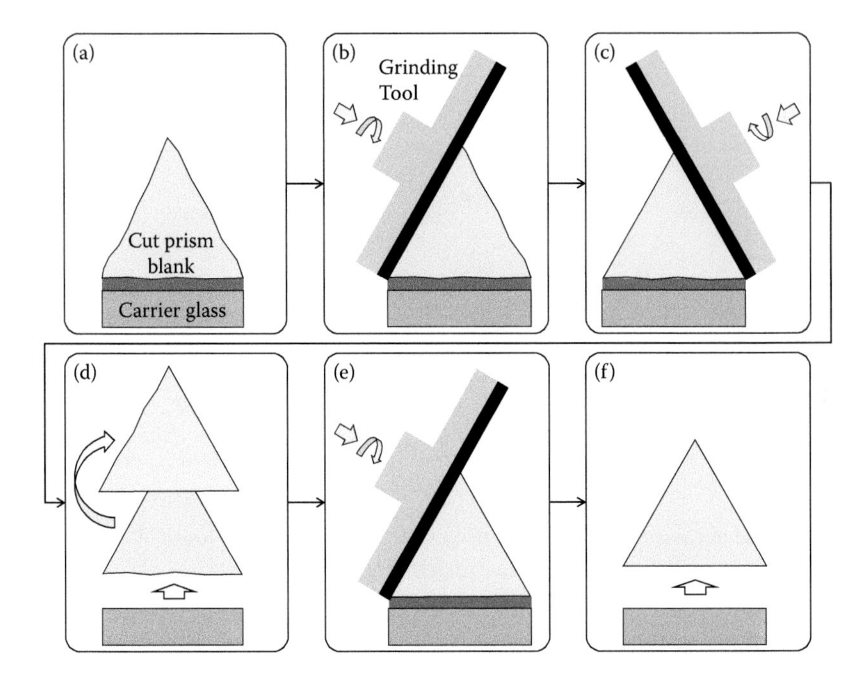


**FIGURE 7.31** Flat grinding of a cut prism blank (a) by successive cementing of the blank on a carrier glass and grinding of each surface (b–e), resulting in a ground prism (f). Note that the deviation in geometry and surface shape of the cut prism blank is displayed exaggeratedly for better visualization and identification of ground and unground surfaces.

Here, the basic geometry of the prism including appropriate allowances is first cut employing circular saw benches. The cut prism blank is then cemented on a carrier glass plate, and the first and second prism surfaces (e.g., the catheti) are ground. The prism is then removed from the carrier glass by dissolving the raw cement, and one ground surface is cemented on another carrier glass. Subsequently, the last remaining cut surface (e.g., the hypotenuse) is ground, and the prism is finally removed from the carrier glass. Instead of carrier glasses, metallic carriers with v-shaped grooves can be used as shown in Figure 7.32.

Finally, flat lens surfaces can be produced in this way. However, a special approach known as *double-sided grinding* is employed for the manufacture of plane-parallel plates such as mirror substrates or laser protection windows. Here, the plate blanks are fixed in a cage and placed on a plane grinding or lapping tool. A second plane grinding or lapping tool is then placed on top of the plates, and both tools are rotated. Moreover, the cage carrying the plates is rotated by an epicyclic gear set. This leads to (1) a chaotic relative motion of the grinding tools on the surfaces of the plates and (2) a homogeneous distribution of pressure or load and material removal, respectively, at each surface point. This technique consequently stands out due to the high achievable parallelism of the ground surfaces and is thus not only applied for flat grinding, but also for flat polishing of windows and mirror or filter substrates.

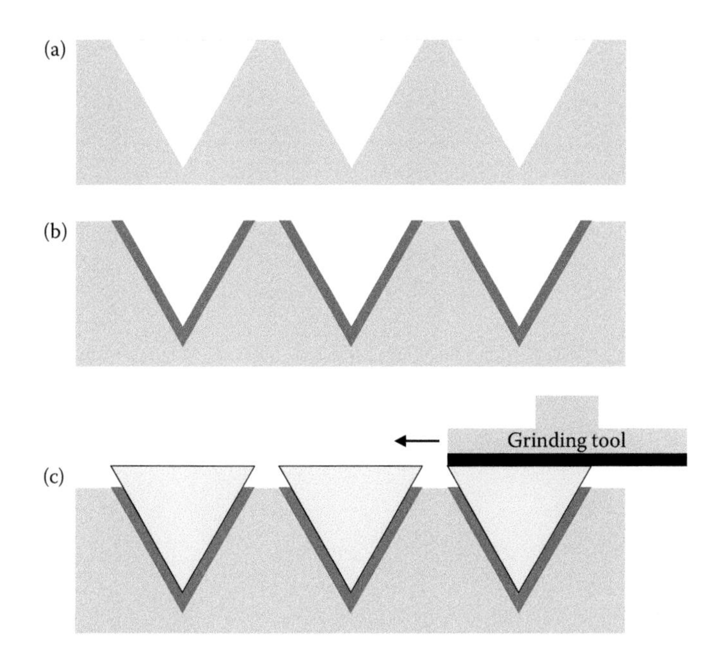


**FIGURE 7.32** Flat grinding of prisms in a v-grooved metallic carrier (a); after applying raw cement to the groove surfaces (b), the prisms are placed in the grooves, and the upturned surfaces are ground (c).

## 7.6 BEVELING

During the entire manufacturing process (cutting, rough grinding, and fine grinding), it should be considered that ground lens or prism surfaces feature sharp edges. In the course of subsequent manufacturing steps, tools could cause flaking at the lens or prism edge, consequently leading to surface damages, if glass chips are quarried out and fall into the gap between the work piece surface and the tool. Against this background, the edges of optical components are usually beveled (a.k.a. chamfered) before and after each manufacturing step as shown in Figure 7.33, where the bevel leg length is usually in the range of some hundreds of microns, and the bevel angle is typically 45°.

In the case of lenses, *beveling* is performed by grinding or lapping the lens edge in spherical cup wheels or lapping tools, respectively, as shown in Figure 7.34. A bevel angle of approximately 45° is achieved, if

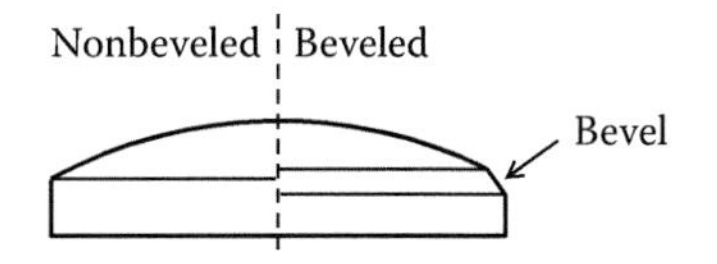


**FIGURE 7.33** Visualization of a nonbeveled (left) and a beveled (right) plano-convex lens surface.

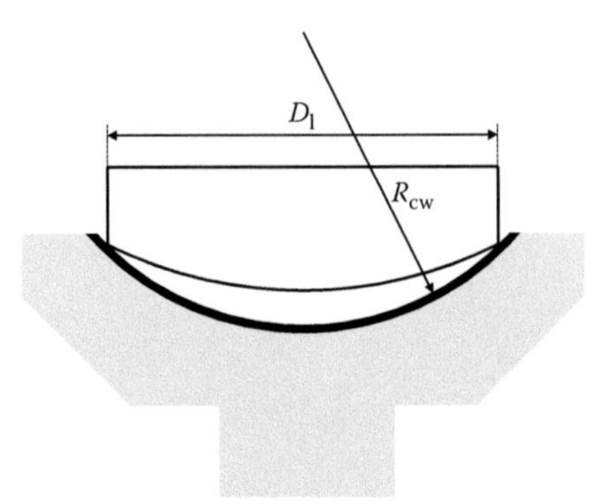


**FIGURE 7.34** Beveling of a lens with the diameter $D_l$ in a spherical cup wheel with the radius of curvature $R_{cw}$.

$$R_{cw} = \frac{D_l}{\sqrt{2}} \approx 0.7 \cdot D_l, \tag{7.9}$$

where $R_{cw}$ is the radius of curvature of the used spherical cup wheel, and $D_l$ is the diameter of the lens.

Bevels are also referred to as chamfers and are of great importance during polishing processes since sharp lens or prism edges may not only give rise to surface damage, but also destroy the comparatively soft tools or polishing pads used in this manufacturing step. Moreover, bevels are applied to finished lenses in order to provide protection against chipping during final mounting by screw connecting or clamping (see Chapter 12). In the context of mounting, bevels can also act as mounting surfaces, which are realized by appropriate tools in the course of the centering process (see Chapter 10). Here, the rotation symmetry of the bevel leg length becomes crucial, since decentered bevels with inconsistent bevel leg lengths gives rise to lens tilting by a tilt angle $\alpha$[9] according to

$$\alpha = \frac{l_{max} - l_{min}}{2 \cdot R}. \tag{7.10}$$

Here, $l_{max}$ is the maximum bevel leg length, $l_{min}$ is the minimum, and $R$ is the radius of curvature of the beveled lens surface. The bevel leg length should thus be constant in order to avoid such tilting.

## 7.7 SURFACE TESTING

During grinding and lapping processes, the dimensions and the surface shape of plane and spherical optics surface are tested at regular intervals with the aid of so-called *spherometers*. In optical manufacturing, spherometers are also known as *lens*

[9] Note that the tilt angle $\alpha$ is given in radians.

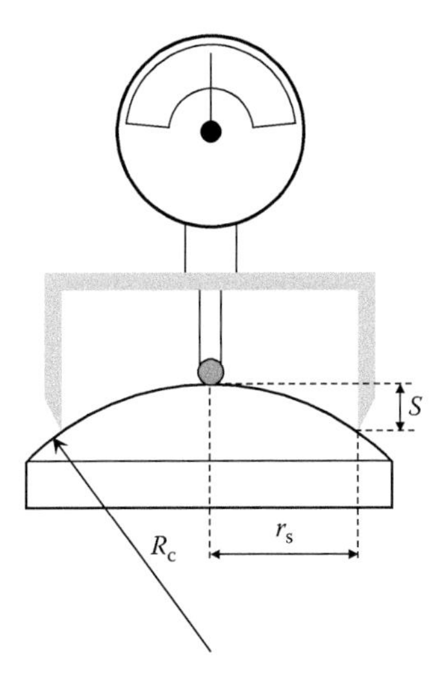


**FIGURE 7.35** Measurement principle for the determination of the sagitta $S$ of a lens surface with a radius of curvature $R_c$ using spherometer rings with a ring-shaped contact zone; $r_s$ is half the diameter of the spherometer ring.

*clocks*, a special form of mechanical dial indicators. This measurement device usually features a resolution of ±50 µm and is used for controlling lens center thicknesses; the work piece thickness is compared to length standards or existing lens samples or prototypes. It is further employed for testing the surface shape (i.e., the deviation in actual radius of curvature from the target radius of curvature). The value measured in this case is the *sagitta S* (i.e., generally the height/depth of a circular arc). Since the lens clock is mounted on a spherometer ring (a.k.a. *measuring bell*), the sagitta results from the diameter $D_s = 2 \cdot r_s$ of this spherometer ring (given by its contact points on the lens surface) and the radius of curvature $R_c$ of the lens surface as shown schematically in Figure 7.35.

The sagitta is generally given by

$$S = \frac{\left(\frac{r_s^2}{R_c}\right)}{1 + \sqrt{\left(1 - \frac{r_s^2}{R_c^2}\right)}}. \quad (7.11)$$

Here, the radius of curvature $R_c$ can be positive (applies for convex lens surfaces) or negative (applies for concave lens surfaces). By measuring the sagitta,[10] the lens' radius of curvature can thus be determined by solving equation (sag) for $R_c$ according to

[10] For a concave lens surface, the measured sagitta is a negative value with respect to a reference flat. For a convex surface, it becomes positive.

$$R_c = \frac{S^2 \cdot r_s^2 + r_s^4}{2 \cdot S \cdot r_s^2}. \tag{7.12}$$

Equations 7.11 and 7.12 can further be simplified when assuming $R_c$ to be a positive value as valid for convex lenses. The sagitta is then given by

$$S = R_c - \sqrt{R_c^2 - r_s^2}, \tag{7.13}$$

and the radius of curvature results from

$$R_c = \frac{r_s^2 + S^2}{2 \cdot S}. \tag{7.14}$$

In practice, a ring spherometer is zeroed to a normal with known reference radius of curvature $R_{ref}$ (i.e., normally the target radius of curvature of the currently machined lens). The form deviation $\Delta f$ is then given by the deviation between the actual surface radius $R_c$ of the device under test and the desired target or reference surface radius $R_{ref}$, assuming both surfaces to be spherical:

$$\Delta f = \left(R_c - \sqrt{R_{ref}^2 - r_s^2}\right) - \left(R_c - \sqrt{R_c^2 - r_s^2}\right). \tag{7.15}$$

The comparison of the surface under test with a reference surface allows a fast and easy evaluation of the lens shape in both a qualitative and a quantitative way. For convex surfaces, a spherometer zeroed to a normal gives negative values if the radius of curvature of the tested surface is higher than the reference surface as shown in Figure 7.36. In contrast, positive values occur if the actual test surface radius is lower.[11]

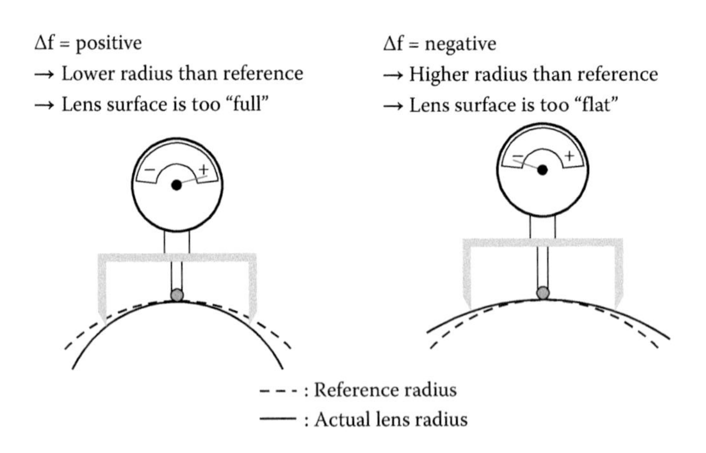


**FIGURE 7.36** Impact of deviation in actual radius of curvature of a lens surface from the target or reference radius on the measurement via spherometers.

[11] If the actual radius of curvature is higher than the reference or target value, the surface is usually called "too flat." In the other case, where the actual radius of curvature is lower than the reference or target value, the surface is referred to as "too full" in optician's slang.

Simple spherometer rings are toroids made of stainless steel or brass, resulting in a ring-shaped contact area with the radius $r_s$ on the lens surface. Alternatively, spherometer rings consisting of three legs can be used, where a statically stable three-point contact is achieved. The actual contact elements can be measuring tips or ball feet (usually made of stainless steel or ruby) as shown in Figure 7.37.

In the latter case, an additional parameter (i.e., the radius $r_b$ of the contact balls) has to be taken into account during testing. Equation 7.11 thus becomes:

$$S = \frac{\left(\frac{r_s^2}{(R_c - r_b)}\right)}{1 + \sqrt{\left(1 - \frac{r_s^2}{R_c - r_b}\right)}}. \tag{7.16}$$

Note that in Equation 7.16, the radius of curvature $R_c$ has to be entered as a positive value for concave lens surfaces and as a negative for convex lens surfaces. Correspondingly, the absolute value of the radius of curvature is then given by

$$|R_c| = \frac{r_s^2}{2 \cdot S} + \frac{S}{2} \pm r_b, \tag{7.17}$$

where $+r_b$ applies for concave lens surfaces and $-r_b$ is valid for convex ones.

Apart from spherometers, other techniques such as tactile surface profilers or measurement systems based on laser triangulation may be employed. However, the use of spherometers represents an easy-to-use and thus well-established solution for surface testing in the surface shaping process, from rough grinding to finish grinding.

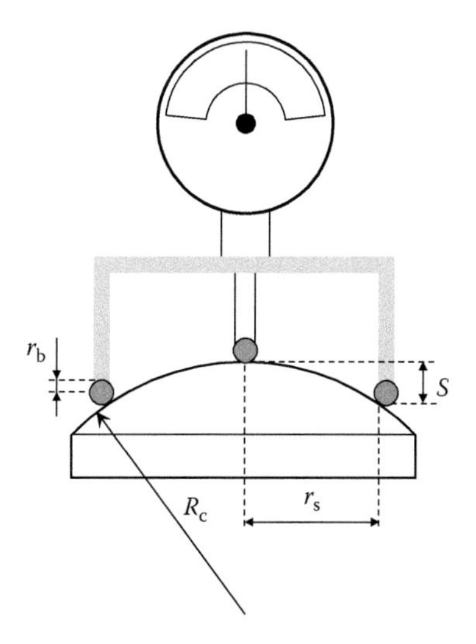


**FIGURE 7.37** Measurement principle for the determination of the sagitta $S$ of a lens surface with a radius of curvature $R$ using spherometer rings with a three-point contact zone; $r_s$ is half the diameter of the spherometer ring, and $r_b$ represents the radius of the contact balls.

## 7.8 SUMMARY

Shaping represents the first step in optics manufacturing; it can be realized in different ways. Preforms of lenses or prisms can be produced by compression molding, where a certain volume of molten glass is formed in adequate molds. Lens or prism blanks can be manufactured by cutting and subsequent rounding or hollow drilling where necessary. In both cases, the blank of the particular optical element with sufficient allowances with respect to the final target geometry is realized.

For the production of lens surfaces, the initial spherical or aspherical surface geometry is then produced by rough grinding. This step is typically performed using cylindrical cup wheels but can also be carried out with the aid of spherical cup wheels (for plane and spherical surfaces) or grinding tools with ball-shaped heads (for aspherical surfaces). In order to reduce surface roughness and to remove microcracks induced by cutting and rough grinding, optics surfaces are subsequently fine or finish ground where the size of the used abrasive grains is successively reduced. Two different types of grinding, bound abrasive grinding and loose abrasive grinding, are applied where the latter is also referred to as lapping.

In the course of shaping processes, the work pieces are usually beveled in order to prevent surface damage by chips that might be quarried out of sharp work piece edges. Further, surface testing during and after each grinding step is essential. Such testing is typically realized by the use of spherometers in optics workshops. The measured variables are the center thickness and the surface shape (e.g., the radius of curvature), determined by comparison to a normal with well-defined dimensions and shape.

## 7.9 FORMULARY AND MAIN SYMBOLS AND ABBREVIATIONS

**Setting angle $\alpha_{CX}$ for shape generation of convex lens surfaces:**

$$\alpha_{CX} = \arcsin \frac{D_{cw}}{2 \cdot (R_c + r)}$$

$D_{cw}$ diameter of used cylindrical cup wheel
$R_c$ target radius of curvature
$r$ radius of cylindrical cup wheel cutting edge

**Setting angle $\alpha_{CC}$ for shape generation of concave lens surfaces:**

$$\alpha_{CC} = \arcsin \frac{D_{cw}}{2 \cdot (R_c - r)}$$

$D_{cw}$ diameter of used cylindrical cup wheel
$R_c$ target radius of curvature
$r$ radius of cylindrical cup wheel cutting edge

**Choice of diameter of cylindrical cup wheel $D_{cw}$:**

$$D_{cw} > (R_c \pm r) \cdot \sin 2 \cdot \alpha$$

$R_c$ target radius of curvature
$r$ radius of cylindrical cup wheel cutting edge
$\alpha$ setting angle
*Note:* $(R + r)$ applies for convex surfaces and $(R - r)$ for concave

**Depth of microcracks $d_{mc}$:**

$$d_{mc} \approx 0.3 \cdot D_g$$

$D_g$ grain grit size

**Degree of coverage $C_p$ of pellets on cup wheel surface:**

$$C_p = \frac{4 \cdot N \cdot \left(\frac{D_p}{2}\right)^2}{D_{cw}^2}$$

$N$ number of pellets
$D_p$ pellet diameter
$D_{cw}$ cup wheel diameter

**Cutting velocity $v_c$ for flat grinding:**

$$v_c = D_t \cdot \pi \cdot n_t$$

$D_t$ grinding tool diameter
$n_t$ grinding tool drive

**Removal volume $V$:**

$$V = \frac{\pi \cdot D_t^2}{4} \cdot \Delta z$$

$D_t$ grinding tool diameter
$\Delta z$ grinding tool feed motion

**Material removal rate $MRR$:**

$$MRR = \frac{\Delta V}{t}$$

$\Delta V$ change in work piece volume
$t$ machining time

**Approximation for generating bevel angles of 45°:**

$$R_{cw} = \frac{D_l}{\sqrt{2}} \approx 0.7 \cdot D_l$$

$R_{cw}$ radius of curvature of used spherical cup wheel
$D_l$ lens diameter

**Lens tilt angle $\alpha$ resulting from decentered bevels:**

$$\alpha = \frac{l_{max} - l_{min}}{2 \cdot R}$$

$l_{max}$ maximum bevel leg length
$l_{min}$ minimum bevel leg length
$R$ radius of curvature of beveled lens surface

**Lens sagitta $S$ (general expression):**

$$S = \frac{\left(\frac{r_s^2}{R_c}\right)}{1 + \sqrt{\left(1 - \frac{r_s^2}{R_c^2}\right)}}$$

$r_s$ half the diameter of used spherometer ring
$R_c$ radius of curvature of tested lens surface
*Notes:* The radius of curvature $R_c$ is positive for convex lens surfaces and negative for concave lens surfaces. This equation applies for spherometer rings with ring-shaped contact zone.

**Radius of curvature $R_c$ of tested lens surface (general expression):**

$$R_c = \frac{S^2 \cdot r_s^2 + r_s^4}{2 \cdot S \cdot r_s^2}$$

$r_s$ half the diameter of used spherometer ring
$S$ sagitta
*Note:* This equation applies for spherometer rings with ring-shaped contact zone.

**Lens sagitta $S$ (for convex surfaces exclusively):**

$$S = R_c - \sqrt{R_c^2 - r_s^2}$$

$R_c$ radius of curvature of tested lens surface
$r_s$ half the diameter of used spherometer ring
*Note:* This equation applies for spherometer rings with ring-shaped contact zone.

**Radius of curvature $R_c$ of tested lens surface (for convex surfaces exclusively):**

$$R_c = \frac{r_s^2 + S^2}{2 \cdot S}.$$

$r_s$ half the diameter of used spherometer ring
$S$ sagitta
*Note:* This equation applies for spherometer rings with ring-shaped contact zone.

**Form deviation $\Delta f$ of a lens surface (with respect to reference):**

$$\Delta f = \left(R_c - \sqrt{R_{ref}^2 - r_s^2}\right) - \left(R_c - \sqrt{R_c^2 - r_s^2}\right).$$

$R_c$ actual surface radius of curvature
$R_{ref}$ reference radius of curvature
$r_s$ half the diameter of used spherometer ring
*Note:* This equation applies for spherometer rings with ring-shaped contact zone.

**Lens sagitta $S$ (general expression):**

$$S = \frac{\left(\dfrac{r_s^2}{(R_c - r_b)}\right)}{1 + \sqrt{\left(1 - \dfrac{r_s^2}{R_c - r_b}\right)}}$$

$r_s$ half the diameter of used spherometer ring
$R_c$ radius of curvature of tested lens surface
$r_b$ radius of contact balls
*Notes:* The radius of curvature $R_c$ is positive for concave lens surfaces and negative for convex lens surfaces. This equation applies for spherometer rings with contact balls.

**Radius of curvature $R_c$ of tested lens surface (general expression for absolute value):**

$$|R_c| = \frac{r_s^2}{2 \cdot S} + \frac{S}{2} \pm r_b$$

$r_s$ half the diameter of used spherometer ring
$S$ sagitta
$r_b$ radius of contact balls
*Note:* $+r_b$ applies for concave lens surfaces and $-r_b$ is valid for convex lens surfaces.

## REFERENCES


Bliedtner, J. and Gräfe, G. 2008. *Optiktechnologie*. München, Germany: Carl Hanser Verlag (in German).

Fynn, G.W. and Powell, W.J.A. 1979. *The Cutting and Polishing of Electro-optic Materials*. New York: John Wiley & Sons.

Karow, H.K. 2004. *Fabrication Methods for Precision Optics*. Hoboken, NJ: John Wiley & Sons.

Neauport, J., Ambard, C., Cormont, P., Darbois, N., Destribats, J., Luitot, C., and Rondeau, O. 2009. Subsurface damage measurement of ground fused silica parts by HF etching techniques. *Optics Express* 22:20448–20456.

Papernov, S. and Schmid, A.W. 2008. Laser-induced surface damage of optical materials: Absorption sources, initiation, growth, and mitigation. *Proceedings of SPIE* 7132:71321J.

Pforte, H. 1995. *Der Optiker.* Homburg, Germany: Verlag Gehlen (in German).

Suratwala, T.I., Miller, P.E., Bude, J.D., Steele, W.A., Shen, N., Monticelli, M.V., Feit, M.D., Laurence, T.A., Norton, M.A., Carr, C.W., and Wong, L.L. 2011. HF-based etching processes for improving laser damage resistance of fused silica optical surfaces. *Journal of the American Ceramic Society* 94:416–428.

Winter Diamantwerkzeuge/Saint-Gobain Abrasives GmbH & Co. KG. 2006. Diamantwerkzeuge zur Bearbeitung feinoptischer, brillenoptischer und technischer Bauelemente. *Technical Information Sheet* (in German).

# 8 Polishing

## 8.1 INTRODUCTION

Polishing of optical components could be termed the supreme discipline in optics manufacturing. The polishing process and its results depend on a large number of interacting and influencing variables and parameters. These variables and parameters arise from the polishing machine used, the applied process parameters, the polishing tool, the polishing agent, and the work piece itself. Further, different variable disturbances have a certain impact on the polishing process and result.

The goal of polishing is not only to realize the final transparent surface of an optical component with the required contour accuracy (see Section 6.3.1) and surface cleanliness (see Section 6.3.3), but also to remove or close digs, pits, and microcracks induced by precedent grinding or lapping and to obtain smooth surfaces with negligible surface roughness and scattering, respectively.

The polishing techniques applied for this challenge can be categorized in different ways, for example, on the basis of the predominant mechanism, the size of the processed area, or the approach/method applied for polishing. Depending on the layout of the polishing machine, different approaches and principles can be applied for polishing. The classical method is the use of lever arm polishing machines. Further, the so-called synchro-speed-polishing method using computerized numerical control machines can be applied, and in some cases, high-precision optical components are produced manually, requiring skilled and experienced crafters.

In any case, the applied process parameters are of essential importance. These parameters are given by the value and distribution of pressure between the polishing tool and the work piece and the relative velocity between the two. Another crucial influencing value for lever arm and manual polishing is the oscillation of the tool or the work piece, respectively, characterized by the zero position, the amplitude, and the frequency of the oscillation.

The polishing tool further significantly impacts the polishing process and results. It consists of a polishing medium carrier, also referred to as a polishing pad, which is applied to a metallic or plastic base body. This polishing pad can be pitch or synthetic material, depending on the material of the work piece to be processed. In addition to the material, its thickness, flexibility, and surface texture are of essential importance. Moreover, the physical dimensions and stiffness of the body material have to be chosen carefully.

The polishing agent brought between the polishing pad and the work piece surface is a suspension consisting of water, polishing grains, and further additives such as defoamers. The size, hardness, chemical properties, and geometry of the polishing grains as well as the concentration of grains within the slurry and the pH-value and temperature of the slurry play important roles in the course of any polishing process.

The chemical properties and pH-value of the slurry have to be specially adapted to the chemical composition of the work piece and its chemical stability (characterized by the resistance classes as introduced in Section 6.2.4.1). In addition, the structure of the work piece material (i.e., amorphous/glassy, crystalline, or glass ceramic) and its mechanical properties such as hardness impact the polishing process. Another important influencing aspect is the work piece geometry, since the process parameters depend on the radius of curvature of a lens surface and its diameter and center thickness. Usually, this fact is expressed by the ratios of the diameter and the center thickness or the diameter and the radius of curvature. As an example, polishing hemispherical lens surfaces is very different from polishing plane surfaces.

In the end, variable disturbances and environmental conditions that cannot be totally controlled impact the dynamics of polishing. Here, vibrations, fluctuations in temperature and humidity, and finally the experience and skills of the worker shall be mentioned.

To summarize, polishing is a complex process that depends on a large number of interacting influencing values. This is also confirmed by the fact that the microscopic description of the underlying mechanism is based on the combination of different hypotheses, as presented hereafter.

## 8.2 HYPOTHESES OF POLISHING PROCESSES

A theoretical description of polishing optical glasses is a complex task, since such polishing processes are based on different interacting mechanisms. These mechanisms are characterized by the so-called removal hypothesis, the flow hypothesis, the chemical hypothesis, and the fretting hypothesis. The weighting of the particular contribution of these phenomena on a polishing process follows from a number of parameters, for example, the mechanical and chemical properties of the work piece material and the polishing slurry and the process parameters (pressure, velocity etc.). Depending on the predominant mechanism, polishing processes are categorized as either chemical-mechanical polishing (CMP), mechanical abrasive polishing, or a mixture of both. The predominance of the particular mechanism mainly depends on ambient conditions during the polishing process (e.g., temperature and humidity) and the composition of the polishing suspension.

### 8.2.1 Removal Hypothesis

The *removal hypothesis* describes a mechanical removal of glass material. During the polishing process, the sharp-edged grains of the used polishing agent roll between the work piece surface and the polishing tool surface. In the course of such rolling, the sharp edges dig into the glass surface and quarry out small glass particles. This process is quite comparable to the lapping process described in Chapter 7. As the grains preferentially affect the roughness peaks of the ground/lapped work piece surface, it is successively smoothed.

### 8.2.2 Flow Hypothesis

In addition to mechanical removal by abrasion, the polishing agent grains perform a ductile displacement of glass material as stated by the *flow hypothesis*. Here, local frictional heat occurs due to the friction between the work piece surface and the polishing pad and grain, respectively, resulting in softening of the glass material. As a result, roughness peaks are dislocated into roughness valleys or digs by material flow. The rough surface is consequently leveled as shown schematically in Figure 8.1.

### 8.2.3 Chemical Hypothesis

The chemical hypothesis describes chemical reactions between the polishing suspension used and the surface of the glass work piece, which is one of the two predominant mechanisms for material removal during CMP (Sabia and Stevens, 2000). Such reactions are supported by water-induced breaking of bonds of the glass network, also known as hydrolytic scission, for example, according to

$$\equiv \mathrm{Si-O-Si} \equiv + \mathrm{H_2O} \leftrightarrow 2 \equiv \mathrm{Si-OH} \tag{8.1}$$

in silicon dioxide-based glasses. The rate and depth of this reaction and the resulting thickness of the chemically modified near-surface layer on the work piece surface are mainly due to the diffusion of water into the glass (Cook, 1990). For a given point in time $t$ (e.g., the machining time during polishing corresponding to the contact time of water from the suspension and the glass surface), the mean diffusion depth $d_{\mathrm{dif}}$ can be estimated by the interrelation

$$d_{\mathrm{dif}} \approx 2 \cdot \sqrt{D \cdot t}\,. \tag{8.2}$$

Here, $D$ is the temperature-dependent diffusion coefficient. In the case of quartz glass or synthetic fused silica, different values for $D$ are reported in the literature. Depending on the process conditions and transport mechanisms of water within the glass, the diffusion coefficient can range from approximately $10^{-15}\,\mathrm{cm^2/s}$ to $10^{-18}\,\mathrm{cm^2/s}$ at a temperature of 90°C (Nogami and Tomozawa, 1984; Lanford et al., 1985).

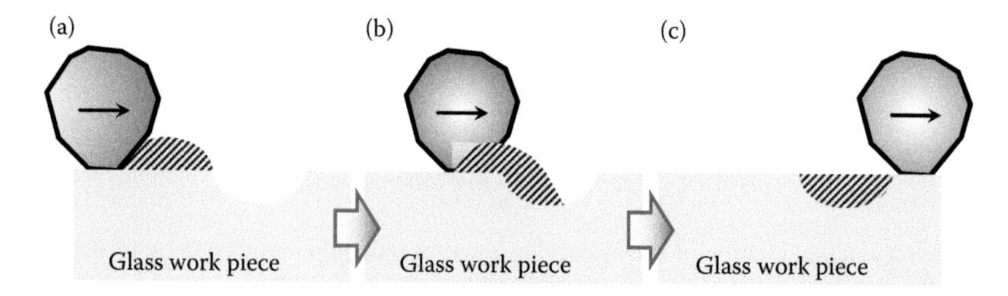


**FIGURE 8.1** Visualization of the flow hypothesis; roughness peaks (dashed structure) are softened by wear in the course of the polishing process (a) and dislocated into roughness valleys or digs (b), consequently resulting in a leveling of the surface (c).

As a result of this phenomenon, silica and/or hydrated silica is redeposited on the work piece surface during polishing, consequently leading to the formation of a so-called *silica gel layer* (Iler, 1979; Cumbo and Jacobs, 1994), according to

$$(SiO_2)_x + 2H_2O \leftrightarrow (SiO_2)_{x-1} + Si(OH)_4. \tag{8.3}$$

This effect contributes to surface smoothing, since silica gel can accumulate within roughness valleys, scratches, and digs and thus fill such geometric surface defects. In some cases, this effect may become problematic since in subsequent cleaning procedures, silica gel can be removed and filled scratches and digs can be excavated. Further, the silica gel layer can selectively dissolve essential constituents such as ions from a thin boundary layer of multicomponent glasses by leaching processes[1] (Evans et al., 2003) as shown schematically in Figure 8.2.

Remaining hydrated silica gel layers can moreover give rise to disturbing effects such as an increase in near-surface absorption. This effect is due to several optically active defects that are induced by hydrogen, for example, the formation of oxygen deficiencies, nonbridging oxygen, and hydrogen centers (also referred to as H(I)- and H(II)-centers). Moreover, the presence of hydrogen may break bonds and generate unpaired electrons in a glass network former oxide, resulting in the formation of so-called E'-centers. Such defects mainly feature high absorption in the ultraviolet wavelength range (Skuja, 1998). Absorption may also be intensified by the presence of contaminations (i.e., residues of hydro-carbonaceous additives such as defoamers or UV-absorbing polishing agents such as zirconium or cerium) within the silica gel layer. Especially in laser optics, this circumstance strongly impacts the laser-induced damage threshold (LIDT). Moreover, the adhesion of coatings or optical cements on polished glass surfaces can be reduced by silica gel layers.

Implanted hydrogen may even lead to severe decomposition of the glass network by water due to hydrolytic scission according to Equation 8.1. Further, silica gel is highly hygroscopic and can thus cause accelerated aging of polished glass surfaces. These effects are also referred to as "glass corrosion" and can be observed in

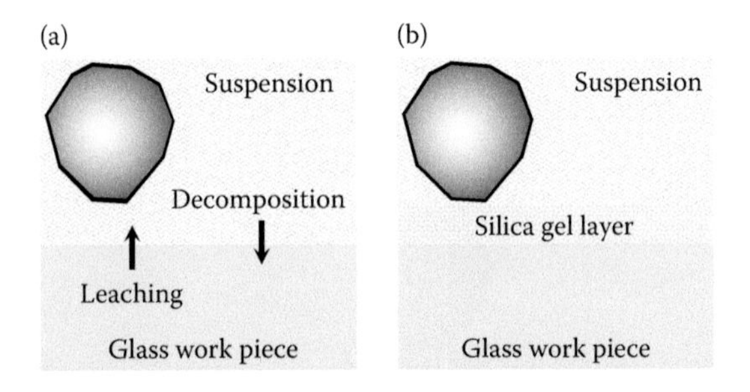


**FIGURE 8.2** Visualization of the effect of glass decomposition and leaching of glass constituents by aqueous polishing suspensions (a) and the formation of a silica gel layer (b).

[1] As a result of selective dissolving of essential glass constituents by leaching, the index of refraction of a near-surface glass layer of silicon dioxide-based glasses can decrease since the glass composition approaches the composition of pure silicon dioxide (when used as glass network former).

window glass panes (here, a white foggy layer can be formed in the course of time) or drinking glasses cleaned in dishwashers (depending on the quality of the water) in everyday life (compare "Climatic Resistance" in Section 6.2.4.1.1).

### 8.2.4 Fretting Hypothesis

The *fretting hypothesis* describes material removal due to fretting, which results from wear at the rough (ground or lapped) glass surface. Such wear arises from the contact pressure of the polishing tool on the glass surface and the relative movement between the two (Dimatteo, 1997). The impact of wear during glass polishing was first described by the English-American engineer *Frank W. Preston* (1896–1989) in the late 1920s (Preston, 1927). According to this description, the polishing time-dependent material removal rate (MRR), that is, the change in height $\Delta h$ over time $\Delta t$, follows from

$$MRR = \frac{\Delta h}{\Delta t} = C_\mathrm{p} \cdot \frac{L}{A} \cdot \frac{\Delta s}{\Delta t}. \tag{8.4}$$

This interrelation is known as the Preston equation. Here, $C_\mathrm{p}$ is *Preston's coefficient* (i.e., a process-specific constant, given in units of area per force, e.g., $\mathrm{m^2/N}$), $L$ is the total load (i.e., the pressure $p$ or the normal force $F_\mathrm{n}$, respectively), $A$ is the area on the glass surface that is exposed to wear (i.e., the contact area between the work piece and the polishing tool), and $\Delta s/\Delta t$ is the relative travel $\Delta s$ between the glass work piece and the polishing tool in a given period of time $\Delta t$. The ratio of $\Delta s$ and $\Delta t$ finally gives the relative velocity $v$. Equation 8.4 can also be expressed as

$$\Delta m = \rho \cdot C_\mathrm{p} \cdot L \cdot \Delta s, \tag{8.5}$$

where $\Delta m$ is the change in mass of the glass work piece and $\rho$ is the glass density. Taking the coefficient of friction $\mu$ between the tool and the work piece surface into account, the area-specific and time-dependent work $W$ required for material removal can be defined as

$$W = \mu \cdot A \cdot L \cdot v \cdot t. \tag{8.6}$$

Here, $t$ is the duration or time of the polishing process. Material removal due to fretting is further supported by chemical reactions between the polishing grain and the glass material since abrasion of chemically modified roughness peaks on the glass surface is enhanced by such reactions.

## 8.3 CLASSICAL POLISHING

Classical or conventional polishing is also known as full lap polishing. The latter denotation clearly indicates the approach applied for such polishing: the whole surface of the work piece is affected by the polishing tool. This method is used for the finishing of most optical components having (comparatively) large-scale plane or curved surfaces, such as lenses or prisms. The principle of classical polishing is shown schematically in Figure 8.3. Here, the work piece (a plano-convex lens in this

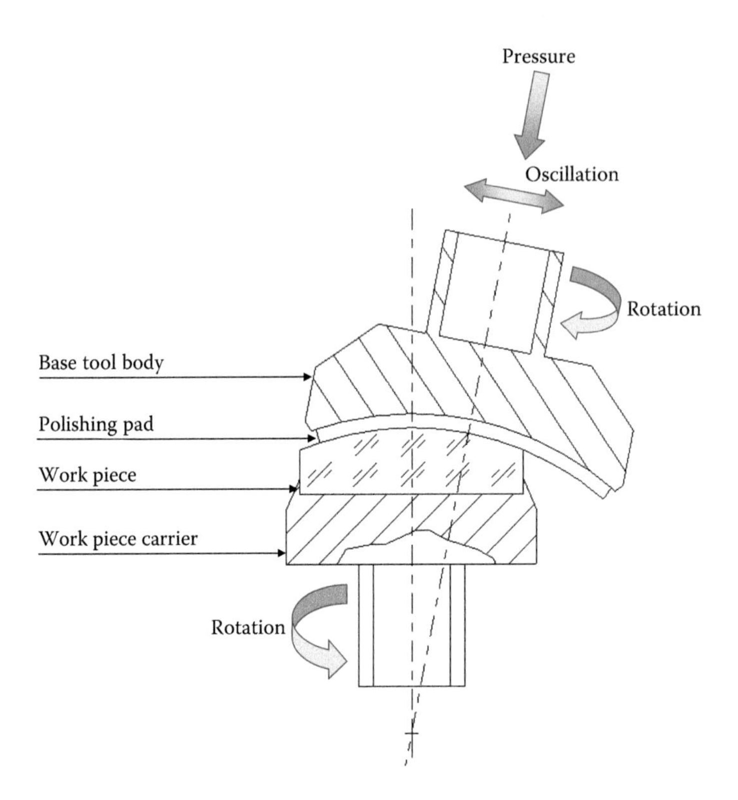


**FIGURE 8.3** Basic principle of classical full lap polishing.

example) is mounted on a work piece carrier, and the polishing tool is put onto the work piece's surface to be processed.

This process is very similar to lapping processes as described in more detail in Section 7.4.2.2. The main difference is given by the polishing tool. It consists of a polishing pad, which is attached to a metallic base tool body.

### 8.3.1 Polishing Pads and Tools

Different types of pads can be applied for polishing: the most classical is *pitch*, which was used for centuries in optical manufacturing. Even though this traditional type of *polishing pad* was replaced by synthetic materials in the last few decades it is still of importance for the production of high-quality precision optics. Pitch is a natural product and is obtained from vacuum distillation of petroleum or tar. After volatilization of oils and other liquids, pasty bitumen, also referred to as earth pitch, remains as a residual product. For optical manufacturing, it is offered in different classes of hardness. These classes, listed in Table 8.1, are defined according to a former in-house standard specification by the German optics manufacturer Zeiss: the particular hardness class corresponds to the pitch temperature (given in centigrade), where a defined test block with a base area of 1 $mm^2$, which is applied to a pitch surface, reaches a penetration depth of 2 mm under predefined conditions (i.e., a load of 1 kg and a fixed measurement time of 10 s).

**TABLE 8.1**
**Classes of Hardness of Polishing Pitches**

| Class of Hardness Name | Standard Hardness Class | Hardness Subclasses |
|---|---|---|
| Soft | 23 | 17–22 |
| Medium | 32 | 24–31 |
| Hard | 42 | 33–41 |

*Source:* Pieplow & Brandt GmbH, *Product Flyer*, 2010.

The stability of shape of a polishing pad made of pitch is thus strongly temperature-dependent. Since high local frictional heat can occur during polishing processes, such tools are deformed in the course of the machining process as shown in Figure 8.4. Against this background, pitch tools can be stabilized by adding different admixtures such as beeswax, colophony, or even wood sawdust (e.g., beech or spruce).[2] As an example, polishing pitch with a hardness of 38 consists of

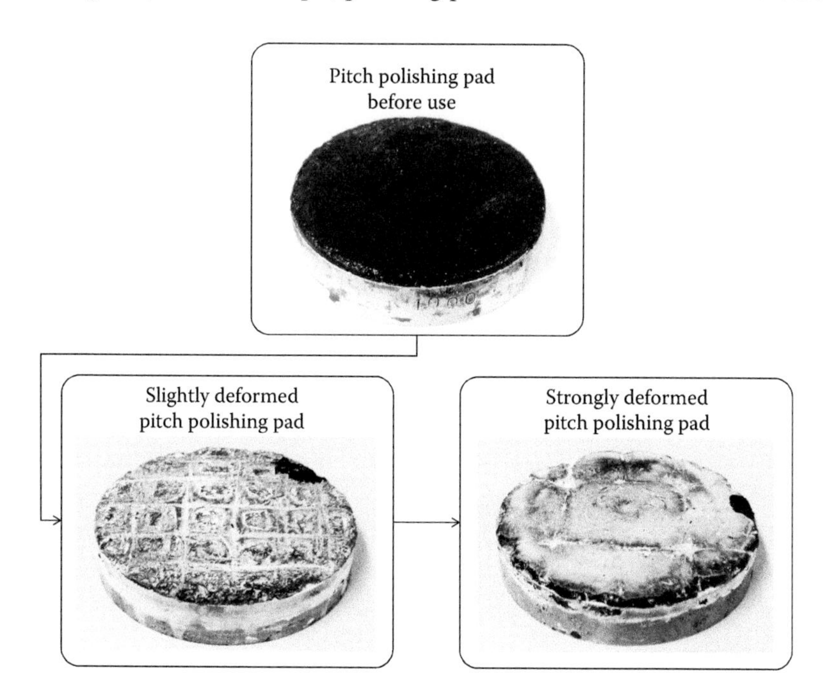


**FIGURE 8.4** Visualization of deformation of a pitch polishing pad due to frictional heat occurring in the course of a polishing process.

[2] In the course of his apprenticeship, the author of the present book has experienced that the preparation of pitch polishing tools is a kind of "witch's kitchen." Indeed, some recipes and selected additives are concealed from others for a long time and sometimes revealed in confidence at the end of a master craftsman's working life ... as a dying sorcerer hands his secret lore down to his apprentice. As a result, a lot of expert knowledge was irretrievably lost in the past.

colophony (30%–40%), bitumen (25%–30%), wax (5%–10%), and resin (20%–25%) (Pforte, 1995).

In addition to the stability of shape, the coefficient of friction $\mu$ between any polishing pad and the glass surface is of special interest according to Equation 8.6. With fused silica, this value amounts to $\mu=0.735$ for polishing pitch with a medium hardness of 26. Pitch features the highest coefficient of friction of any materials used as a polishing pad, as shown by the comparison in Table 8.2.

For the preparation of pitch polishing tools, a metallic base body is preheated. At the same time, a certain volume of pitch is heated to a semifluid. This semifluid is then deposited onto the warm base body, resulting in a thermal gluing of the pitch on the base body's surface. Subsequently, a fine ground work piece, for example a lens, is greased and then pressed into the warm and deformable pitch surface where the grease inhibits the adhesion of the work piece on the pitch surface. As a result, the pitch polishing tool is preshaped by the work piece itself. The preshaped pitch tool is then cooled to ambient temperature and further textured by cutting grooves into its surface in order to realize reticulated or spiral-grooved patterns as shown in Figure 8.5.

**TABLE 8.2**
**Comparison of Coefficients of Friction $\mu$ of Different Polishing Pad Materials When Polishing Fused Silica Including the Preferential Application of the Particular Polishing Pad Medium**

| Polishing Pad Material | Coefficient of Friction $\mu$ | Application |
|---|---|---|
| Pitch (medium hardness=26) | 0.735 | Precision polishing |
| Synthetic or natural felt | 0.685 | Rough polishing |
| Polishing cloth | 0.625 | Rough polishing |
| Polyurethane | 0.622 | Rough and precision polishing |

*Source:* Bliedtner, J., and Gräfe, G., *Optiktechnologie*, Carl Hanser Verlag, München, Germany, 2008 (in German).

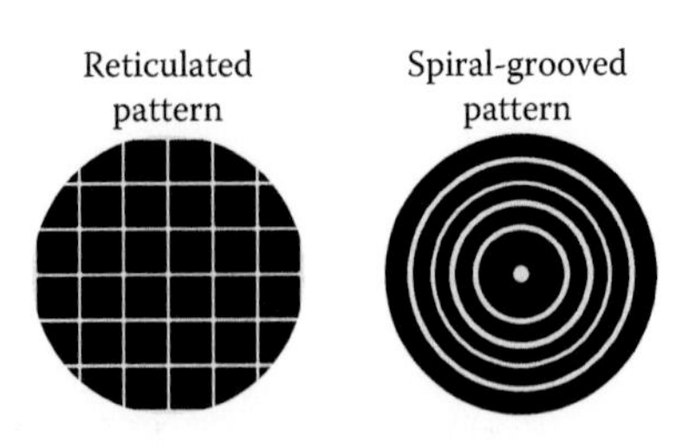


**FIGURE 8.5** Different textures for polishing pads: reticulated pattern (left) and spiral-grooved pattern (right). The gray background represents the base body; the effective polishing pad surface is given by the black structures.

Such pattering is performed for adjusting the contact time and local strength of interaction, respectively, of the pitch tool and work piece during polishing. Additionally, the grooves act as a reservoir for the polishing suspension.

By now, synthetic polishing pads are established in optical manufacturing. In comparison to pitch pads, this type of polishing pad stands out due to high long-term form stability; it is thus appropriate for large-volume production. It is further less sensitive to changes or rises in temperature and allows high process velocities. The most common material for synthetic polishing pads is polyurethane foil. Such foils are available in different thicknesses, ranging from approximately 0.5–25 mm. For some applications, polyurethane pads may be foamed and provided with specific polishing agent fillers such as cerium or zirconium. The coefficient of friction of this type of polishing pad amounts to $\mu = 0.622$ when polishing fused silica. In order to optimize the effective contact zone between the polishing pad and the work piece synthetic, polishing pads are usually cut into blossom-like structures as shown in Figure 8.6.

Instead of pitch or plastics, synthetic or natural felt or polishing cloth may be used in some cases for rough polishing. When polishing fused silica, the coefficient of friction of these pad materials amounts to $\mu_{felt} = 0.685$ and $\mu_{cloth} = 0.625$, respectively; see Table 8.2. Moreover, foils with embedded diamond grains of very small size can be used for some polishing tasks. The main advantage is then that polishing can be performed using pure water as lubricant instead of polishing suspensions as required for polishing with pads made of pitch, plastics, felt, or cloth.

**FIGURE 8.6** Polishing tool with blossom-like structured polishing pad made of polyurethane.

### 8.3.2 Polishing Suspension

In the course of the polishing process, the *polishing suspension* (a.k.a., polishing slurry) is brought in between the polishing pad and the work piece surface. Polishing suspensions are a mixture of water, a certain polishing agent, and (eventually) further additives such as defoamers or alkaline/acid additives in order to adapt the pH-value of the suspension to the chemical properties of the particular glass of the work piece. The polishing agent could be compared to fine lapping powders, where the size of polishing grains ranges from some hundreds of nanometers to some microns (Bliedtner and Gräfe, 2008). These grains initiate polishing by removal or material flow (compare Sections 8.2.1 and 8.2.2). A selection of commonly used polishing agents is listed in Table 8.3.

These polishing agents differ significantly in terms of hardness as well as surface geometry and sharpness of the grains. In order to obtain the optimum polishing result, the polishing agents listed in Table 8.3 may be mixed, and other trace elements may be added. As an example, cerium polishing suspensions contain a considerable portion of lanthanum, and zirconium-based polishing suspensions are a mixture of zirconium and hafnium in practice (Neauport et al., 2005).

The choice of the particular polishing agent depends on the material of the work piece. For instance, cerium oxide has turned out to be very appropriate for polishing fused silica. The ideal combination of the polishing grain material and the work piece material is mainly dominated by the particular hardness. Another influencing factor is the crystalline shape and size of the polishing grains, which strongly impact the polishing efficiency. Both parameters can be adapted by appropriate pretreatment of polishing grains, for example, by calcination processes (Kirk and Wood, 1995).

The interaction of the polishing suspension and the work piece surface unexclusively depends on the material properties of the polishing agent, but also on the concentration $C_s$ of the polishing suspension. This value is given by

$$C_s = \frac{m_{pa}}{m_s} \cdot 100\%, \quad (8.7)$$

where $m_{pa}$ is the mass of the polishing agent within the suspension, and $m_s$ is its total mass. The polishing suspension concentration usually amounts to $C_s = 5\%–30\%$ (Bliedtner and Gräfe, 2008) and should be adapted dynamically in

**TABLE 8.3**
**Selection of Commonly Used Polishing Agents**

| Polishing Agent | Pseudonym |
|---|---|
| Cerium oxide ($CeO_2$) | Ceria/opaline |
| Aluminum oxide ($Al_2O_3$) | Alumina |
| Zirconium oxide ($ZrO_2$) | Zirconia |
| Iron oxide ($Fe_2O_3$) | Polishing rouge |

the course of a polishing process: for high-efficient material removal at the beginning of the process, a quite viscous suspension with a high share of polishing agent grains is usually applied. At the end of the process, the suspension should be attenuated to a thinner fluid in order to achieve high surface cleanliness by avoiding the so-called *wiedergrau-effect* (i.e., the formation of a thin polishing agent film on the polished work piece surface). For very high surface cleanliness, the size of the polishing agent grains within the slurry can be reduced additionally.[3] Generally, the concentration of the polishing suspension impacts not only surface roughness, but also the surface shape. Polishing with a suspension of low concentration leads to the formation of a slightly concave surface with respect to the target shape, whereas the use of suspensions of high concentration gives rise to slight convexity.

### 8.3.3 Polishing Process

As already summarized in the introduction of the present chapter, the goal of polishing includes several aspects: the most important aspect is the realization of the final surface shape within the specified tolerances for surface accuracy as indicated by code number "3" according to DIN ISO 10110 and the generation of a glass surface with marginal surface defects as specified by the surface cleanliness (code number "5" according to DIN ISO 10110). Further, final surface smoothing is performed by polishing in order to achieve the required grade of polishing and to produce an even and transparent surface without any or with only marginal scattering characteristics. An example of such surface smoothing is shown by the comparison of a lapped and a polished glass surface in Figure 8.7.

One has to notice that surface smoothing by polishing is not only due to material removal, but also obtained by closing digs and micro cracks (induced in the course of previous rough and fine grinding) by material flow as presented in Section 8.2.1 and the formation of silica gel, compare Section 8.2.3.

Polishing processes can be classified into two different main categories: full face polishing, for spherical, plane, or cylindrical surfaces, and zonal polishing, used for aspherical or free-form surfaces. Full face polishing can be performed employing different polishing machines. First, CNC machines with two (or more) axes and at least two motorized spindles for rough and fine grinding (compare Section 7.3.1) can be used for the so-called *synchro speed polishing* approach. Here, the work piece spindle and the polishing tool spindle have the same sense of rotation and are rotated at high and almost identical rotational speed. Further, the work piece diameter is half the polishing tool diameter. As a result, the cutting velocity on the work piece

[3] During classical polishing, the process is interrupted several times in order to test the work piece surface and to survey its actual shape and cleanliness. For this purpose, the polishing suspension is usually wiped away from the glass surface using a sponge. This sponge is then cleaned in a wash pan. After some time, the polishing agent washed out of the sponge redeposits on the bottom of the wash pan. This redeposited polishing agent is highly suitable for precision polishing since the grains contained therein were broken and crushed and consequently reduced in size in the course of the polishing process.

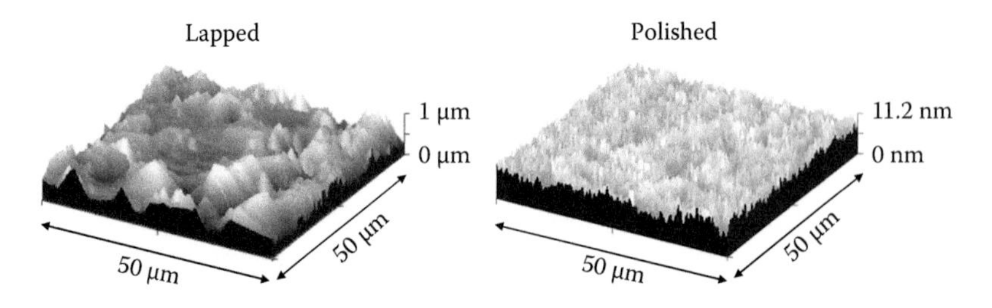


**FIGURE 8.7** Qualitative comparison of the surface roughness of fused silica surfaces after lapping (left) and polishing (right) measured via atomic force microscopy. Please note the different scaling of the height-axes.

surface is constant, resulting in constant material removal. For this approach, polishing tools with synthetic polishing pads are employed.

Second, polishing can be performed using lever arm machines with eccentric spindle drives as are applied for lapping (Compare Section 7.4.1). Here, polishing tools with pitch polishing pads are mostly used. The tool is oscillated over the lens surface by a lever arm, driven by an eccentric tappet. The final surface shape and accuracy can thus be influenced by different parameters or variables: (1) the polishing pad material used and the composition of the polishing suspension, (2) the amplitude of oscillation, (3) the pressure of the tool on the work piece, (4) the pattern of the polishing tool, and (5) the velocity or drive of the tool spindle and the work piece spindle, respectively. As shown in Figure 8.8, the latter parameter directly depends on the work piece diameter.

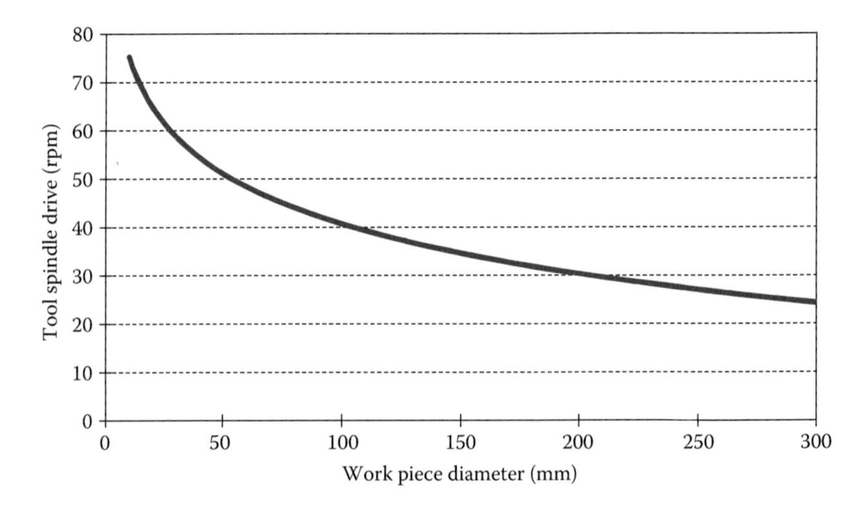


**FIGURE 8.8** Recommended polishing tool spindle drive vs. work piece diameter for lever arm machine polishing of spherical surfaces. (From Bliedtner, J., Gräfe, G., *Optiktechnologie*, Carl Hanser Verlag, München, Germany, 2008 (in German).)

Moreover, the ratio of the work piece spindle drive and the tool spindle drive is of essential importance: for simple surface geometries (i.e., standard radii of curvature and work piece diameters), high surface accuracy is achieved if this ratio is close to 1.

The polishing process on lever arm machines is quite comparable to lapping but differs significantly in terms of the tools, working materials, and basic underlying mechanisms for material removal. For polishing plane-parallel plates (e.g., windows or mirror substrates), the approach of *double-sided polishing* (compare Section 7.5) can be applied. This method allows for the production of polished plane-parallel components with high parallelism and marginal wave front deformation.

For polishing aspherical surfaces, zonal polishing using polishing tools with spherical (ball-shaped) heads is used (compare Section 7.3.2). The accuracy of this method directly depends on the stability and precision of the tool spindle and the resulting tool path. Against this background, sophisticated machines with several axes and air bearings are usually employed for zonal polishing. However, final correction of polished surfaces may be required after the actual polishing process in order to obtain the specified form accuracy. Such correction or finishing is referred to as subaperture correction; it is realized by unconventional polishing techniques as presented in the following sections.

## 8.4 UNCONVENTIONAL POLISHING TECHNIQUES

Classical polishing as described above represents the standard method for the last production step of optics surfaces. However, modern high-quality optics may require the development and application of novel and unconventional techniques. This especially applies for the polishing of free-form surfaces as well as for aspherization and local subaperture precision correction where zonal machining becomes necessary. Moreover, increasing surface quality of UV optics has gained importance in the last decades where now surface roughness values in the angstrom range are required and realized by super polishing in order to reduce scattering. Such high precision can be achieved by different techniques and methods as introduced hereafter.

### 8.4.1 MAGNETO-RHEOLOGICAL FINISHING

For subaperture correction, the method of *magneto-rheological finishing* (MRF) was invented in the mid-1980s (Kordonski and Jacobs, 1996). Here, the polishing pad is given by a magneto-rheological fluid, abbreviated MR fluid, consisting of the main components water, carbonyl iron particles, and abrasive grains, as well as further additives. When exposed to an external magnetic field, such a liquid becomes solid. Based on this effect, the nanohardness of the fluid is regulated locally by the magnetic field strength (Shorey et al., 2001). In practice, the MR fluid is transported onto the surface of a rotating wheel as shown in Figure 8.9.

The work piece is mounted on an inclinable axis and brought in contact with the fluid film formed on the rotating wheel. The local material removal on the work

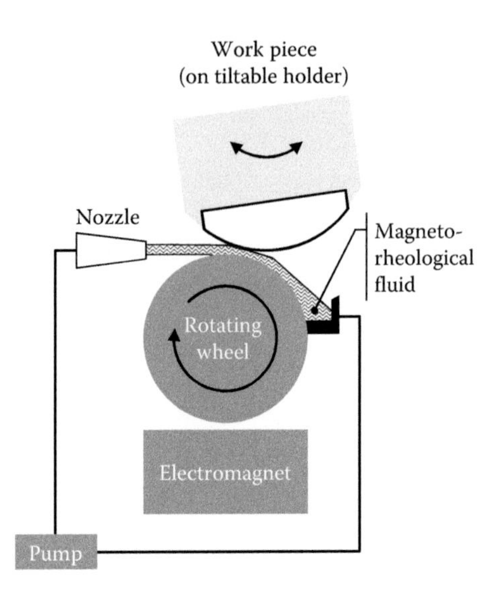


**FIGURE 8.9** Working principle of magneto-rheological finishing (MRF). (Adapted from Shorey, A.B. et al., *Applied Optics*, 40, 20–33, 2001.)

piece surface finally results from the MR fluid hardness and the dwell time of the work piece surface on the fluid film. Applying this technique, high surface accuracy in the range of some hundredths of the test wavelength can be realized.

### 8.4.2 Plasma Polishing

The application of plasmas represents a relatively novel approach for etching and polishing surfaces of optical components. Here, material removal is achieved by applying reactive process gases or gas mixtures. Within plasma volumes, such molecular gases are dissociated by collisions of electrons with the gas molecules. As a result, atomic species with a high chemical reactivity are generated. This technique is thus usually referred to as reactive atomic plasma technology (RAPT) (Fanara et al., 2006). The process gas generally used consists of fluorochemical compounds ($MF_x$). After electron-induced dissociation, fluorine (F) reacts with the network former silicon dioxide ($SiO_2$) to volatile gaseous silicon tetrafluoride ($SiF_4$) and gaseous dioxygen ($O_2$), according to

$$SiO_2(s) + 4F(g) \rightarrow SiF_4(g) + O_2(g), \tag{8.8}$$

consequently resulting in the removal of near-surface glass material as shown schematically in Figure 8.10.

Since this method works without any mechanical interaction at the surface, the formation of microcracks (as occurring during classical grinding, lapping, and polishing) is inhibited. However, special equipment for operational safety, such as extraction systems and filters, is required due to the emergence of the hazardous volatile

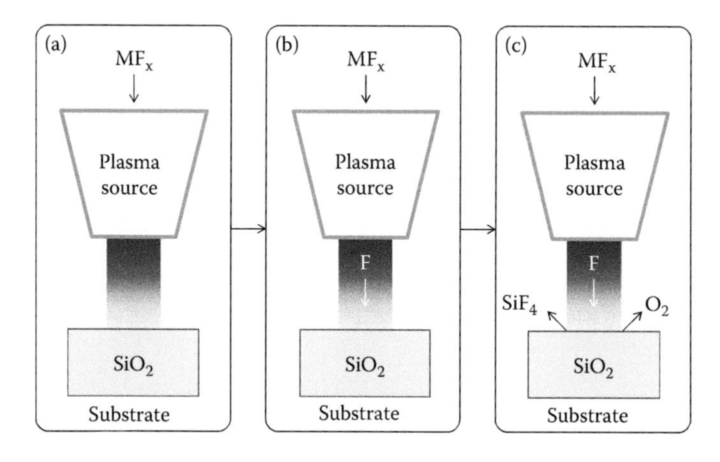


**FIGURE 8.10** Principle of glass polishing by RAPT. Fluorochemical compounds ($MF_x$) are fed into a plasma discharge (a) and dissociated to fluorine (F), which is blown onto the glass surface (b). Polishing is then due to the reaction of fluorine and the glass network former silicon dioxide ($SiO_2$) to silicon tetrafluoride ($SiF_4$) and dioxygen ($O_2$) in gaseous and volatile form (c).

reaction product silicon tetrafluoride. Different types of plasmas can be applied for reactive atomic plasma polishing where the process efficiency depends on the plasma source used (i.e., the discharge principle or geometry) and the composition of the process gas (Fanara et al., 2006; Oh et al., 2010). For example, volume etch rates up to 10 mm$^3$/min can be achieved when applying microwave-excited plasma jets and a process gas mixture of argon (Ar), sulfur hexafluoride ($SF_6$), and nitrogen ($N_2$) (Schindler et al., 2005). Finally, RAPT can be used for polishing or correcting other silicon-based optical materials commonly used in the manufacture of high precision optical components such as glass ceramics (Yao et al., 2010), silicon carbide (SiC) (Wang et al., 2006; Arnold and Böhm, 2012), or pure silicon (Si) (Zhang et al., 2008). Apart from polishing, RAPT can be employed for the realization of free-form surfaces (Arnold et al., 2010, 2016).

Another novel approach is plasma polishing using inert process gases. Here, plasma-physical mechanisms are applied in order to achieve surface smoothing (Gerhard et al., 2012, 2013). For this purpose, the work piece is arranged directly between the high-voltage electrode and the ground electrode of a plasma generator, which is given by a capacitor setup as shown in Figure 8.11.

The work piece itself consequently acts as dielectric separation of the two electrodes, resulting in the formation of a so-called dielectric barrier discharge plasma. This type of plasma is usually abbreviated DBD and stands out due to very low gas temperatures in the range of ambient temperature; the heating of the glass material is thus negligible and well below its softening point. Since the plasma is ignited directly on the work piece surface, several surface-specific plasma phenomena arise from this configuration. First, a so-called plasma sheath layer is formed. Within this layer, the plasma condition of quasi-neutrality is not valid, and both ions and electrons are accelerated, resulting in a certain particle bombardment on the surface. Second,

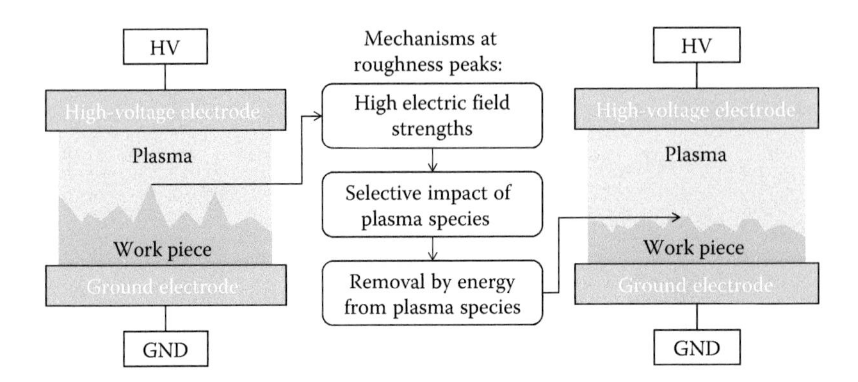


**FIGURE 8.11** Principle of inert gas plasma polishing. A glass work piece is brought in a direct plasma discharge, and polishing is due to the formation of high electric field strengths and a selective impact of plasma species at roughness peaks.

metastable species are formed within the plasma volume and successively de-excited due to collisions at the surface. Such de-excitation comes along with a considerable transfer of energy into the surface, which contributes to material removal. Third, high electric field strengths occur at roughness peaks of the work piece. This effect leads to a selective removal of such peaks due to the abovementioned plasma phenomena of the plasma (bombardment and de-excitation). Due to the use of plasma-physical instead of plasma-chemical mechanisms, this approach can be applied to quite different optical media (amorphous and partially crystalline) with different chemical compositions. It allows not only a further polishing of prepolished surfaces, but also a certain smoothing of ground/lapped optical media (Gerhard et al., 2013). Since plasma-induced smoothing comes along with the removal of surface-adherent UV-absorbing residues of working materials used during classical optical manufacturing (polishing agents, lubricants, etc.), applying this approach additionally results in a notable increase in LIDT (Gerhard et al., 2017).

Finally, polishing by atmospheric pressure plasmas can be achieved by plasma-induced surface melting (Donskoi et al., 1976), where the glass surface is heated far above its melting temperature, for example, to approximately 1900 K (Paetzelt et al., 2013). Comparable to fire polishing,[4] surface smoothing is then due to self-arrangement of the molten glass layer by surface tension.

### 8.4.3 Laser Polishing

The effect of surface smoothing by means of surface tension of molten glass is also the underlying mechanism for *laser polishing*. Here, the glass surface is heated and molten by incident laser irradiation where laser wavelengths in the infrared or ultraviolet wavelength range—which are well absorbed at the glass surface—are chosen.

[4] Fire polishing describes the effect of self-arrangement of glass surfaces exposed to high temperatures. For optical bulk articles such as single-use camera lenses or simple video projector condenser lenses, this is achieved by heating the surface of compression molded lens surfaces using burners.

This can be achieved by the use of carbon dioxide ($CO_2$) lasers with an emission wavelength of $\lambda = 10.6\,\mu m$. The heat input $Q$ into the glass surface is then given by

$$Q = (1 - R) \cdot I_0 \cdot \alpha \cdot e^{-\alpha \cdot t}, \tag{8.9}$$

where $R$ is the reflectance of the glass surface, $I_0$ is the intensity of the incident laser beam, $\alpha$ is the absorption coefficient of the glass, and $t$ is the thickness of the near-surface glass layer where laser irradiation is absorbed[5] (Laguarta et al., 1994). Alternatively, UV lasers such as frequency-tripled Nd:YAG-lasers ($\lambda = 355\,nm$) can be used for laser polishing (Wei et al., 2012). The achievable surface roughness is in the range of some hundreds of picometers.

Finally, it should be mentioned that laser polishing by local melting allows for the healing of damaged sites on glass surfaces, such as digs or cracks, as well as the mitigation of laser damage probability of UV optics by $CO_2$-laser annealing of polished or super polished surfaces (Doualle et al., 2016, 2017).

### 8.4.4 Fluid Jet Polishing

Subaperture correction and surface finishing can also be performed by *fluid jet polishing*, where a suspension of water and abrasive grains such as silicon carbide (SiC) is guided onto the glass surface. Material removal is then accomplished through the high pressure of the fluid jet, that is, approximately $6 \cdot 10^5$ Pa (Fähnle et al., 1998) and the resulting kinetic energy of the suspension. Here, the type and hardness of the used abrasives, the abrasive particle concentration, the hardness of the glass, and the angle of incidence of the fluid jet on the glass surface are essential paramters (Booij et al., 2002; Fang et al., 2006). When optimizing the footprint of the fluid jet on the surface to be machined, low surface roughness in the range of $\lambda/10$ can be realized (Booij et al., 2004).

## 8.5 SUMMARY

In the course of polishing, the final smooth and transparent surface of an optical component with the required shape accuracy and surface cleanliness is realized. The mechanisms for material removal and surface smoothing during polishing are described by different hypotheses: first, mechanical removal of glass material is induced by polishing grains comparable to lapping. Second, a dislocation of material from roughness peaks into roughness valleys due to material flow occurs. Third, digs and scratches are filled and closed by silica gel, which is formed due to chemical reactions of the aqueous polishing suspension with the glass surface. This effect comes along with a decomposition of glass by hydrolytic scission as well as leaching and diffusion of water into a near-surface glass layer. Fourth, material removal

---

[5] The thickness of the near-surface glass layer where laser irradiation is absorbed is found at the depth within the glass bulk material where the initial intensity of incoming laser light is reduced by $1/e$ (i.e., the reciprocal of Euler's number $e \approx 2.7183$). It thus corresponds to the optical penetration depth $d_{opt}$ which is given by the reciprocal of the absorption coefficient $\alpha$ ($d_{opt} = 1/\alpha$).

is accomplished by fretting, which results from wear of the polishing tool on the ground glass surface as described by the Preston equation.

Different tool materials can be employed for polishing, where commonly used polishing pads are made of synthetics such as polyurethane or pitch. Such polishing pads are brought in direct contact with the work piece surface, and polishing suspension is simultaneously supplied to the polishing zone. The polishing suspension is a mixture of water and abrasive grains made of cerium oxide, zirconium oxide, etc., but may also contain further alkaline or acid additives and defoamers. Polishing can be performed on computerized numerical control machines that are used for synchro speed polishing or on classical lever arm machines. In the latter case, the process is comparable to lapping in terms of the machining strategy and process control.

Moreover, several unconventional polishing methods are available for subaperture correction or aspherization by zonal machining. In the case of MRF, local polishing is realized by the use of fluids with adjustable nanohardness. Plasma polishing can be categorized in three different types on the basis of the underlying mechanism for material removal: plasma-chemical etching of glass, plasma-physical smoothing of roughness peaks, and plasma-induced heating and melting of glass surfaces. The latter mechanism and the accompanying self-arrangement of molten near-surface glass layers due to surface tension are also applied for laser polishing. Finally, local polishing can be realized by abrasive fluid jets at high fluid pressure.

## 8.6 FORMULARY AND MAIN SYMBOLS AND ABBREVIATIONS

**Mean diffusion depth $d_{\text{dif}}$:**

$$d_{\text{dif}} \approx 2 \cdot \sqrt{D \cdot t}$$

$D$ diffusion coefficient
$t$ time

***MRR* during polishing (Preston equation):**

$$MRR = \frac{\Delta h}{\Delta t} = C_{\text{p}} \cdot \frac{L}{A} \cdot \frac{\Delta s}{\Delta t}$$

$\Delta h$ change in height
$\Delta t$ time (duration of polishing process)
$C_{\text{p}}$ Preston's coefficient
$L$ total load (pressure $p$ or the normal force $F_{\text{n}}$)
$A$ work piece surface area exposed to wear
$\Delta s$ relative travel between glass work piece and polishing tool

**Change in work piece mass $\Delta M$:**

$$\Delta M = \rho \cdot C_{\text{p}} \cdot L \cdot \Delta s$$

$\rho$ glass density
$C_{\text{p}}$ Preston's coefficient

$L$ total load (pressure $p$ or the normal force $F_n$)
$\Delta s$ relative travel between glass work piece and polishing tool

**Work $W$ required for material removal:**

$$W = \mu \cdot A \cdot L \cdot v \cdot t$$

$\mu$ coefficient of friction between tool and work piece surface
$A$ work piece surface area
$L$ total load (pressure $p$ or the normal force $F_n$)
$v$ velocity
$t$ time

**Concentration $C_s$ of polishing suspension:**

$$C_s = \frac{m_{pa}}{m_s} \cdot 100\%$$

$m_{pa}$ mass of polishing agent within suspension
$m_s$ total mass of polishing suspension

**Heat input $Q$ into glass surface during laser polishing:**

$$Q = (1 - R) \cdot I_0 \cdot \alpha \cdot e^{-\alpha \cdot t}$$

$R$ reflectance of glass suerface
$I_0$ intensity of incident laser beam
$\alpha$ absorption coefficient of glass
$t$ thickness of near-surface glass layer

## REFERENCES

Arnold, T., and Böhm, G. 2012. Application of atmospheric plasma jet machining (PJM) for effective surface figuring of SiC. *Precision Engineering* 36:546–553.

Arnold, T., Boehm, G., and Paetzelt, H. 2016. New freeform manufacturing chains based on atmospheric plasma jet machining. *Journal of the European Optical Society—Rapid Publications* 11:16002.

Arnold, T., Boehm, G., Eichentopf, I.-M., Janietz, M., Meister, J., and Schindler, A. 2010. Plasma Jet Machining—A novel technology for precision machining of optical elements. *Vakuum in Forschung und Praxis* 22:10–16.

Bliedtner, J., and Gräfe, G. 2008. *Optiktechnologie*. München, Germany: Carl Hanser Verlag (in German).

Booij, S.M., Fähnle, O.W., and Braat, J.J.M. 2004. Shaping with fluid jet polishing by footprint optimization. *Applied Optics* 43:67–69.

Booij, S.M., van Brug, H., Braat, J.J.M, and Fähnle, O.W. 2002. Nanometer deep shaping with fluid jet polishing. *Optical Engineering* 41:1926–1931.

Cook, L.M. 1990. Chemical processes in glass polishing. *Journal of Non-Crystalline Solids* 120:152–171.

Cumbo, M.J., and Jacobs, S.D. 1994. Determination of near-surface forces in optical polishing using atomic force microscopy. *Nanotechnology* 5:70–79.

Dimatteo, N.D. 1997. *ASM Handbook Volume 19: Fatigue and Fracture*. Novelty, OH: ASM International.

Donskoi, A.V., Dresvin, S.V., Orlova, M.A., Osovskii, B.B., Khait, O.D., and Paushkin E.V. 1976. Plasma polishing of surface of wares made of silicate glass of any composition. *Glass and Ceramics* 33:162–165.

Doualle, T., Gallais, L., Cormont, P., Donval, T., Lamaignère, L., and Rullier, J.L. 2016. Effect of annealing on the laser induced damage of polished and $CO_2$ laser-processed fused silica surfaces. *Journal of Applied Physics* 119:213106.

Doualle, T., Gallais, L., Monneret, S., Bouillet, S., Bourgeade, A., Ameil, C., Lamaignère, L., and Cormont, P. 2017. $CO_2$ laser microprocessing for laser damage growth mitigation of fused silica optics. *Optical Engineering* 56:011022.

Evans, C.J., Paul, E., Dornfeld, D., Lucca, D.A., Byrne, G., Tricard, M., Klocke, F., Dambon, O., and Mullany, B.A. 2003. Material removal mechanisms in lapping and polishing. *CIRP Annals* 52:611–633.

Fähnle, O.W., van Brug, H., and Frankena, H.J. 1998. Fluid jet polishing of optical surfaces. *Applied Optics* 37:6771–6773.

Fanara, C., Shore, P., Nicholls, J.R., Lyford, N., Kelley, J. Carr, J., and Sommer, P. 2006. A new Reactive Atom Plasma Technology (RAPT) for precision machining: The etching of ULE® surfaces. *Advanced Engineering Materials* 8:933–939.

Fang, H., Guo, P., and Yu, J. 2006. Surface roughness and material removal in fluid jet polishing. *Applied Optics* 45:4012–4019.

Gerhard, C., Roux, S., Brückner, S., Wieneke, S., and Viöl, W. 2012. Low-temperature atmospheric pressure argon plasma treatment and hybrid laser-plasma ablation of barite crown and heavy flint glass. *Applied Optics* 51:3847–3852.

Gerhard, C., Tasche, D., Munser, N., and Dyck, H. 2017. Increase in nanosecond laser-induced damage threshold of sapphire windows by means of direct dielectric barrier discharge plasma treatment. *Optics Letters* 42:49–52.

Gerhard, C., Weihs, T., Luca, A., Wieneke, S., and Viöl, W. 2013. Polishing of optical media by dielectric barrier discharge inert gas plasma at atmospheric pressure. *Journal of the European Optical Society—Rapid Publications* 8:13081.

Iler, R.K. 1979. *The Chemistry of Silica: Solubility, Polymerization, Colloid and Surface Properties and Biochemistry of Silica*. New York: Wiley.

Kirk, N.B., and Wood J.V. 1995. The effect of the calcination process on the crystallite shape of sol-gel cerium oxide used for glass polishing. *Journal of Materials Science* 30:2171–2175.

Kordonski, W.I., and Jacobs, S.D. 1996. Magnetorheological finishing. *International Journal of Modern Physics B* 10:2837.

Laguarta, F., Lupon, N., and Armengol, J. 1994. Optical glass polishing by controlled laser surface-heat treatment. *Applied Optics* 33:6508–6513.

Lanford, W., Burman, C., and Doremus, R. 1985. Diffusion of water in $SiO_2$ at low temperature. In *Advances in Materials Characterization II*, eds. R. Snyder, R. Condrate, and P. Johnson, 203–208. New York: Plenum.

Neauport, J., Lamaignere, L., Bercegol, H., Pilon, F., and Birolleau, J.-C. 2005. Polishing-induced contamination of fused silica optics and laser induced damage density at 351 nm. *Optics Express* 13:10163–10171.

Nogami, M., and Tomozawa, M. 1984. Diffusion of water in high silica glasses at low temperature. *Physics and Chemistry of Glasses* 25:82–85.

Oh, J.S., Park, J.B., Gil, E., and Yeom, G.Y. 2010. High-speed etching of $SiO_2$ using a remote-type pin-to-plate dielectric barrier discharge at atmospheric pressure. *Journal of Physics D: Applied Physics* 43:425207.

Paetzelt, H., Böhm, G., and Arnold, T. 2013. Plasma jet polishing of rough fused silica surfaces. *Proceedings of the 13th International Conference of the European Society for Precision Engineering and Nanotechnology*, Berlin, May 27–31, 2013.

Pforte, H. 1995. *Der Optiker*. Homburg, Germany: Verlag Gehlen.

Pieplow & Brandt GmbH. 2010. *Product Flyer*. http://www.pieplow-brandt.de/downloads/lpeng.pdf.

Preston, F.W. 1927. The theory and design of plate glass polishing machines. *Journal of the Society of Glass Technology* 11:214–256.

Sabia, R., and Stevens, H.J. 2000. Performance characterization of cerium oxide abrasives for chemical-mechanical polishing of glass. *Machining Science and Technology* 4:235–251.

Schindler, A., Hänsel, T., Frost, F., and Rauschenbach, B. 2005. Modern methods of highly precise figuring and polishing. *Glass Science and Technology* 78:111.

Shorey, A.B., Jacobs, S.D., Kordonski, W.I., and Gans, R.F. 2001. Experiments and observations regarding the mechanisms of glass removal in magnetorheological finishing. *Applied Optics* 40:20–33.

Skuja, L. 1998. Optically active oxygen-deficiency-related centers in amorphous silicon dioxide. *Journal of Non-Crystalline Solids* 239:16–48.

Wang, B., Zhao, Q., Wang, L., and Dong, S. 2006. Application of atmospheric pressure plasma in the ultrasmooth polishing of SiC optics. *Materials Science Forum* 532–533:504–507.

Wei, X., Xie, X.Z., Hu, W., and Huang, J.F. 2012. Polishing sapphire substrates by 355 nm ultraviolet laser. *International Journal of Optics* 2012:238367.

Yao, Y.X., Wang, B., Wang, J.H., Jin, H.L., Zhang, Y.F., and Dong, S. 2010. Chemical machining of Zerodur material with atmospheric pressure plasma jet. *CIRP Annals—Manufacturing Technology* 59:337–340.

Zhang, J., Wang, B., and Dong, S. 2008. Application of atmospheric pressure plasma polishing method in machining of silicon ultra-smooth surfaces. *Frontiers of Electrical and Electronic Engineering in China* 3:480–487.

# 9 Cementing

## 9.1 INTRODUCTION

Cementing plays an important role in optics manufacturing, since for many applications, single optical components are cemented. Probably, the best-known example is an achromatic lens. Cementing is moreover applied to realize lens triplets or prism groups such as Amici prisms. Generally, the aim of cementing is to realize easy-to-handle simple optical systems with stable and long-term connections. In cemented lens groups, precise matching of the particular optical axes of the involved single lenses is required. For this purpose, not only the alignment of the lenses during cementing, in terms of the lens positions to each other, but also a constant thickness of the cement layer becomes crucial.

Generally, the term *cementing* describes bonding of surfaces with weak binding forces and adhesion. In contrast to glues, where bonding results from chemical reactions of the glue and the surface, cements do not react with the surface. As a consequence, the surface is not chemically modified. Avoiding surface modification is of great importance in optics manufacturing, since an alteration in stoichiometry of near-surface glass layers by reduction,[1] leaching, or implantation of atoms or molecules can give rise to a modification in index of refraction and reflectance, respectively (Williams, 1965; Gerhard et al., 2012, 2013). In addition to cementing, other bonding methods are used in some cases.

In this chapter, classical cementing using fine cements as well as novel approaches for lens bonding are presented. Further, possible errors that can occur in the course of cementing and the impact of such errors on the imaging quality of cemented optical groups are introduced.

## 9.2 CLASSICAL CEMENTING

### 9.2.1 Cementing Procedure

Cementing is performed using so-called *fine cement* or optical cement. This type of cement differs significantly from the raw cement used for fixing optical elements on holders or carriers as presented in Section 7.2.2.1. Fine cements generally feature high transmittance in the visible wavelength range and low viscosity. The latter characteristic allows (1) realizing thin cement layers with negligible impact on the total optical power of a cemented lens group, (2) easy adjustment of the single optical elements to be cemented, and (3) filling residual surface errors such as scratches or digs at the same time.[2]

---

[1] That is, the removal of oxygen from the glass oxides.

[2] Actually, optical fine cements are sometimes referred to as "fillers."

The most classical type of optical fine cement is the so-called *Canada balsam* (i.e., natural resin of North American fir). This resin is the oldest optical cement known and was the only available material for cementing in former times. Even though different synthetic cements are in hand nowadays, it is still used for some special applications due to its comparatively high elasticity.

Modern optical fine cements mainly consist of copolymers, and two different main types can be identified: UV-curing cement and two-component cement curing by polymerization due to heating (e.g., for 20 min up to 1 h, depending on the cement and catalyst ratio) or by storing at ambient conditions (e.g., for 24 h up to 6 days). Those cements feature comparable indices of refraction as standard optical glasses (i.e., in the range from approximately 1.45–1.6). In order to adjust this value, thinners and refractive index adjusters are available and can be added to the fine cement. Against this background, it should be considered that the index of refraction may change in the course of a curing process. As an example, cross-linked vinyl copolymer features an index of refraction of 1.53 in the liquid state and 1.55 in the solid state after curing at a temperature of 25°C. Fine cements should not contain any air bubbles, which would finally lead to pores and optically active microball lenses, respectively, within cured cement layers.

The actual cementing procedure is performed on a *cementing workstation*. This station consists of a work piece holder mounted on a work piece spindle, a pin edge guide, and an optical metrology device such as an autocollimation telescope for alignment of the single components to be cemented.

An example for a cementing procedure of a lens doublet is shown in Figure 9.1. Here, the first lens (lens 1) is mounted in the work piece holder, usually a vacuum chuck. Its optical axis is then aligned to the mechanical axis of the work piece spindle. Subsequently, a drop of fine cement is applied to the surface to be cemented, which is—if applicable—a concave surface in order to concentrate the applied cement at the lens center (Figure 9.1a). The volume of the cement drop is determined on the basis of different basic considerations. First, it results from the lens surface geometry (radius of curvature and area) and the target cement layer thickness. Second, possible shrinkage of the cement during curing is considered. To give an example, the volume of cross-linked vinyl copolymer cements is reduced by 4% after curing with respect to the liquid volume before curing.

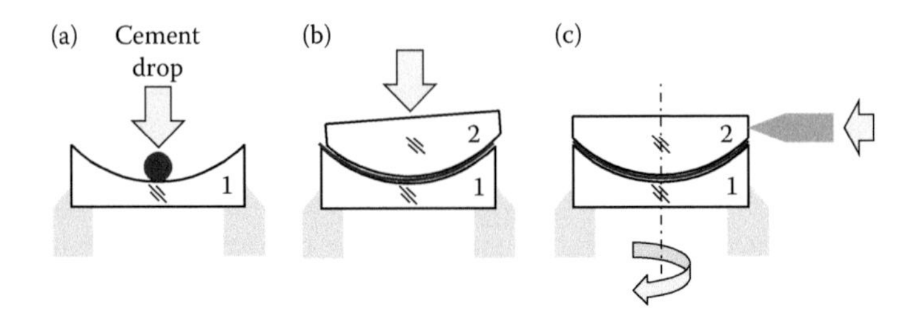


**FIGURE 9.1** Process steps during cementing; application of a cement drop on lens 1 (a), pressing of the applied cement drop to a thin and closed cement layer between lens 1 and lens 2 (b), and adjustment of lens 2 with respect to lens 1 (c).

After applying the fine cement to the lens surface, the second lens (lens 2) is placed on the fine cement, which is then pressed to a thin closed layer with constant thickness (Figure 9.1b). The alignment of the optical axes of both lenses is then performed by rotating the work piece spindle and the whole lens group, respectively, and displacing the second lens laterally by a mechanical stop, that is, the abovementioned pin edge guide (Figure 9.1c).

As mentioned above, the integration of an optical metrology device is mandatory in the cementing workstation for adjusting the optical axis of the second lens to the mechanical axis of the cementing workstation and the optical axis of the first lens, respectively. Alignment thus means to reduce the tilt between the lenses, which is determined by sending a laser beam through the lens group and measuring the *wobble circle* radius of the laser focal point. In the course of this measurement process, the second lens is displaced laterally as shown in Figure 9.2 until the minimum wobble circle radius is found or until the residual centering error, which directly follows from the wobble circle radius (see Section 6.3.2), meets the required specifications.

The required accuracy of a cementation in terms of the tilt between the particular optical axes is usually in the range of some angular minutes but may even amount to some angular seconds for systems of high precision. Once the alignment is finished, the cement is provisionally fixed by irradiating the cemented doublet with UV-light or by heating for some seconds. After fixing, the cemented lens group is removed from the cementing workstation and cured by long-term UV irradiation in an UV oven, tempering at some tens of centigrade in an oven or simple storing at ambient conditions for several days, depending on the type of the used cement.

In some cases, shrinking of the cement during final curing may lead to a misalignment and higher tilts of the optical axes than acceptable. The knowledge of the behavior of the used fine cement during curing is thus of great importance in order to avoid waste. Moreover, extensive cleaning of the surfaces to be cemented prior to the cementing procedure is necessary in order to avoid inclusions such as dust particles.

It should finally be noted that the thickness of cement layers must exceed the coherence length of white light[3] in order to avoid constructive or destructive

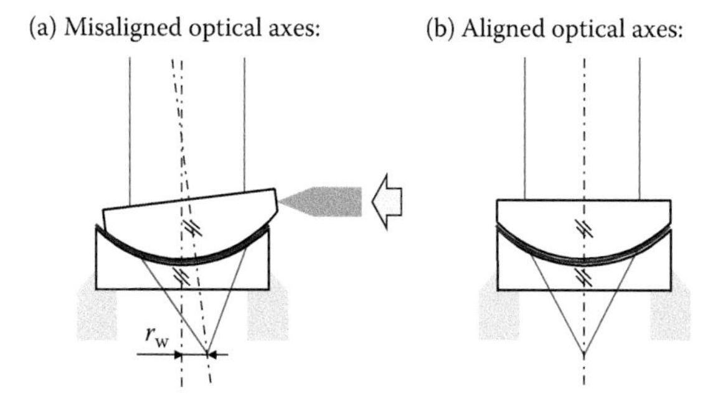


**FIGURE 9.2** Principle of alignment of the optical axes of single lenses for the production of a cemented doublet by measuring (a) and minimizing (b) the wobble circle radius $r_w$.

[3] The coherence length of white light amounts to approximately 50 µm.

interference within the cement layer. Otherwise, such a layer could act as a reflective or antireflective coating (see Chapter 11), which may support the formation of disturbing phenomena such as ghost images.

### 9.2.2 Cementing Errors

During cementing of lens doublets or triplets, different errors can occur due to inappropriate alignment of the optical axes of the involved single lenses. A selection of possible cementing errors is shown in the example of a simple lens doublet in Figure 9.3.

First, the optical axis of one lens might be tilted with respect to the optical axis of the other lens (Figure 9.3a) where the thickness of the cement layer is constant. In this case, incoming light is deviated from its intended path, leading to a deformation of the focal point, for example, due to the formation of coma. Second, the optical axes of both involved lenses can be shifted or displaced in a parallel fashion (Figure 9.3b). Such lateral displacement or offset leads to the formation of a *cement wedge* with position-dependent thickness. This kind of cementing error may cause a shift of incoming light, an offset of its focal point from the optical axis, and a decrease in image quality, respectively.

The combination of both cementing errors, tilt and lateral offset, gives rise to the formation of a severe cement wedge and may thus strongly deviate light passing through such a cemented lens group and finally reduce imaging quality. The latter aspect becomes obvious when one looks at the simulation of the impact of centering errors on the wave front deformation as shown in Figure 9.4. Here, wave front

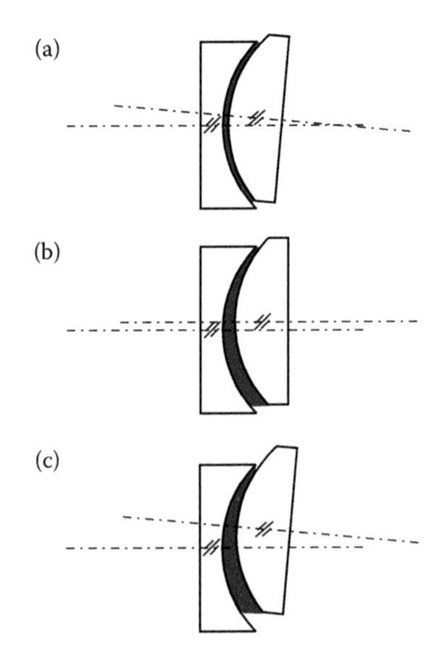


**FIGURE 9.3** Visualization of different possible cementing errors: tilt of optical axes (a), lateral offset of optical axes (b), and the combination of both (c). In (b) and (c), the so-called cement wedge is formed.

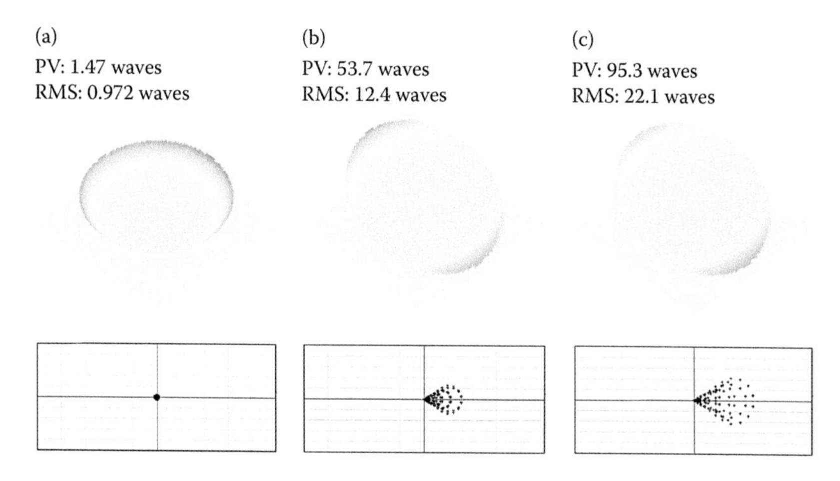


**FIGURE 9.4** Comparison of wave front plots including the peak-to-valley (PV) and root-mean-squared (RMS) wavefront deformation (top) and spot diagrams (bottom) of an achromatic lens without any cementing errors (a), with a tilt of the optical axes of both involved single lenses (b), and with both tilt and decenter (c) as described in more detail in the running text. (Figure was generated using the software WinLens3D Basic from Qioptiq Photonics GmbH & Co. KG.)

analysis was performed for a standard achromatic doublet with a focal length of 100 mm. Figure 9.4a shows the wave front of light after passing such an achromatic doublet without any manufacturing errors. The peak-to-valley value of the wave front is 1.47 waves (at a test wavelength of 546 nm) in this case. After simulating a tilt of 1° between the optical axes of the two lenses (i.e., lens 2 is tilted by 1° with respect to lens 1), the peak-to-valley value has increased to 53.7 waves. An additional decentering of the second lens by 500 μm with respect to the optical axis of the first lens gives rise to a further increase in peak-to-valley value up to 95.3 waves. Moreover, comatic aberration results from such a tilt and decenter of the second lens as shown by the spot diagrams in Figure 9.4 (bottom).

In addition to tilt and decenter, the thickness of the cement layer and the difference in radius of curvature of the surfaces to be cemented are important. If well chosen and accurately mixed, the used cement features the same index of refraction as one of the single lenses of the lens group. The cement thus merely gives rise to a simple increase in center thickness of this lens, since the interface from the lens to the cement does not exhibit any difference in index of refraction. In comparison to tilt and decenter, an increase in center thickness does not significantly impact the imaging performance and can be considered during the design and tolerancing of the lens group.

Another behavior is found if the index of refraction of the cement differs from both indices of refraction of the involved lenses, since the cement layer then represents a third optical element in between the two cemented lenses. In this case, the fitting of the surfaces to be cemented gains essential importance. For large differences in radii of curvature, the cement layer can act as a meniscus (i.e., either a concave-convex or convex-concave lens, compare Section 4.2.1) between the two actual lenses. This effect is shown by the simulation of cement wedges with different indices of refraction in Figure 9.5.

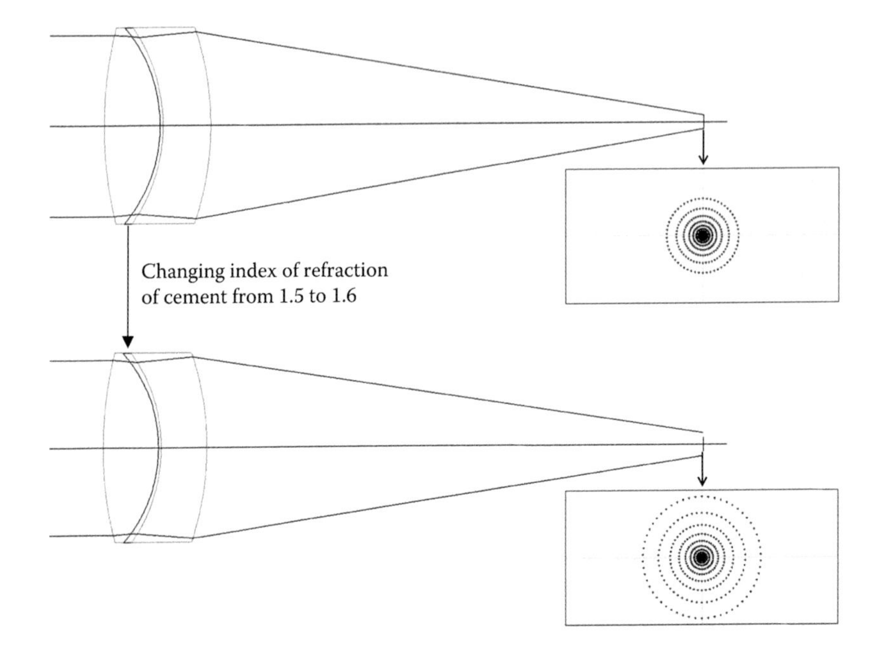


**FIGURE 9.5** Visualization of the impact of changes in index of refraction of a meniscus-shaped cement layer on the spot diameter and focal plane, respectively, of a cemented lens doublet. (For more information, see running text below.) (Figure was generated using the software WinLens3D Basic from Qioptiq Photonics GmbH & Co. KG.)

Here, the meniscus-like shape of the cement layer can be seen. For better visualization, an exaggerated difference in radii of curvature of the lens surfaces to be centered of 2 mm was simulated, whereas the chosen cement layer center thickness of 300 μm is a quite realistic value. The indices of refraction of the involved lenses were 1.5 for the first lens and 1.9 for the second. First, the index of refraction of the cement layer was set to 1.5, thus corresponding to the index of refraction of the first lens. In this case, the total effective focal length of the lens doublet was 52.82 mm, according to the simulation. After changing the index of refraction of the cement layer or meniscus, respectively, from 1.5 to 1.6, the effective focal length was 54.85 mm. The absolute change in focal length due to the different index of refraction of the cement wedge is thus 2.03 mm.

Even though cementation is a well-established and cost-efficient technique for bonding optical components, fine cements have some disadvantageous properties. First, shrinkage during curing can lead not only to centering errors as mentioned above; it can also cause large-scale or local mechanical tension and thus induce stress birefringence. Second, the long-term stability of cementations is another task since adhesion of cement with one or even both glass surfaces may decrease over time. This effect is mainly due to the different coefficients of thermal expansion of the cemented glasses and the accompanying strain of the glass–cement interfaces resulting from variations in temperature. Consequently, a thin, wedged air gap is formed at the border of a cemented surface where water from the ambient air can penetrate and induce further separation of the cement–glass bond. Such separation becomes visible in the form of interference patterns, since the air gap acts as a thin optically active layer. In order to avoid this effect, the border cylinder of cemented lens groups is usually sealed with lacquer.[4]

In addition to shrinkage and redeemableness, two aspects limit the fields of application for bonding via cementation: fine cements feature a low laser-induced damage threshold (LIDT) in comparison to optical glasses (see Table 6.11 in Section 6.4). A cement layer is thus the critical element of any optical system used in laser technology at medium and high laser power. Moreover, most cements—especially UV-curing ones—are highly UV-absorbing, so initially transparent cement layers become gray when exposed to UV irradiation for a longer time. Against this background, other approaches and methods for bonding laser and UV optics were developed in the past, as presented hereafter.

## 9.3 UNCONVENTIONAL BONDING METHODS

### 9.3.1 Optical Contact Bonding

In addition to classical cementing using fine cements as described above, different approaches for direct bonding of glass surfaces are in hand and applied for lens or prism groups of high precision used for laser or UV-optics. The basic approach is the so-called *optical contact bonding* (Haisma and Spierings, 2002). Here, optical components with high surface accuracy and surface smoothness are cleaned and directly contacted without any cement or filler between, as shown in Figure 9.6. As long as the surface form deviation and the air gap between the components to be bonded are lower than one nanometer, bonding occurs due to pure physical adhesion arising from intermolecular phenomena (Smith, 1965), such as dipole–dipole or van der Waals interactions[5] of the involved glass surfaces (Greco et al., 2001). In order to avoid the separation of these bonds in the course of time, the border cylinder of the bonded group is finally sealed with lacquer.

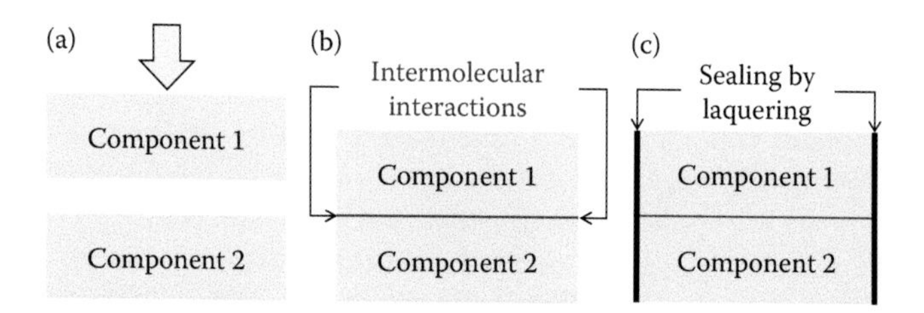


**FIGURE 9.6** Principle of optical contact bonding: two optical components are directly contacted (a), and adhesion occurs due to intermolecular interactions between the surfaces (b). Finally, the border cylinder of the bonded group is sealed with lacquer (c).

[4] Here, black, absorbing lacquer is used, since its secondary function is to absorb scattered and vagabonding light in order to prevent the formation of ghost images.

[5] Named after the Dutch physicist and Nobel Prize Laureate (physics, 1910) *Johannes Diderik van der Waals* (1837–1923).

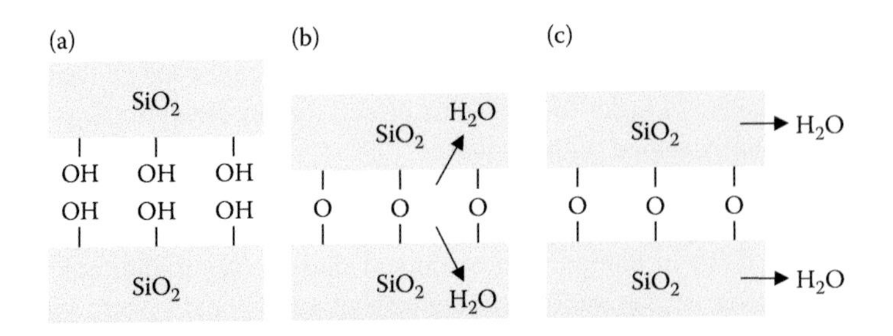


**FIGURE 9.7** Principle of activated covalent bonding; hydroxyl (OH) groups are attached to the glass surfaces to be bonded (a), leading to the formation of covalent bonding by oxygen atoms (c), after diffusion of water (b) into the glass where the water is finally removed from the glass by tempering (c). (Adapted from Turner, T., and Casnedi, P., *EuroPhotonics*, 27–29, 2013.)

The adhesion strength and long-term stability of assemblies produced by optical contact bonding can even be improved by an activation of the involved glass surfaces prior to contacting. This method is known as *activated covalent bonding* where hydroxyl (OH) groups are attached on the glass surfaces by chemical processes (Turner and Casnedi, 2013). Water from such hydroxyl groups then diffuses into the glass, resulting in covalent bonding of remaining oxygen atoms as shown in Figure 9.7. Finally, the water can be removed from the glass bulk material by tempering.

Surface activation for improved optical contact bonding can also be realized by plasma treatment with nitrogenous or oxygen-containing process gases (Kalkowski et al., 2011). Moreover, improved adhesion can be obtained when introducing silica nanoparticles to the bonding process. Such nanoparticles are applied to the contact zone between the involved glass surfaces and subsequently polymerize in the course of a sol-gel chemistry process. In this way, a branched network that connects the surfaces is formed (Sivasankar and Chu, 2007).

### 9.3.2 Laser Welding

Bonding of optical components can also be achieved by the comparatively novel approach of *laser welding*, where chiefly laser sources with pulse durations in the femtosecond range are applied (Tamaki et al., 2006; Richter et al., 2011; Hélie et al., 2012; Zimmermann et al., 2013). Such laser sources stand out due to the fact that the laser irradiation is absorbed in transparent media on the basis of multiphoton absorption even though the medium exhibits marginal absorbance at the used laser wavelength.

Comparable to optical contact bonding, the single optical components are brought in direct contact for laser welding where high surface accuracy is required. More precisely, the fit of both involved surfaces should be as fine as possible in order to prevent the formation of interference patterns within the resulting air gap. Then, incident laser irradiation is focused onto the contact zone, as shown in Figure 9.8, and glass material of both components is molten and mixed.

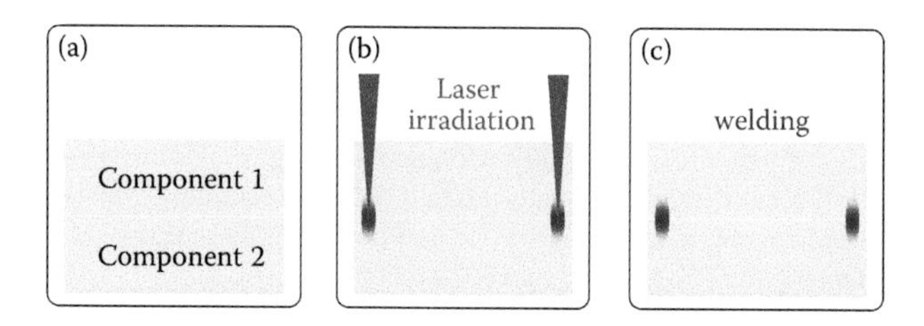


**FIGURE 9.8** Principle of laser welding: two optical components are directly contacted (a), and focused laser irradiation is applied to the contact zone (b), resulting in a welded joint by melting and mixing of glass material from both components (c).

As a result of such mixing, the components are permanently welded, and this type of connection exhibits high resistance to age. It is obvious that the control and choice of the laser process parameters become essential, since overheating and uncontrolled cooling of the molten glass might give rise to stress birefringence on the one hand and deformation of the contact surfaces on the other hand.

## 9.4 SUMMARY

Cementing or optical contact bonding is applied in order to connect single components to lens or prism groups. The most classical approach is cementation, where optical fine cement or filler is used as adhesive. Different types of cements, UV-curing or cured by tempering or storing, are in hand, and the index of refraction of such cements can be adjusted by additives within certain limits. In matters of lens groups such as achromatic doublets or triplets, the alignment of the particular axes of the involved single components becomes of essential importance in order to avoid cementing errors such as tilts and decenters and a reduction in imaging quality, respectively. This is achieved by the integration of optical metrology devices in the used cementing workstations.

Optical cements feature comparatively low LIDT and poor transmittance in the ultraviolet wavelength range. For optics used in these contexts, unconventional bonding methods are thus applied: optical contact bonding is performed without any cement or filler; adhesion is exclusively due to intermolecular forces between the glass surfaces. For this purpose, high surface accuracy and fit are required. The activation of the surfaces to be bonded by chemical or plasma treatment allows increasing the bond strength and the long-term stability of groups connected via optical contact bonding. Finally, laser welding is an alternative and novel solution for stable bonding of optics surfaces with high resistance to age.

## REFERENCES

Gerhard, C., Tasche, D., Brückner, S., Wieneke, S., and Viöl, W. 2012. Near-surface modification of optical properties of fused silica by low-temperature hydrogenous atmospheric pressure plasma. *Optics Letters* 37:566–568.

Gerhard, C., Weihs, T., Tasche, D., Brückner, S., Wieneke, S., and Viöl, W. 2013. Atmospheric pressure plasma treatment of fused silica, related surface and near-surface effects and applications. *Plasma Chemistry and Plasma Processing* 33:895–905.

Greco, V., Marchesini, F., and Molesini, G. 2001. Optical contact and van der Waals interactions: The role of the surface topography in determining the bonding strength of thick glass plates. *Journal of Optics A: Pure and Applied Optics* 3:85–88.

Haisma, J., and Spierings, G.A.C.M. 2002. Contact bonding, including direct-bonding in a historical and recent context of materials science and technology, physics and chemistry: Historical review in a broader scope and comparative outlook. *Materials Science and Engineering R* 37:1–60.

Hélie, D., Bégin, M., Lacroix, F., and Vallée, R. 2012. Reinforced direct bonding of optical materials by femtosecond laser welding. *Applied Optics* 51:2098–2106.

Kalkowski, G., Risse, S., Rothhardt, C., Rohde, M., and Eberhardt, R. 2011. Optical contacting of low-expansion materials. *Proceedings of SPIE* 8126:81261F-1.

Richter, S., Döring, S., Tünnermann, A., and Nolte, S. 2011. Bonding of glass with femtosecond laser pulses at high repetition rates. *Applied Physics A* 103:257–261.

Sivasankar, S., and Chu, S. 2007. Optical bonding using silica nanoparticle sol-gel chemistry. *Nano Letters* 7:3031–3034.

Smith, H.I. 1965. Optical contact bonding. *The Journal of the Acoustical Society of America* 37:928–929.

Tamaki, T., Watanabe, W., and Itoh, K. 2006. Laser micro-welding of transparent materials by a localized heat accumulation effect using a femtosecond fiber laser at 1558 nm. *Optics Express* 14:10460–10468.

Turner, T., and Casnedi, P. 2013. Novel bonding technology improves optical assemblies. *EuroPhotonics* 2013:27–29.

Williams, E.L. 1965. The diffusion of oxygen in fused silica. *Journal of the American Ceramic Society* 48:190–194.

Zimmermann, F., Richter, S., Döring, S., Tünnermann, A., and Nolte, S. 2013. Ultrastable bonding of glass with femtosecond laser bursts. *Applied Optics* 52:1149–1154.

# 10 Centering

## 10.1 INTRODUCTION

After the manufacturing steps of preshaping, rough grinding, fine grinding, and polishing, a lens border cylinder is usually tilted with respect to the optical axis of the lens. Such tilt is referred to as lens decenter or centering error. It occurs due to the fact that a lens is fixed to and removed from different holders and work piece carriers several times in the course of the manufacturing process, leading to variations in lens alignment in the particular machine. The centering error preferentially arises for lenses that were placed at the rim of a carrier body for lapping (see Section 7.4.2.2) and subsequent polishing, since in this case the interaction of the tool and the lens surface is strongly asymmetric and not rotational-symmetric.

The centering error can contribute to a reduction in imaging quality of stacked optomechanical systems (compare Section 12.2.1), where lenses are stacked into tubes and mounts and fixed by screw connections. This procedure is based on a certain self-alignment of the lens border cylinder with respect to the mechanical surfaces of the tubes and mounts. Lens decenter consequently results in a tilt of a lens within a stacked optical system as shown in Figure 10.1. Such possible tilt can have severe impact on the imaging quality as shown by the comparison of astigmatism resulting from the tilt of a single lens in Figure 10.2. Here, the initial image plane features certain astigmatism without any lens tilt due to inclined incidence of light. When tilting the lens, the initial image plane becomes asymmetrical, and astigmatism is increased.

As an idealized description, the goal of centering is to grind the lens border cylinder in that vein that its mechanical axis is parallel to its optical axis as visualized in Figure 10.3. However, perfect parallelism of both axes is nearly impossible to achieve, and centering is performed in order to fit the requirements as determined in the course of optical design (see Section 5.5) and defined in the manufacturing drawing in practice. As already discussed in Section 6.3.2, the example shown in Figure 6.6 specifies a maximum centering error of 10 arc minutes. This is indicated by the expression "4/10′" according to DIN ISO 10110.

Centering can be performed for single lenses or cemented lens groups such as doublets or triplets. In the latter case, the lenses are cemented prior to centering due to the fact that cementing can lead to decenter of lens groups even though the used single lenses are well centered. In addition to a correction of the tilt between the optical and the mechanical axes, centering allows realizing bevels as protection chamfers or mounting surfaces at the same time when using appropriate centering tools. The commonly used machines, tools, and methods for centering are presented in this chapter.

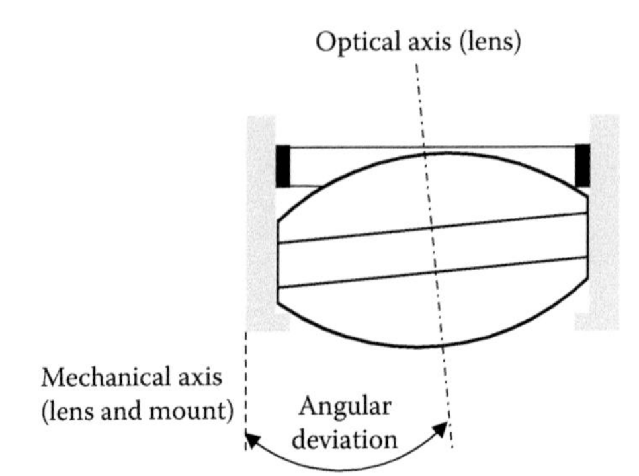


**FIGURE 10.1** Visualization of the impact of lens centering error on the orientation of a mounted lens, leading to an angular deviation between the optical axis of the lens and the mechanical axis of the lens and the mount.

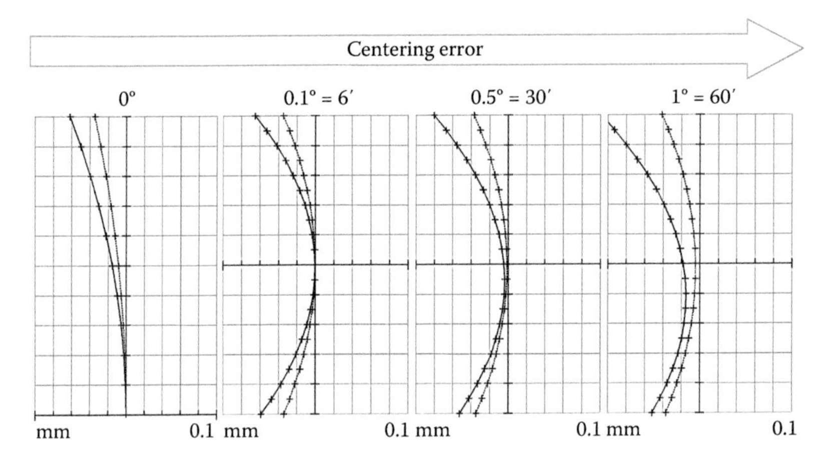


**FIGURE 10.2** Field diagram for astigmatism of a single lens without tilt (0°) and with tilt (0.1°–1°). (Figure was generated using the software WinLens3D Basic from Qioptiq Photonics GmbH & Co. KG.)

## 10.2 CENTERING METHODS

### 10.2.1 Classical Centering

Classical centering is performed on turning machines (a.k.a. *centering machines*). The principal setup of these machines is comparable to turning lathes, thus consisting of two opposite work piece spindles with coaxially arranged mechanical axes. Both work piece spindles are rotated at the same rotation velocity $\omega$. The mechanical alignment of these spindles and the control of the rotation velocity are of essential importance, since slight variations in alignment and rotation can cause severe damage of the lens surfaces. In addition to the work piece spindles, a centering machine features a third spindle, the tool spindle, which is arranged parallel to the work piece spindles.

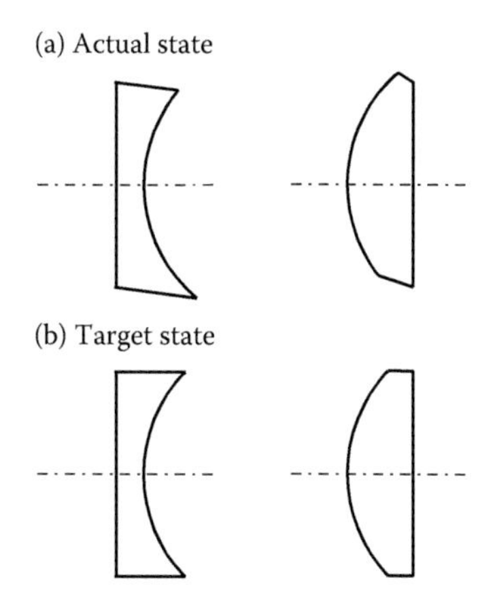


**FIGURE 10.3** Visualization of the actual state of a lens after rough grinding, finish grinding, and polishing where the mechanical axis of the lens border cylinder is tilted with respect to the optical axis (a) and the idealized target state without any tilt of both axes (b).

In the course of a lens centering process, this tool spindle is slowly brought close to the work piece spindles, and the lens border cylinder is successively ground by the centering tool. Simultaneously, cooling lubricant, as used for rough grinding or fine grinding with bound abrasive grains, is applied.

For fixing a lens in a centering machine, different approaches are available: first, it can be cemented on a holder using raw cement. This method is preferentially used for special lens geometries. Second, a lens can be held by *clamping bells*, see Figure 10.4 (i.e., cylindrical cup points made of stainless steel, copper, or brass mounted on the work piece spindles).

The advantage of this technique is that the lens is not only held but also aligned by the used clamping bells as follows: For fixing a lens between two clamping bells, the bells are pressed together with moderate pressure applying oil to the lens surface for greasing. As a result, the optical axis of the lens is self-orientated parallel to the mechanical axes of the clamping bells and work piece spindles, respectively, as shown in Figure 10.5.

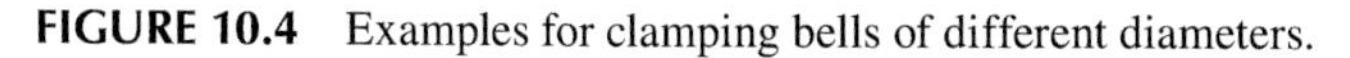

**FIGURE 10.4** Examples for clamping bells of different diameters.

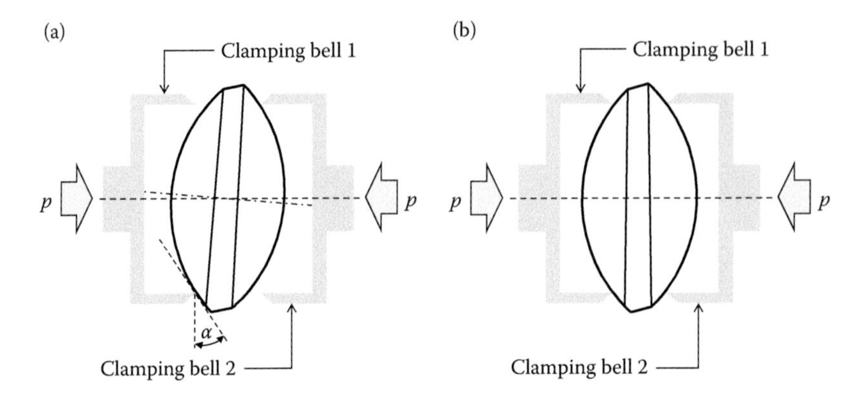


**FIGURE 10.5** Principle of fixing and aligning lenses by clamping bells in a centering machine; the lens is placed between two clamping bells that are pressed together (a). The optical axis of the lens is consequently aligned with the mechanical axes of the clamping bells and the work piece spindles, respectively (b).

Such self-alignment is due to the fact that the optical axis of a lens is given by the straight line through its extreme surface points, the so-called vertices. The optical axis thus lies at the thickest position of a convex and at the thinnest position of a concave lens where the nominal lens center thickness is found. Lens self-alignment thus occurs when pressing the ring-shaped clamping bells on a spherical lens surface, which automatically slides into the centered position with respect to the bells and the work piece spindles of the centering machine. Due to this advantage, this method is usually applied for fixing lenses during centering. However, self-alignment is only achieved as long as the so-called *sliding angle* $\alpha$ (see Figure 10.5), which is found between the tangent of the lens surface and the perpendicular to the cylinder axis of the clamping bells, is higher than approximately 7°. This angle can be calculated by

$$\alpha = \arcsin\left(\frac{D_1}{2 \cdot R_1}\right) + \arcsin\left(\frac{D_2}{2 \cdot R_2}\right). \tag{10.1}$$

Here, $D_1$ is the diameter of the clamping bell applied to the first lens surface with the radius of curvature $R_1$, and $D_2$ is the diameter[1] of the clamping bell used for holding the second lens surface with $R_2$. The sliding angle $\alpha$ finally gives the frictional force $F$ between the clamping bell and the lens surface according to

$$F = \mu \cdot p \cdot \cos\alpha, \tag{10.2}$$

where $\mu$ is the coefficient of friction[2], and $p$ is the pressure of the clamping bell on the lens surface.

[1] If possible, clamping bells with the same diameter are used for holding the lens surfaces. However, the use of clamping bells of different diameters may become necessary for special lens geometries, for example, for lenses with high difference in radii of curvature.

[2] For instance, the coefficient of friction of polished glass on polished steel is approximately 0.14 (Karow, 2004).

For siding angles lower than 7°, the lens has to be cemented on a holder and aligned optically, for example, by measuring the wobble circle radius[3] as already discussed in Section 6.3.2. Alternatively, the *centering runout* can be measured with the aid of a lens clock as introduced in Section 7.7. For this purpose, the lens is placed on a clamping bell and rotated. The runout $\Delta z$, given in millimeters, is then measured close to the lens edge and the centering error follows from

$$CE = 3434 \cdot \left( \frac{\Delta z}{R_1} + \frac{\Delta z}{R_2} \right), \tag{10.3}$$

taking the radii of curvature of the lens surfaces, $R_1$ and $R_2$, into account. This interrelationship gives the centering error in arc minutes. For a description in radians, the factor 3437 is left out.

After fixing, the optical axis of the lens is congruent with the mechanical axes of the used clamping bells or holders and the work piece spindles. The lens is then rotated, and its border cylinder is successively ground by the *centering tool* as shown in Figure 10.6 until the target lens diameter is obtained. For this purpose, plane grinding tools as shown in Figure 10.6 or grinding wheels as shown in Figure 10.7 are employed. The surface of these tools features comparable abrasive coatings as saw blades (see Section 7.2.2.1) or grinding tools (see Section 7.3.1), where the choice of the size of the embedded abrasive grains depends on the required surface roughness of the lens border cylinder as specified in the manufacturing drawing. As an example, this parameter is indicated by the element "G2" (or by two chevrons according to DIN 3140) in Figure 6.16 in Section 6.3.4. The lens border cylinder should thus be medium ground, where the target residual surface roughness amounts to 5 ± 1 µm (4–6 µm, compare Table 6.9).

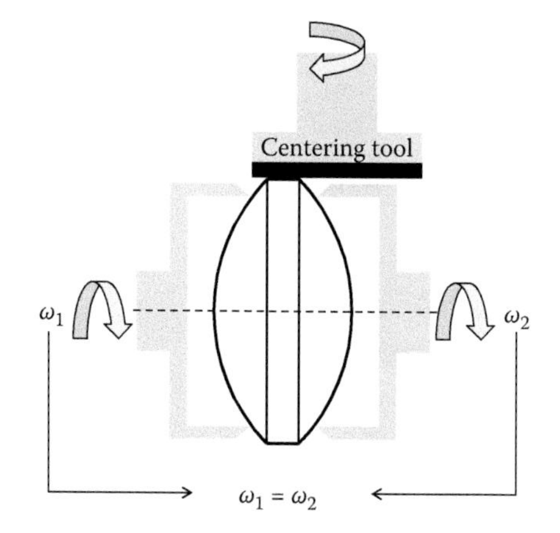


**FIGURE 10.6** Centering of a lens using a plane grinding tool.

[3] Measurement of the wobble circle radius by an integrated optical metrology device may also be carried out during fixing lenses with clamping bells in order to control proper alignment, especially in the case of lenses with low radii of curvature and low sliding angles, respectively.

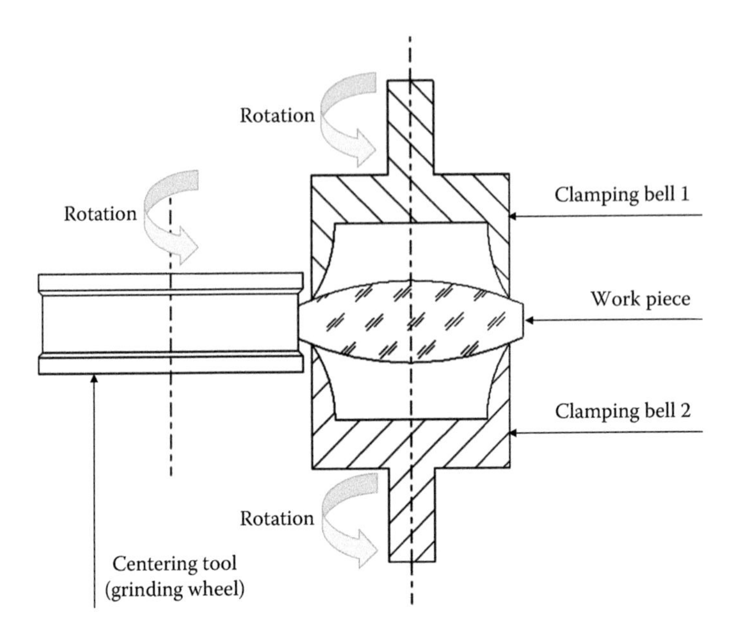


**FIGURE 10.7** Centering of a lens using a grinding wheel.

Depending on the centering tool geometry, edging, grooving, and beveling can be realized simultaneously or subsequent to the actual centering process where necessary. A selection of possible grinding wheel geometries used for this purpose is shown in Figure 10.8.

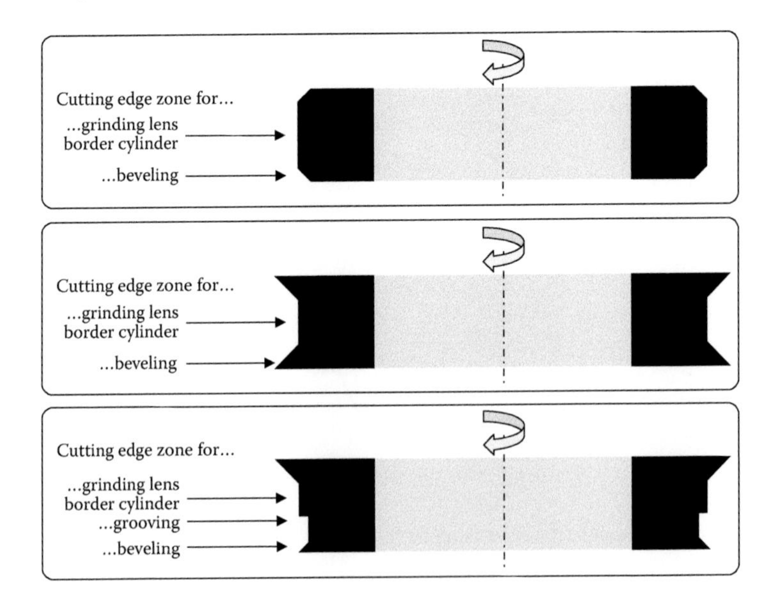


**FIGURE 10.8** Examples for grinding wheels used for centering and eventual beveling and grooving of lens border cylinders.

Moreover, the tool spindle can perform a relative lateral motion with respect to the work piece spindle. This allows generating other lens border geometries than cylinders, for example, ellipses[4] or even two-dimensional free forms, which may become necessary for subsequent mounting and assembly. As described in Chapter 12, different approaches can be chosen for this last step in manufacturing optomechanical systems. The particular centering accuracy is also defined on the basis of the used method for mounting and assembly and the theoretically and practically achievable precision.

### 10.2.2 Precision Centering of Cemented Optics

For optical systems of high performance, the approach of *precision centering* allows a reduction in position errors that may arise from mounting and assembly by classical stacking and screw connecting (see Section 12.2.1). Here, a lens or cemented lens group is glued into a mount, and less attention is paid to the deviation between the optical axis and the mechanical axis of the mount. As shown in Figure 10.9, the resulting glued optomechanical group is subsequently centered by grinding the mount border cylinder instead of the lens border cylinder.

## 10.3 SUMMARY

In the course of optics manufacturing steps such as rough grinding, finish grinding, and polishing, lens decenter may occur. Such decenter is also referred to as centering error and is defined as the deviation between the border cylinder axis and the optical axis of a lens. When mounting decentered lenses in mounts or tubes, a decentered lens is tilted, since it is aligned along its border cylinder due to the direct contact with the inner mount cylinder. The optical axis thus features a certain deviation from the mechanical axis of an optomechanical system, leading to the formation of astigmatism or distortion.

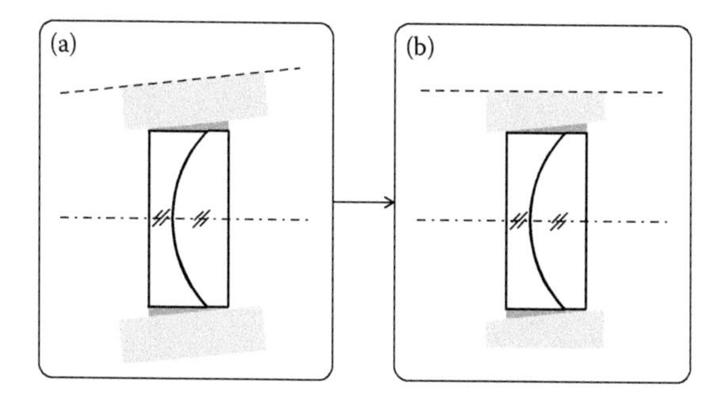


**FIGURE 10.9** Principle of precision centering; a lens is glued into a mount (a), and the entire optomechanical group is finally centered (b).

[4] For example, 45°; deflecting mirrors are often elliptical for optomechanical laboratory equipment.

The aim of centering is to reduce this effect by grinding the lens border cylinder parallel and concentric to the optical axis. For this purpose, a lens is placed between two clamping bells that are pressed together. Self-alignment of the optical axis with respect to the axes of the clamping bells and the work piece spindles, respectively, is achieved if the sliding angle is higher than 7° as valid for the bigger part of standard lenses. After clamping, the lens border cylinder is ground by a centering tool. This tool is mounted on the tool spindle parallel to the work piece spindle. During centering, the lens edges may simultaneously be beveled and shaped, depending on the geometry of the used centering tool.

## 10.4 FORMULARY AND MAIN SYMBOLS AND ABBREVIATIONS

**Sliding angle $\alpha$:**

$$\alpha = \arcsin\left(\frac{D_1}{2 \cdot R_1}\right) + \arcsin\left(\frac{D_2}{2 \cdot R_2}\right)$$

$D_1$ diameter of the clamping bell applied to first lens surface
$D_2$ diameter of the clamping bell applied to second lens surface
$R_1$ radius of curvature of first lens surface
$R_2$ radius of curvature of second lens surface

**Frictional force $F$ between clamping bell and lens surface:**

$$F = \mu \cdot p \cdot \cos\alpha$$

$\mu$ coefficient of friction
$p$ pressure of clamping bell on lens surface
$\alpha$ sliding angle

**Centering error $CE$ (in arc minutes):**

$$CE = 3434 \cdot \left(\frac{\Delta z}{R_1} + \frac{\Delta z}{R_2}\right)$$

$\Delta z$ runout
$R_1$ radius of curvature of first lens surface
$R_2$ radius of curvature of second lens surface

## REFERENCE

Karow, H.K. 2004. *Fabrication Methods for Precision Optics.* Hoboken, NJ: John Wiley & Sons.

# 11 Coating

## 11.1 INTRODUCTION

After the actual manufacturing, the optically active surfaces of optical components are usually coated with functional layers. For instance, the application of antireflective coatings allows a considerable increase in total transmission of optical systems by decreasing losses due to reflection. Moreover, such coatings reduce the amount of parasitic and vagabonding light in optical systems and consequently prevent the formation of ghost images.

Simple reflecting components or systems can be generated by metallic mirror coatings, and sophisticated dielectric reflective layers allow the realization of high surface reflectance for broad wave bands or discrete wavelengths as, for example, required for laser mirrors with a high laser-induced damage threshold. In addition, beam splitters, color filters, or polarizers produced by coating processes are of high relevance. The functionality of dielectric coatings is principally based on interference phenomena where the particular behavior of the coating strongly depends on the coating material's index of refraction and the thickness of the coating layer. For the design of such a layer, one has to consider that its reflectance and transmission depend not only on wavelength, but also on polarization of light and its angle of incidence according to the Fresnel equations (see Section 2.3).

In this chapter, the basic functionality and mode of operation of optical coatings in general and especially of reflective and antireflective coatings are presented. Further, underlying mechanisms of layer growth and the well-established coating techniques CVD and PVD are introduced.

## 11.2 BASICS OF OPTICAL COATINGS

Depending on the final function of an optical coating, different coating materials can be used. There are two main categories of optical coatings: metallic and dielectric; the latter type is applied in most cases. This type of coating consists of at least one but usually more thin layers made of transparent dielectric media. A selection of commonly used dielectric coating media is listed in Table 11.1.

The basic principle of dielectric coatings can be described at the example of a single layer coating as follows: Incident light is partially reflected at the first interface given by the interface of the ambient medium and the coating layer. The nonreflected light then passes through the layer, and another fraction is reflected at its back. Both reflected fractions feature a certain *optical path difference* $\delta$, also referred to as *optical retardation*, given by

$$\delta = \frac{2 \cdot \pi}{\lambda} \cdot n_1 \cdot t_1. \tag{11.1}$$

**TABLE 11.1**
**Selected Coating Materials Including the Particular Refractive Index**

| Coating Material | Refractive Index at $\lambda = 589.29$ nm |
|---|---|
| Magnesium fluoride ($MgF_2$) | 1.38 |
| Silicon dioxide ($SiO_2$) | 1.46 |
| Aluminum oxide ($Al_2O_3$) | 1.76 |
| Zirconium dioxide ($ZrO_2$) | 2.18 |
| Zinc sulfide (ZnS) | 2.37 |
| Titanium dioxide ($TiO_2$) | 2.52 |

Here, $\lambda$ is the wavelength of the incident light, $n_1$ is the index of refraction of the coating layer material, and $t_1$ is its thickness. The reflectance $R$ of such a coated surface is directly related to the resulting path difference between the two reflected fractions of light according to

$$R = \frac{r_1^2 + r_2^2 + 2 \cdot r_1 \cdot r_2 \cdot \cos(2 \cdot \delta)}{1 + r_1^2 \cdot r_2^2 + 2 \cdot r_1 \cdot r_2 \cdot \cos(2 \cdot \delta)}, \tag{11.2}$$

where $r_1$ is the reflectivity at the first interface (ambient medium–coating layer), and $r_2$ is the reflectivity at the second one (coating layer–substrate). Since the optical path difference depends on both the index of refraction of the coating layer and its thickness (see Equation 11.1), these two parameters finally determine the function and type of the coating, either reflective or antireflective, as introduced in more detail in the following sections.

### 11.2.1 Reflective Coatings

Optical *reflective coatings* or *mirror coatings* can be realized by metallic or dielectric layers. For metallic coatings, incoming light is reflected on the surface of the metal where the coating can be applied to a substrate's front or back. In the first case, an additional dielectric protection coating such as silicon dioxide is applied onto the metal surface in order to prevent oxidation and an accompanying decrease in reflectance.

Commonly used metals for metallic mirror coatings are aluminum (Al), silver (Ag), and gold (Au). As shown in Figure 11.1, aluminum features quite high reflectance in the ultraviolet wavelength range and can be used for broad wave bands. In contrast, silver and gold exhibit comparatively poor UV reflectance but higher reflectance in the visible and near-infrared wavelength ranges in comparison to aluminum.

In contrast to metallic mirror coatings, dielectric reflective coatings are based on the principle of constructive interference of the fractions of light reflected at the front and back of the coating layer as visualized in Figure 11.2.

High reflectance then results from the fact that both reflected fractions superimpose in phase. The particular amplitudes of the light waves consequently sum up, and

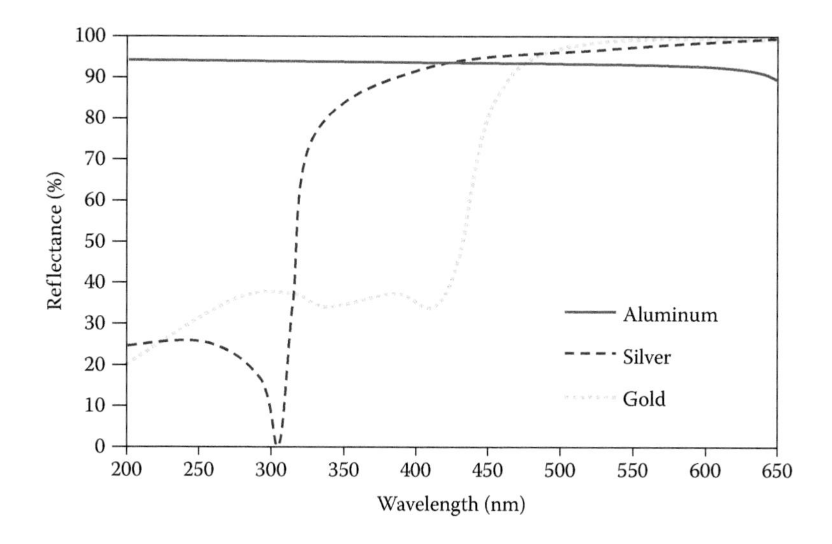


**FIGURE 11.1** Qualitative representation of the reflectance of the metallic coating materials aluminum, silver, and gold vs. wavelength with focus on the ultraviolet wavelength range.

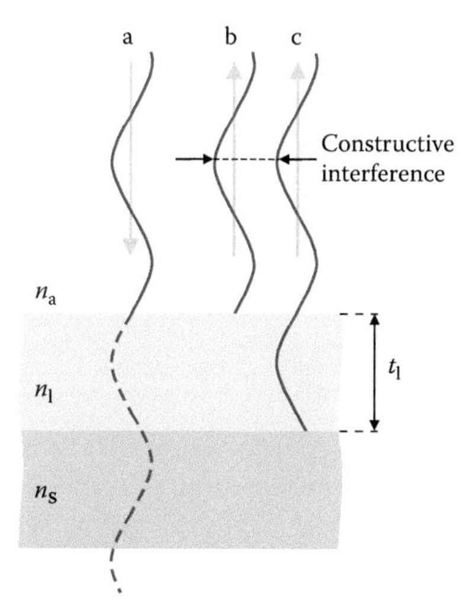


**FIGURE 11.2** Visualization of the formation of constructive interference at dielectric transparent coatings by partial reflection of incident light (a) on the front (b) and the back (c) of the coating layer, leading to increased reflectance.

the total amplitude of reflected light is increased. For such *constructive interference* at normal incidence, the wavelength-dependent thickness of the coating layer $t_1$ is

$$t_1 = \frac{\lambda}{2 \cdot n_1}, \tag{11.3}$$

with $n_l$ being the wavelength-dependent refractive index of the layer material. In optical coating technology, the goal is to achieve high reflection for a broad wavelength band. Since the reflectance of an optics or coating surface depends on the angle of incidence of light as well as its wavelength and its polarization, this cannot be realized by single layers. In this case, multilayer systems, also referred to as film systems, consisting of several layers of two different coating materials are chosen. The total reflectance of such film systems is given by

$$R_{\text{tot}} = \left( \frac{1 - \left( \frac{n_{\text{m1}}}{n_{\text{m2}}} \right)^{2 \cdot N}}{1 + \left( \frac{n_{\text{m1}}}{n_{\text{m2}}} \right)^{2 \cdot N}} \right)^2 . \tag{11.4}$$

Here, $N$ is the number of layers, $n_{\text{m1}}$ is the index of refraction of the first coating material, and $n_{\text{m2}}$ is the index of refraction of the second coating material.

Taking the abovementioned dependencies of reflectance into account, the performance of any reflective coating is not only described by its reflectance, but also by the wavelength range (or discrete wavelength), angle of incidence (or range of angles of incidence), and polarization direction or even directions where the indicated reflectance is valid. These parameters are thus of essential importance for a proper specification of coating reflectance and quality.

Finally, it should be mentioned that sometimes the reflectance of a coated optical component is expressed by its finesse $F$, given by

$$F = \frac{\pi \cdot \sqrt{R}}{(1 - R)} . \tag{11.5}$$

This way of specification is commonly used for the description of Fabry-Pérot resonators or interferometers.[1]

### 11.2.2 Antireflective Coatings

The goal of the application of *antireflective coatings* is to reduce the amount of reflected light at optical interfaces and to increase transmission.[2] Thus, transparent dielectric materials are necessarily used for this type of coating. As shown in the example of dielectric reflective coatings in Section 11.2.1, the coating layer thickness is of essential importance for the type of resulting interference from the reflected fractions of incoming light (see Equation 11.3). By adjusting this thickness, the effect of destructive interference can be realized where the resulting total amplitude of the superimposed fractions of reflected light waves is minimized or annihilated as shown in Figure 11.3.

[1] This special type of interferometer was invented by the French physicists *Maurice Paul Auguste Charles Fabry* (1867–1945) and *Jean-Baptiste Alfred Pérot* (1863–1925); it is used for the manipulation of the wavelength of light in different fields of applications such as optical telecommunications, laser technology, and spectroscopy.

[2] The reduction of reflection by antireflection coatings is also referred to as blooming. Such coatings are also applied to eyeglasses in order to reduce disturbing reflections and ghost images in the visual field of spectacle wearers.

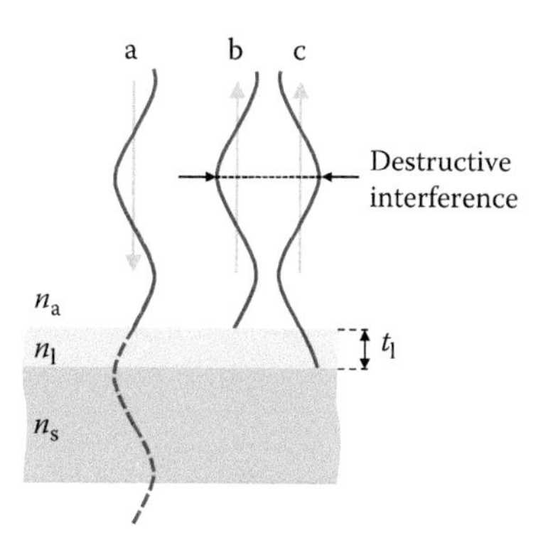


**FIGURE 11.3** Visualization of the formation of destructive interference at dielectric transparent coatings by partial reflection of incident light (a) on the front (b) and the back (c) of the coating layer, leading to decreased reflectance and increased transmittance, respectively.

For *destructive interference*, the wavelength-dependent thickness of the dielectric coating layer $t_l$ follows from

$$t_l = \frac{\lambda}{4 \cdot n_l}, \tag{11.6}$$

where $n_l$ is the wavelength-dependent refractive index of the coating layer material. This interrelationship is valid for perpendicular incidence, where the angle of incidence amounts to $\varepsilon = 0°$. In case of angular incident beams, where $0° < \varepsilon < 90°$, the angle of incidence has to be taken into account since the geometric and the optical path length, respectively, are extended. Equation 11.6 has then to be rewritten as

$$t_l = \frac{\lambda}{4 \cdot \sqrt{n_l^2 - n_a^2 \cdot \sin^2 \varepsilon}}, \tag{11.7}$$

with $n_a$ being the index of refraction of the ambient medium.

Considerable reduction in residual reflectance at glass surfaces can be achieved by merely one coating layer. The reflectance $R$ of such a single-layer antireflective coating at normal incidence is given by

$$R = \left( \frac{n_s \cdot n_a - n_l^2}{n_s \cdot n_a + n_l^2} \right)^2. \tag{11.8}$$

Here, $n_s$ is the index of refraction of the substrate, $n_a$ the one of the ambient medium, and $n_l$ the one of the coating layer material. The impact of such a coating on surface reflectance is shown in Figure 11.4 at the example of an uncoated fused silica surface and a fused silica surface with a single-layer antireflective coating made of magnesium fluoride ($MgF_2$).

However, a number of coating layers made of different dielectric media is required for the realization of antireflective coatings with high performance, that is, low residual reflectance and high transmission for a broad range of wavelengths and angles of incidence. The importance of increasing transmission through optical elements by

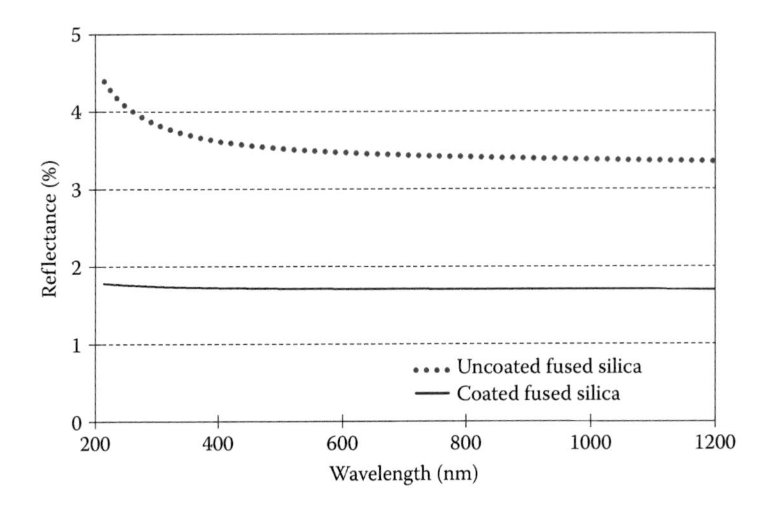


**FIGURE 11.4** Comparison of the reflectance of a fused silica surface with and without a single antireflective layer made of magnesium fluoride. (Data from Dodge, M.J., *Appl. Optics*, 23, 1980–1985, 1984; Malitson, I.H., *J. Optical Soc. Am.*, 55, 1205–1208, 1965; Tan, C.Z., *J. Non-Crystalline Solids*, 223, 158–163, 1998.)

appropriate coatings becomes obvious when considering that the total transmittance $T_{total}$ of an optical system is given by the product of the transmittances of each optical interface (indicated by the indices 1–$n$) according to

$$T_{total} = T_1 \cdot T_2 \cdot T_3,\ldots,T_n. \tag{11.9}$$

For systems where all the involved optical interfaces feature the same transmittance, this interrelationship can be rewritten as

$$T_{total} = T^n, \tag{11.10}$$

where $n$ is the number of interfaces.[3]

### 11.2.3 Filter Coatings

In addition to reflective and antireflective coatings, a large variety of different dielectric filter coatings such as color filters[4] or polarizers can be realized. An overview on the most important types and particular functions of optical filters is given in Table 11.2.

[3] In practice, optical systems are usually an assembly of components with differing dimensions and thicknesses that are made of different glasses and consequently feature quite different transmittances. However, Equation 11.10 allows a first estimation of the total transmittance of optical systems.

[4] Color filters can also be realized using colored glass bulk material as introduced in Section 3.2.1.5. These filters work as absorption filters where the main part of incoming light is absorbed, and merely the wavelength or wavelength range of interest is transmitted.

**TABLE 11.2**
**Overview on Different Types and Particular Functions of Optical Filters**

| Type of Filter | Function |
|---|---|
| Dichroic filter | Realization of high reflectance for one wavelength of interest and high transmittance for another wavelength of interest |
| Monochromatic filter | Realization of high transmission for one selected wavelength |
| Notch filter | Blocking of selected single wavelength or wavelength range (e.g., UV or IR) |
| Polarizer | Conversion of nonpolarized light to polarized light with well-defined polarization |
| Beam splitter | Separation of incoming light into two fractions with defined intensity or polarization, respectively |
| Neutral density filter | Blocking of broad wavelength range (transmitted light intensity continuously adjustable when using gray wedges) |

The last type listed in Table 11.2, neutral density filter, is specified by its *optical density* (OD), which indicates the grade of blocking of light according to

$$OD = \log_{10} T^{-1}. \tag{11.11}$$

Here, $T$ is the transmittance of the filter.

## 11.3 MECHANISMS OF LAYER GROWTH

In the course of a coating process where dielectric media are deposited on the surface of optical components, different mechanisms of *layer growth* (2D layer growth, 3D layer growth, and a combination of both) can occur as shown in Figure 11.5.

In the case of pure 2D layer growth,[5] a monolayer of coating material is initially formed on the substrate surface. Once the entire surface is covered, the next monolayer is formed on top of the first one, and so on; the optical coating thus grows monolayer by monolayer. This behavior is due to the fact that the interaction of the coating and the substrate atoms is much higher than the interaction between the coating atoms[6] (Frank and van der Merwe, 1949a,b,c).

In contrast, a higher interaction between coating atoms than between coating atoms and substrate atoms is found in the case of 3D layer growth.[7] This leads to the formation of separated clusters or nanoparticles with a thickness of several atom layers on the substrate surface as shown in Figure 11.5b. After a certain time, these clusters grow together in the course of the coating process, finally forming a closed coating layer (Volmer and Weber, 1926).

[5] This type of layer growth is also known as Frank-van-der-Merwe growth.

[6] High interaction of coating and substrate atoms is referred to as adhesion, whereas cohesion is a high interaction between coating atoms.

[7] 3D layer growth is also referred to as Volmer-Weber growth.

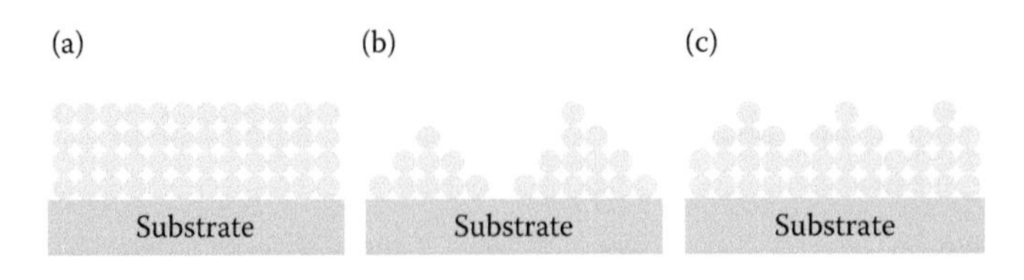


**FIGURE 11.5** Schematic of different mechanisms of layer growth; 2D layer growth (a), 3D layer growth (b), and the combination of 2D and 3D layer growth (c) The circles represent coating material molecules.

The third type of layer growth, a combination of both abovementioned mechanisms, is the so-called Stranski-Krastanow growth where, first, some monolayers are grown on the substrate surface by 2D layer growth. These monolayers form an adhesion layer for subsequent 3D layer growth since adhesion of the coating atoms at the formed layer is higher than at the pure substrate surface (Stranski and Krastanov, 1938).

## 11.4 COATING TECHNIQUES

Even though several coating processes that work at atmospheric pressure were developed in the past, the deposition of optical coatings is mainly performed in a vacuum or low-pressure environment. In such conditions, a high *free length of path l* (see also Section 13.4.1.1) is found. This parameter depends on the pressure $p$ according to

$$l = \frac{k_B \cdot T}{\sqrt{2} \cdot p \cdot \sigma}, \tag{11.12}$$

where $k_B$ is the Boltzmann constant,[8] $T$ is temperature, and $\sigma$ is the collision diameter.[9] Hence, the coating material can efficiently be brought onto the substrate surface, which is usually placed at a certain distance from the coating material inlet (for chemical vapor deposition, see Section 11.4.1) or generation zone (for physical vapor deposition, see Section 11.4.2) since the higher the free length of path, the lower the number or density of potential collision partners.

### 11.4.1 Chemical Vapor Deposition

The deposition of solid coatings from gaseous raw material, the so-called precursor gases, is referred to as *chemical vapor deposition* (CVD). This coating technique is performed in heated process chambers where the precursor gas is inserted and dissociated by heat. Hence, volatile reactants such as atoms or molecules are generated. These reactants are adsorbed at the substrate surface and arrange a closed surface layer due to surface diffusion, that is, lateral diffusion parallel to the substrate surface.

[8] The Boltzmann constant amounts to $k_B \approx 1.381 \cdot 10^{-23}$ J/K.

[9] The collision diameter is defined as the distance between two colliding molecules. For molecules of the same kind, it is given by the product of the molecule diameters and pi. For molecules of different kind, the particular molecule diameters are considered by their arithmetic means.

The diffusion of precursor gases can also be realized by collisions with free electrons in a plasma environment where low-temperature plasmas with gas temperatures in the range of some hundreds centigrade are applied. Within the plasma, radicals and ions are formed and the deposition of the coating layer is achieved by reactions of such plasma species at the substrate surface. This technique is referred to as *plasma-enhanced CVD*; it allows the deposition of temperature-sensitive coating materials and the treatment of temperature-sensitive substrates.

Generally, the method of CVD is employed for a number of different applications such as the deposition of crystalline coatings on wafers or the manufacture of fused silica (compare Section 3.2.2.1). It is also applied for the generation of protection layers on metallic mirror coatings as, for example, silicon dioxide in order to prevent oxidation of the metallic coating and to avoid the accompanying alteration of its optical properties and reflectance, respectively. As an example, gas mixtures containing silane ($SiH_4$) and oxygen ($O_2$) can be used for this task. In the course of such deposition process, solid silicon dioxide ($SiO_2$, the actual coating material) and volatile molecular hydrogen ($H_2$) are finally formed according to

$$SiH_4(g)+O_2(g)\rightarrow SiO_2(s)+2H_2(g). \qquad (11.13)$$

Deposition of silicon dioxide layers can also be achieved by a mixture of dichlorosilane ($SiCl_2H_2$) and nitrous oxide ($N_2O$), or tetraethylorthosilicate ($Si(OC_2H_5)_4$), where the latter gas mixture is also known as and commonly abbreviated TEOS.

### 11.4.2 Physical Vapor Deposition

For *physical vapor deposition* (PVD), the coating material is provided in solid state and vaporized by physical methods. Deposition of such vaporized material is then due to condensation on the substrate surface within a heated process chamber as shown in Figure 11.6.

The vaporization of solid coating material, commonly called *target*, can be realized by a number of different techniques as, for example, thermal evaporation, electron beam evaporation by electron guns, laser beam evaporation,[10] cathodic arc deposition, or sputter deposition. In the latter case, the coating material is pulverized by incident primary ions from an ion gun as also used for physical etching (see Section 13.4.1).

In practice, the substrates to be coated by PVD are usually placed in a holder, which is driven by a planetary gear and rotated in order to achieve high coating homogeneity and uniformity over the entire substrate surface. High homogeneity and uniformity of deposited coatings in terms of chemical composition, index of refraction, and thickness is of essential importance since the final functionality and quality of any dielectric coating is directly related to these parameters according to Equations 11.3 and 11.6. As an example, uniformity in reflectance over the entire surface of an optical component with an antireflective coating of 0.1%–0.5% is usually required. The control of both the coating material composition and layer thickness is thus an important issue, where the latter parameter can be measured with the aid of sensors based on oscillating crystals.

[10] The deposition of coating material evaporated by laser irradiation is referred to as pulsed laser deposition (PLD).

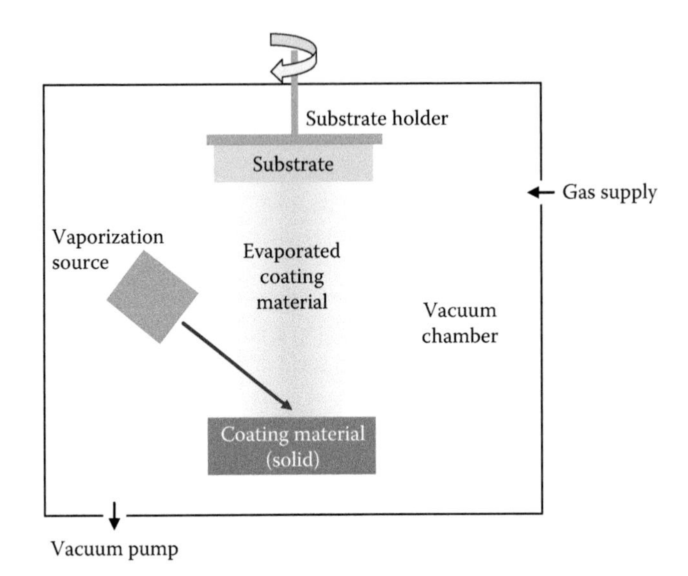


**FIGURE 11.6** Schematic of a PVD process chamber including the coating material, vaporization source, and rotating substrate holder.

## 11.5 SUMMARY

In optics manufacturing, metallic or dielectric and transparent coating materials can be used. In the latter case, the functionality, either reflective or antireflective, depends on the index of refraction and the thickness of the coating layer and the resulting type of interference. Coating is usually performed at low pressure or in a vacuum in order to ensure high free lengths of path. The coating material can be provided in gaseous or solid state. In the first case, the CVD process, so-called coating precursors are used and the deposition of the actual coating is due to adsorption of atoms or molecules from the precursor gas on the substrate surface within a process chamber. The second case is referred to as PVD. Here, the solid coating material, the so-called target, is vaporized by thermal evaporation, electron or ion beam evaporation, laser beam evaporation, or cathodic arcs. The gaseous coating material generated in this vein then condensates on the substrate surface where the substrates are placed in a rotating holder in order to ensure high coating homogeneity and uniformity.

## 11.6 FORMULARY AND MAIN SYMBOLS AND ABBREVIATIONS

**Optical path difference $\delta$ (optical retardation):**

$$\delta = \frac{2 \cdot \pi}{\lambda} \cdot n_1 \cdot t_1$$

$\lambda$ wavelength of incident light
$n_1$ index of refraction of coating layer material
$t_1$ thickness of coating layer

**Reflectance $R$ of a surface with single-layer dielectric coating:**

$$R = \frac{r_1^2 + r_2^2 + 2 \cdot r_1 \cdot r_2 \cdot \cos(2 \cdot \delta)}{1 + r_1^2 \cdot r_2^2 + 2 \cdot r_1 \cdot r_2 \cdot \cos(2 \cdot \delta)}$$

$r_1$ reflectivity at first interface (ambient medium–coating layer)
$r_2$ reflectivity at second interface (coating layer–substrate)
$\delta$ optical path difference

**Condition for constructive interference at normal incidence:**

$$t_1 = \frac{\lambda}{2 \cdot n_1}$$

$t_1$ thickness of coating layer
$\lambda$ wavelength of incident light
$n_1$ index of refraction of coating layer material

**Total reflectance of multilayer coating:**

$$R_{tot} = \left( \frac{1 - \left( \frac{n_{m1}}{n_{m2}} \right)^{2 \cdot N}}{1 + \left( \frac{n_{m1}}{n_{m2}} \right)^{2 \cdot N}} \right)^2$$

$n_{m1}$ index of refraction of first coating material
$n_{m2}$ index of refraction of second coating material
$N$ number of layers

**Finesse $F$:**

$$F = \frac{\pi \cdot \sqrt{R}}{(1 - R)}$$

$R$ reflectance of coating

**Condition for destructive interference at normal incidence:**

$$t_1 = \frac{\lambda}{4 \cdot n_1}$$

$t_1$ thickness of coating layer
$\lambda$ wavelength of incident light
$n_1$ index of refraction of coating layer material

**Condition for destructive interference at inclined incidence of light:**

$$t_1 = \frac{\lambda}{4 \cdot \sqrt{n_1^2 - n_a^2 \cdot \sin^2 \varepsilon}}$$

$t_1$ thickness of coating layer
$\lambda$ wavelength of incident light

$n_1$ index of refraction of coating layer material
$n_a$ index of refraction of ambient medium
$\varepsilon$ angle of incidence

**Reflectance *R* of single-layer antireflective coating at normal incidence:**

$$R = \left( \frac{n_s \cdot n_a - n_1^2}{n_s \cdot n_a + n_1^2} \right)^2$$

$n_s$ index of refraction of substrate
$n_a$ index of refraction of ambient medium
$n_1$ index of refraction of coating layer material

**Total transmittance $T_{total}$:**

$$T_{total} = T_1 \cdot T_2 \cdot T_3, \ldots, T_n$$

$T_{1-n}$ partial transmittance of involved interfaces

**Optical density *OD* of neutral density filters:**

$$OD = \log_{10} T^{-1}$$

$T$ filter transmittance

**Free length of path *l*:**

$$l = \frac{k_B \cdot T}{\sqrt{2} \cdot p \cdot \sigma}$$

$k_B$ Boltzmann constant
$T$ temperature
$p$ pressure
$\sigma$ collision diameter

## REFERENCES

Dodge, M.J. 1984. Refractive properties of magnesium fluoride. *Applied Optics* 23:1980–1985.

Frank, F.C., and van der Merwe, J.H. 1949a. One-dimensional dislocations. I. Static theory. *Proceedings of the Royal Society of London A* 198:205–216.

Frank, F.C., and van der Merwe, J.H. 1949b. One-dimensional dislocations. II. Misfitting monolayers and oriented overgrowth. *Proceedings of the Royal Society of London A* 198:216–225.

Frank, F.C., and van der Merwe, J.H. 1949c. One-dimensional dislocations. III. Influence of the second harmonic term in the potential representation on the properties of the model. *Proceedings of the Royal Society of London A* 200:125–134.

Malitson, I.H. 1965. Interspecimen comparison of the refractive index of fused silica. *Journal of the Optical Society of America* 55:1205–1208.

Stranski, I.N., and Krastanov, L. 1938. Zur Theorie der orientierten Ausscheidung von Ionenkristallen aufeinander. *Sitzungsberichte der Kaiserlichen Akademie der Wissenschaften in Wien—mathematisch-naturwissenschaftliche Classe* 146:797–810 (in German).

Tan, C.Z. 1998. Determination of refractive index of silica glass for infrared wavelengths by IR spectroscopy. *Journal of Non-Crystalline Solids* 223:158–163.

Volmer, M., and Weber, A. 1926. Keimbildung in übersättigten Gebilden. *Zeitschrift für Physikalische Chemie* 119:277–301 (in German).

# 12 Assembly of Optomechanical Systems

## 12.1 INTRODUCTION

As described in previous chapters, optical components such as lenses or prisms feature high accuracy and precision in terms of geometry and surface shape. However, the best optics is quite useless without mechanics. In order to provide utilizable systems (e.g., microscope lenses, camera objectives, or binoculars) optics and mechanics have to be brought together by assembly. It is essential that the used mounts, tubes, and holders feature high precision, as introduced in Section 6.5, since mounting errors that result from inappropriate accuracy of these mechanical elements can cause severe decreases in imaging quality of optomechanical systems.

Depending on the geometry of optical elements, the arrangement of a final optical system (either on-axis or folded), and the materials used, different techniques and approaches for mounting and assembly are applied. The commonly used and most important are presented in this chapter. Further, possible assembly or mounting errors as well as the impact of such errors on the imaging quality of optical systems are introduced.

## 12.2 MOUNTING TECHNIQUES

Depending on the final application and further requirements, where pricing may play an important role, different approaches can be applied for mounting optical components and assembling optomechanical systems, respectively. The commonly used methods (i.e., screw connecting, gluing, and clamping) are presented hereafter using the example of one or more single lenses. However, gluing and clamping are also applied for fixing prisms.

### 12.2.1 Screw Connecting

For mounting a single lens by *screw connecting*, the lens is placed on the bearing surface of a mount with internal thread. As shown in Figure 12.1, the lens is then fixed by screwing a threaded ring (a.k.a. ring nut) with a proper external thread into the internal thread of the mount.

This method can also be used for mounting several lenses into one mount in order to set up systems consisting of several lenses such as microscope objectives. Figure 12.2 shows an example for the application of this technique, referred to as *stacking*. Here, the first optical component such as a single lens, doublet, or triplet is placed on the bearing surface of the mount (Figure 12.2a and b). Subsequently, a spacer ring is placed on top of this optical component (Figure 12.2c), and a second

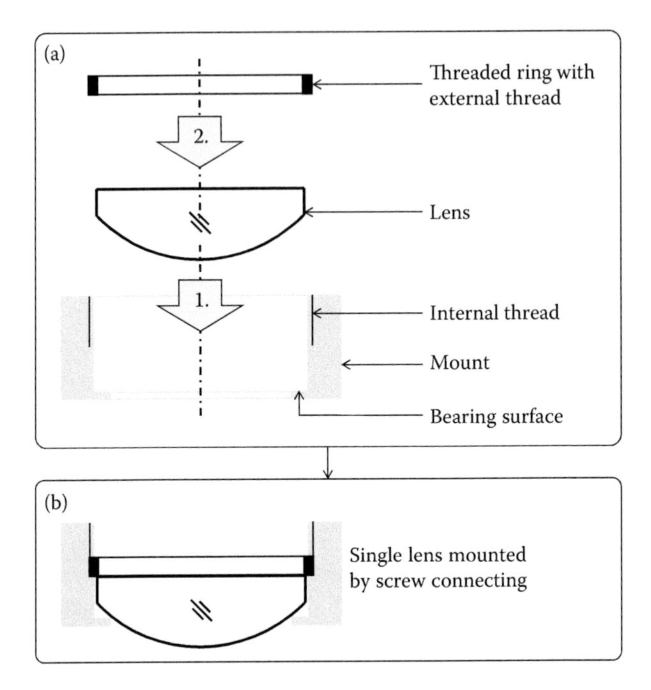


**FIGURE 12.1** Principle of screw connecting; first, a lens is placed in a mount with bearing surface and internal thread. Second, a threaded ring is screwed into the mount (a), resulting in stable fixing of the lens within the mount (b).

component is placed on the spacer ring (Figure 12.2d). In this way, the required air gap between the optical components is realized. Finally, assembly is finished by screw connecting (Figure 12.2e and f).

Stacking is a well-established standard technique for the assembly of on-axis optomechanical systems including several single optical elements, for example,

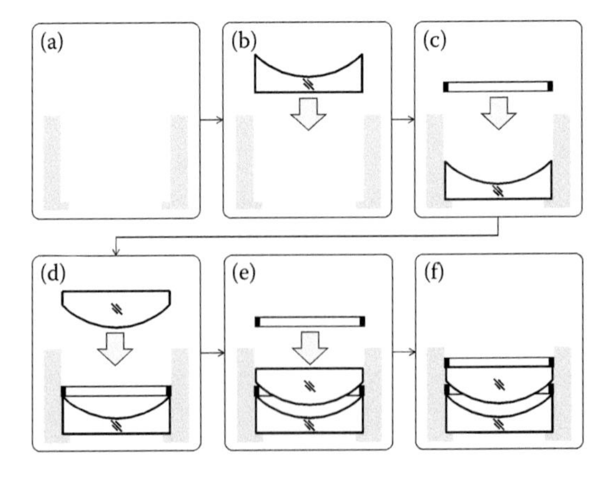


**FIGURE 12.2** Assembly of optomechanical systems by stacking optical and mechanical components in a mount. For a detailed description of each particular step, see running text.

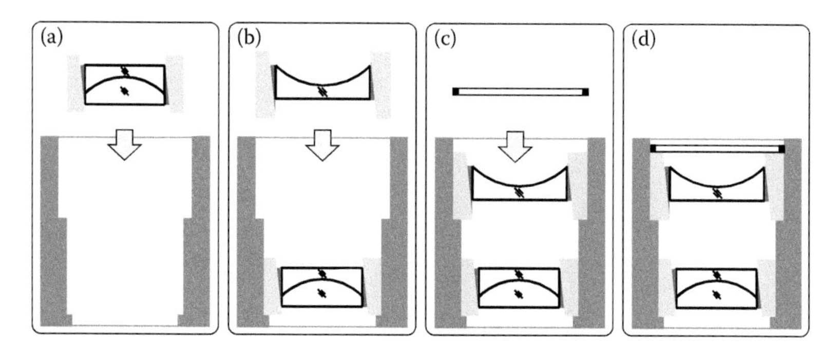


**FIGURE 12.3** Assembly of an optomechanical system consisting of precision centered optomechanical groups by stacking (a–c) and final screw connecting (d).

air-gapped achromatic doublets or triplets. It is also applied for mounting precision-centered optomechanical groups (compare Section 10.2.2) as shown in Figure 12.3.

### 12.2.2 Gluing

Assembly by *gluing* is quite comparable to screw connecting: first, a lens is placed in a mount with a bearing surface. Second, the lens is fixed, in this case with two-component glue or UV-curing glue[1], where two different main types of gluing are distinguished, circumference gluing and ring gluing. For circumference gluing, the glue is applied to the lens border, and it thus fills the gap between the lens border cylinder and the inner mount cylinder as shown in Figure 12.4a. In contrast, putting a ring-shaped glue seam on the lens surface is referred to as ring gluing, see Figure 12.4b.

Ring gluing is thus quite comparable to screw connecting, but it is much more economic and cost-efficient since threaded mounts and threaded rings are not required. This also applies for circumference gluing, which additionally gives the possibility of fine adjustment of lenses (or other optical components) in mounts before the final curing of the glue. Due to these advantages, gluing has also been established as a standard technique for mounting optical systems. Figure 12.5 shows an example of such an optical system, a microscope objective assembled by gluing.

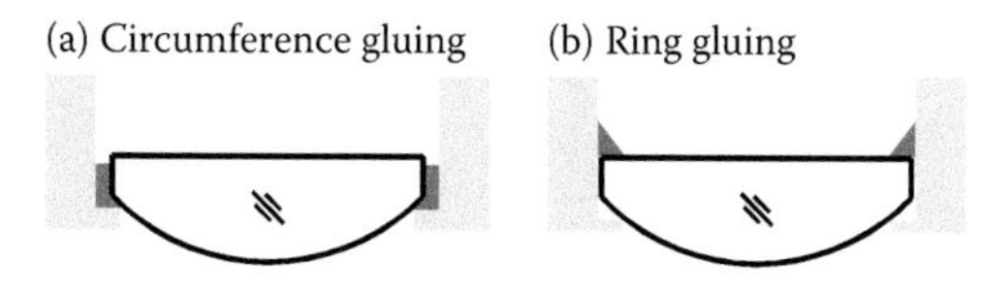


**FIGURE 12.4** Different types of gluing: circumference gluing, where glue is applied to the lens border cylinder (a) and ring gluing, where a glue seam is applied to the lens surface (b).

[1] Note that UV-curing glue is not UV-curing fine cement.

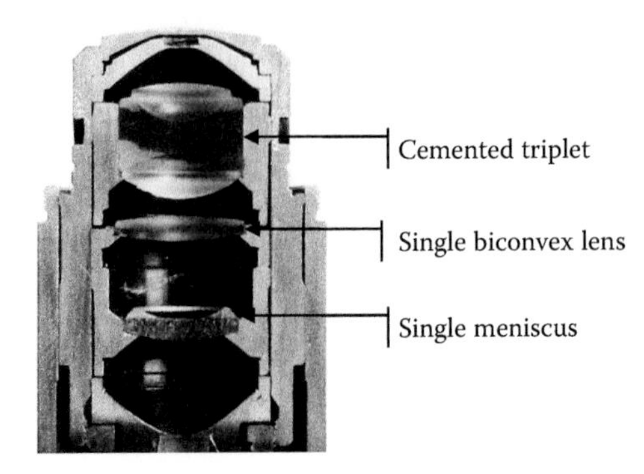


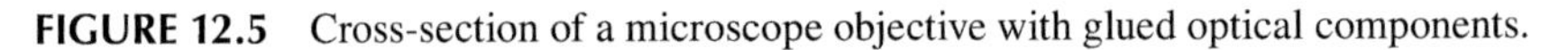

**FIGURE 12.5** Cross-section of a microscope objective with glued optical components.

### 12.2.3 Clamping

*Clamping* is another standard technique for mounting optical components. One possibility is to use tension springs as shown in Figure 12.6. In this example, a lens is placed on the bearing surface of a mount with at least two threaded holes at the end face. Mounting of the lens is then realized by screwing tension springs to the threaded holes, resulting in fixing due to the contact pressure of the tension springs on the lens edge.

Since the tension springs used for this method are elastic, vibrations can easily be compensated. This fact might be of advantage for some special applications. Second only to gluing, clamping with tension springs is the main approach for mounting prisms on mechanical holders.

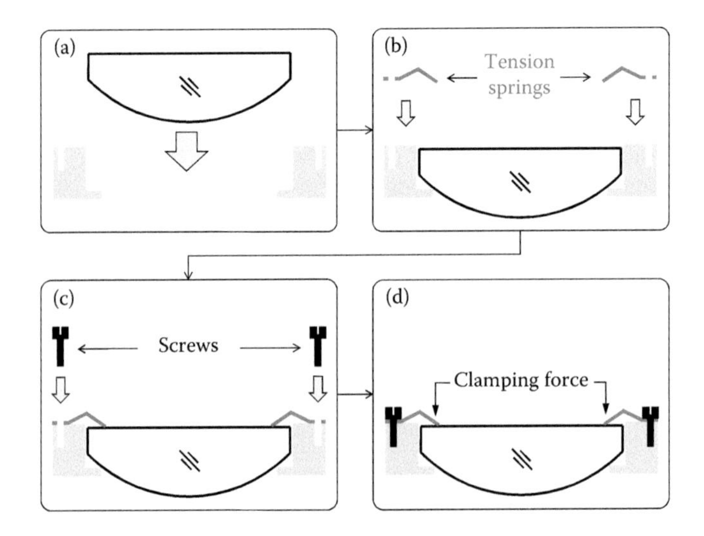


**FIGURE 12.6** Principle of clamping by tension springs: a lens is placed in a mount with threaded holes (a), where tension springs are screwed in (b, c). Fixing of the lens is then due to the clamping force of the screwed tension springs (d).

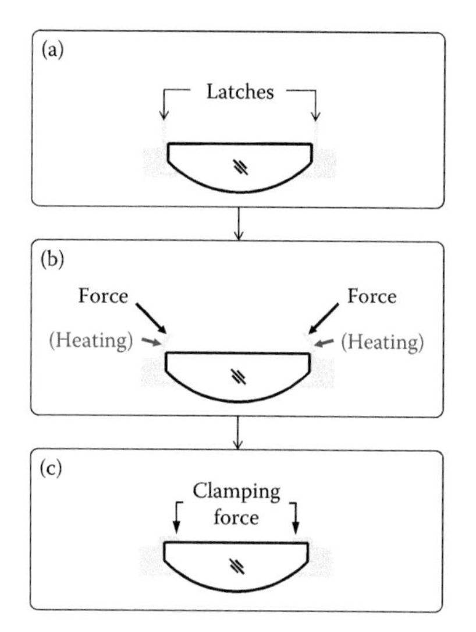


**FIGURE 12.7** Principle of clamping by bending; a lens is placed in a mount with latches (a) that are heated and bent (b). Fixing of the lens is then due to the clamping force of the bent latches (c).

Clamping of optical components can also be carried out by bending or press forming. In this case, the component is placed in a mount with a bearing surface and at least two latches at the end face as shown in Figure 12.7a. These latches are then bent or press formed toward the lens surface, where heating of the latches is necessary in some cases, depending on the material of the mount (Figure 12.7b). In doing so, the lens is finally fixed by the contact pressure of the bent latches (Figure 12.7c).

### 12.2.4 Glass-Metal Soldering

Mounting of optomechanical systems by *glass-metal soldering* is a comparatively novel technique. The main issue with this approach is that brazing solders do not adhere very well on glass surfaces. In order to overcome this restriction, the border cylinder or side face of optical components is initially coated with a thin metal layer. This layer then acts as an adhesive or primer for the actual brazing solder. After applying such a metallic layer, the coated optical component is placed in a mount as shown in Figure 12.8. Subsequently, solid brazing solder is added and molten, for example, by focused laser irradiation. The molten solder finally fills the gap between the metallic coating on the lens border cylinder and the inner cylinder of the mount.

The method of glass-metal soldering has turned out to be an appropriate mounting technique for optomechanical vacuum equipment. It is leak proof and hermetically sealed (Cheng et al., 2001), and no disturbing and—sometimes—destroying outgazing of substances occurs, in contrast to glues. Glass-metal soldering is furthermore applied for the assembly of miniaturized optomechanical devices (Stauffer et al., 2005).

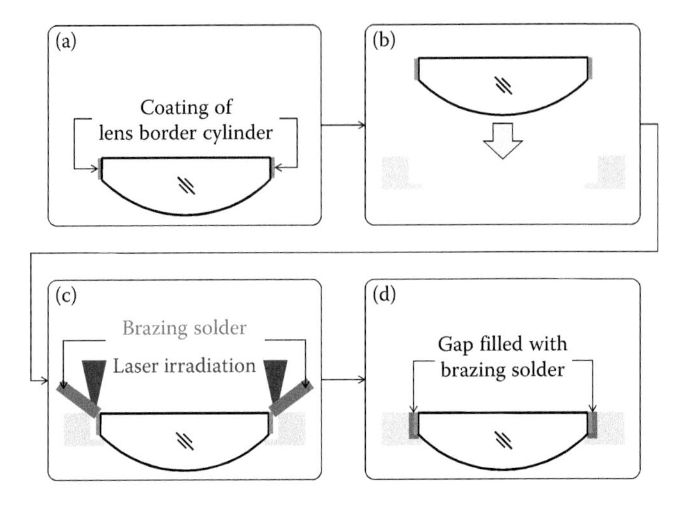


**FIGURE 12.8** Principle of glass-metal soldering; the border cylinder of a lens is coated with a metallic adhesive layer (a) and placed in a mount (b). Brazing solder is then added and molten by laser irradiation (c), and the gap between the mount and the coated lens border cylinder is filled with brazing solder (d).

## 12.3 MOUNTING ERRORS

It turns out that not only precision optical components, but also mechanical elements such as mounts, tubes, or holders of high precision, are required for the assembly of optomechanical systems with high imaging quality. However, different mounting errors can result from inaccuracies of mechanical elements.

Generally, the relation between the outer lens diameter and the inner mount diameter is of essential importance. A high difference in diameters may lead to the tilt of mounted lenses due to canting, which can occur in the course of assembly as shown in Figure 12.9a. Another possible effect is a lateral decenter of the lens with respect to the mechanical axis of the mount. This error is even increased by inaccurate or tilted bearing surfaces, as shown in Figure 12.9b, which may also cause additional lens tilt. Finally, inappropriate tolerancing of spacer rings used for stacking may lead to distance errors between mounted lenses or lens groups.

The method of assembly by gluing can give rise to additional disturbing effects. During curing, the used glue may shrink and induce tensile and compressive stress. Inhomogeneous shrinking of the glue could thus not only cause displacement or tilt of glued optical components, but also initiate a slight deformation of components and a decrease in surface accuracy, respectively, as well as (local) stress birefringence.

Local compressive stress and stress birefringence may also result from assembly by clamping or bending where the tension springs (clamping) or lathes (bending) can potentially feature different contact pressures. In optical components made of comparatively soft optical media, such as calcium fluoride ($CaF_2$), this effect can induce local stress birefringence and gradients in index of refraction, respectively.

Tilt and decenter of optical components as well as the formation of stress birefringence might also occur during assembly by glass-metal soldering because of

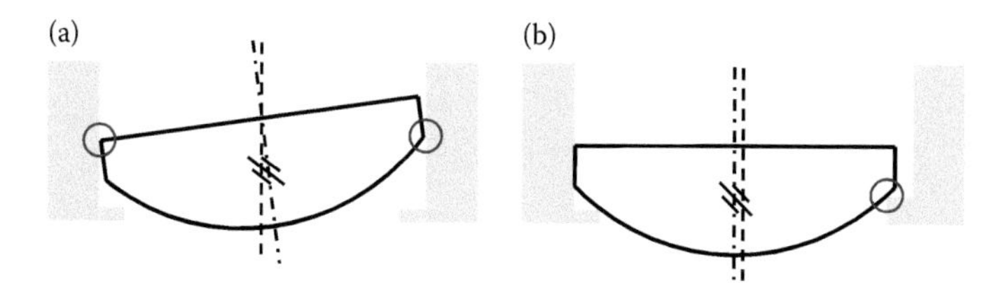


**FIGURE 12.9** Visualization of two possible lens mounting errors: tilt due to canting (a) and lateral decenter due to high difference in outer lens and inner mount diameter and an inaccurate or tilted bearing surface (b). Circles indicate the trouble spots.

inhomogeneous (i.e., nonrotational-symmetric) filling of the gap between the coated lens border cylinder and the mount by the used brazing solder. One has to consider that this method is based on the use of different materials (glass, metallic coating, brazing solder, and mount material) with different coefficients of thermal expansion. This fact can cause severe strain (Hull and Burger, 1934) and irreversible deformation of the lens or mount. Further, inhomogeneous heating by the incident focused laser irradiation has to be avoided. Control of the process parameters during glass-metal soldering is thus of essential importance.

## 12.4 SUMMARY

For the assembly of optomechanical groups and systems, different techniques are in hand. Single lenses and lens groups can be mounted by screw connecting where the optical component is placed in a mount with a bearing surface and an internal thread and fixed by a threaded ring. This method is also applied in the assembly of air-gapped optical systems: several lenses are compiled, and the air gap thickness results from spacer rings arranged between the lenses. In this case, the assembly procedure is referred to as stacking, which may also be applied for precision-centered groups. Optical components can furthermore be mounted by gluing or clamping with tension springs or bent latches. Finally, the novel approach of glass-metal soldering allows the setup of optomechanical assemblies suitable for vacuum applications.

Inappropriate tolerances and form deviations of the used mechanical elements, such as tilted bearing surfaces or deviations in diameter, may lead to tilt and decenter of mounted optical elements. Such position or distance errors may have a severe impact on the imaging quality of optomechanical systems. The contour accuracy and precision of mounts, spacer rings, and tubes is thus of essential importance.

## REFERENCES

Cheng, Y.T., Lin, L., and Najafi, K. 2001. A hermetic glass-silicon package formed using localized aluminum/silicon-glass bonding. *Journal of Microelectromechanical Systems* 10:392–399.

Hull, A.W., and Burger, E.E. 1934. Glass-to-metal seals. *Physics* 5:384–405.

Stauffer, L., Würsch, A., Gächter, B., Siercks, K., Verettas, I., Rossopoulos, S., and Clavel, R. 2005. A surface-mounted device assembly technique for small optics based on laser reflow soldering. *Optics and Lasers in Engineering* 43:365–372.

# 13 Microoptics

## 13.1 INTRODUCTION

In the last decades, microoptical components and systems have gained in importance in a number of different fields of applications in which pure microoptical, optoelectronic, or optomechanical systems are required. For example, microoptical elements are essential key devices of daily-used convenience goods such as multimedia terminals, safety devices, and medical equipment. Microlenses are used in fiber couplers for telecommunication, in CD, DVD, and blue-ray players, in micro cameras for safety monitoring or park assistance devices, as well as in endoscopes for diagnosis and surgery applications. Moreover, microoptics is of essential relevance in both industrial manufacturing and vision, as well as research. Here, microlens arrays allow the homogenization of laser beams and the legalization of light-sections for scanners or microscopic analysis systems.

By definition, a three-dimensional component is referred to as a micro component if at least two of its dimensions (length, width, or height) are smaller than 1 mm. Relating to optical elements, this definition can be extended by considering the proportion of the geometrical dimension and the wavelength of the used light. The size of microoptical components is merely some orders of magnitudes higher than the wavelength, resulting in a notable impact of the wave character of light on its propagation characteristics when passing through such devices and finally the imaging properties. The effect of diffraction, which is usually neglected in classical optical imaging models such as the paraxial or geometric-optical imaging model, has thus to be taken into account.

The manufacture of micro-optical components can be performed by different techniques such as laser machining or reactive ion etching. In this chapter, the basics of microoptical components as well as suitable manufacturing methods and systems are presented.

## 13.2 BASICS OF MICROOPTICS

### 13.2.1 Formation of Airy Discs

*Diffraction* is one effect due to the wave character of light. A well-known example of this effect is diffraction at apertures such as slits, stops, or diaphragms. The particular type of diffraction follows from the aperture widths or diameters where Fresnel-diffraction occurs for apertures that are some orders of magnitudes larger than the wavelength of the light passing through. According to the Huygens-Fresnel principle, diffraction of light at an aperture edge gives rise to the emergence of new circular fundamental waves. During propagation, these waves are subject to constructive and destructive interference, leading to the formation of a diffraction pattern (a.k.a. *Airy disc*[1]) as shown in Figure 13.1.

[1] The Airy disc is named after the English mathematician and astronomer *Sir George Bidell Airy* (1801–1892).

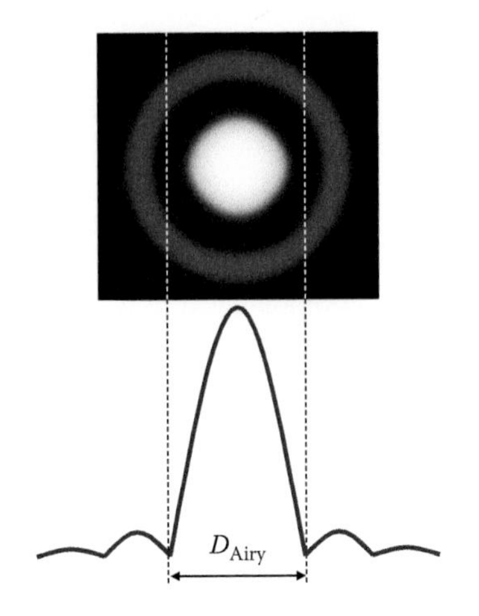


**FIGURE 13.1** Diffraction pattern (top) and corresponding Airy disc with the Airy disc diameter $D_{Airy}$ (bottom), representing the distribution of light energy in an image spot.

When focusing a parallel bundle of light by a lens with a given focal length *f*, the diameter of this Airy disc follows from

$$D_{Airy} = 2.4392 \cdot \frac{\lambda \cdot f}{2 \cdot a}, \tag{13.1}$$

with $\lambda$ being the wavelength and $a$ being half the diameter of the aperture stop (i.e., the aperture stop radius). In microoptics, the diameter of the aperture stop is given by the diameter of a microlens. Considering this fact, it turns out that the focus diameter is not infinitely small, as expected from geometrical imaging models, but features a certain actual focus diameter in reality as shown in Figure 13.2.

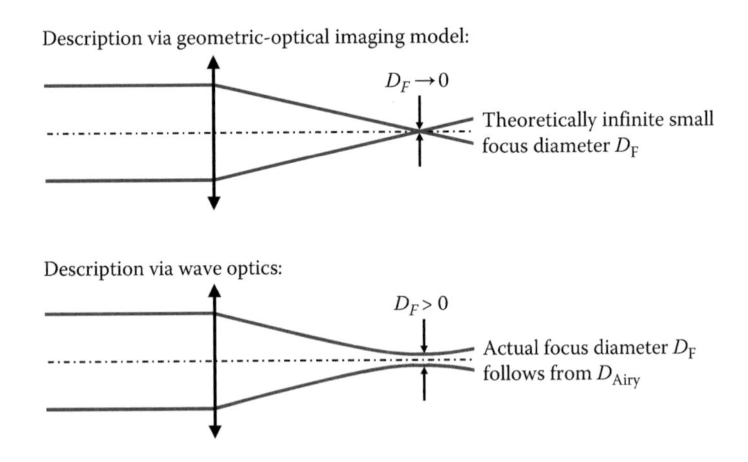


**FIGURE 13.2** Comparison of propagation of light through a focusing optical element and resulting focus diameter $D_F$ as described by the geometric-optical imaging model (top) and wave optics (bottom).

This behavior is well known and taken into account by specifying the diffraction limit of an optical component or system in macrooptics. Here, focus diameters higher than the theoretical diffraction-limited diameter are usually found, due to optical aberrations such as spherical aberration. Diffraction can thus be neglected in many cases.

In the case of microoptics, the superposition of classical aberrations and additional diffraction effects notably influence the imgaing performance since the smaller an optical component, the higher the impact of diffraction. For example, microlenses with low Fresnel numbers (for definition of the Fresnel number see Section 13.2.3.1) feature dramatic defocus. Diffraction thus contributes to the actual focal length where this value is usually reduced.

### 13.2.2 Talbot Self-Images

Apart from single microlenses, microlens arrays are widely used microoptics. As a first approximation, such microlens arrays can be described as a grating of multiple slits or multiple stops. In this case, the so-called *Talbot-effect*, named after the English scientist *William Henry Fox Talbot* (1800–1877) can be observed. This effect describes the distribution of light intensity behind optical arrays and gratings (Talbot, 1836). Talbot observed that for a particular distance from the grating, the distribution of light intensity corresponds to the grating structure itself, forming a so-called Talbot self-image. This distance is the Talbot length $z_T$, given by

$$z_T = \frac{\lambda}{1 - \sqrt{1 - \frac{\lambda^2}{(2 \cdot a)^2}}}. \quad (13.2)$$

Here, $\lambda$ is the wavelength, and $2 \cdot a$ is the free diameter of the grating elements (i.e., the lens diameter of the particular microlenses in microoptical arrays). For apertures that are much higher than the wavelength, Equation 13.2 becomes

$$z_T \approx \frac{2 \cdot (2 \cdot a)^2}{\lambda}. \quad (13.3)$$

At half the Talbot length (and even-number multiples of half the Talbot length), the formed Talbot self-image is laterally displaced by half the aperture diameter. Between such self-images, further images with increased periods with respect to the grating period can be found. This pattern is referred to as *Talbot carpet*, which might result in the formation of ghost images or internal images in microoptical systems.

Talbot self-images of gratings or arrays can be observed in near-field diffraction (a.k.a. Fresnel diffraction), which is defined by the Fresnel number. This number is an essential parameter for the specification of microoptical elements as described hereafter.

### 13.2.3 Specification of Microoptical Components

Microoptical components can be specified by different parameters and quality scores based on physical properties or geometric parameters as described hereafter. These parameters are the Fresnel number, the Strehl-ratio, the modulation transfer function, the point spread function (PSF), the Rayleigh- and Maréchal-criteria, and different geometrical quality scores.

#### 13.2.3.1 Fresnel Number

The *Fresnel number F* allows the definition of the predominant type of diffraction of any stop, array, or grating on the basis of geometric considerations. It is given by

$$F = \frac{a^2}{d \cdot \lambda}, \tag{13.4}$$

where $d$ is the distance between the grating and the detector. For $F \ll 1$, the type of Fraunhofer diffraction is found; it represents a far field approximation. The distance between the aperture and the detector is thus much higher than the aperture diameter. In contrast, the near-field approximation follows from $F \approx 1$, where the type of diffraction is Fresnel diffraction. The distance to the aperture is then in the order of magnitude of the aperture diameter. As an example, for aperture diameters corresponding to the wavelength, $F$ becomes 1 if $d = 8 \cdot a$. Finally, the case of geometrical optics where diffraction can be neglected is found for $F \gg 1$. Since the aperture diameter is much higher than the distance to the detector, this case can be described as downscaled classical imaging, which allows the application of common ray optical calculations.

#### 13.2.3.2 Strehl-Ratio

The *Strehl-ratio S*, named after the German physicist and mathematician *Karl Strehl* (1864–1940), specifies the ratio of the particular intensities of a real image disc of an object point imaged by an optical component or optical system, respectively, and the theoretical maximum intensity, which follows from the diffraction pattern formed by the appropriate Airy disc. Strictly speaking, the Strehl-ratio should be calculated on the basis of the surface integrals of both considered intensity patterns but is usually determined on the basis of the maximum intensity values as shown in Figure 13.3.

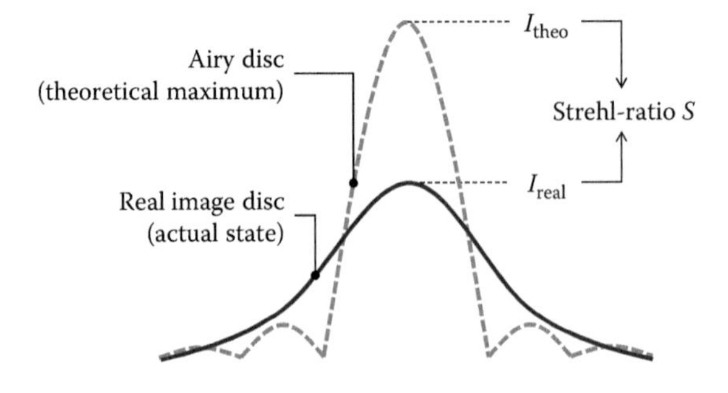


**FIGURE 13.3** Definition of the Strehl-ratio on the basis of the intensities $I$ of a theoretical Airy disc and a real image disc.

It is thus given by

$$S = \frac{I_{\text{real}}}{I_{\text{theo}}}, \tag{13.5}$$

where $I_{\text{real}}$ is the maximum intensity of the real image disc, and $I_{\text{theo}}$ is the maximum intensity of the theoretical Airy disc intensity pattern. An ideal optical system consequently features the maximum Strehl-ratio of $S=1$. In practice, the real image disc differs significantly from the Airy disc due to deformation of an image point by diffraction and optical aberrations, resulting in Strehl-ratios lower than 1.

#### 13.2.3.3 Modulation Transfer Function

The quality of optical imaging and optical resolution power of microoptical components and systems (and optics in general) also results from the contrast transfer. Any object features a certain contrast in intensity, quantitatively given by the Michelson contrast $C_{\text{M}}$, named after the American physicist *Albert Abraham Michelson* (1852–1931). The *Michelson contrast* is also known as *modulation M* and follows from

$$C_{\text{M}} = M = \frac{I_{\max} - I_{\min}}{I_{\max} + I_{\min}}. \tag{13.6}$$

Here, $I_{\max}$ is the intensity of bright object areas and $I_{\min}$ the intensity of dark object areas, for example, the intensities of black and white lines of a black-and-white stripe pattern. The ratio of the modulation of an image $M_{\text{image}}$ and the modulation of an object $M_{\text{object}}$ gives the so-called *modulation transfer function MTF* according to

$$MTF(R) = \frac{M_{\text{image}}(R)}{M_{\text{object}}(R)}. \tag{13.7}$$

The fineness or structure of the object is considered by its *spatial frequency R* (i.e., the reciprocal value of its spatial period length, given in line pairs or cycles per millimeter). Referring to the abovementioned example of a black-and-white stripe pattern, this period length is twice the distance between black and white lines in case of equidistant structures. Both the particular Michelson contrasts or modulations and the modular transfer function are thus given for a particular spatial frequency as considered in Equation 13.7. The spatial frequency, where $MTF(R)=0$, is defined as *cut-off-frequency* $f_{\text{cut-off}}$, given by

$$f_{\text{cut-off}} = \frac{1}{\arctan\left(\dfrac{\lambda}{2 \cdot a}\right)}. \tag{13.8}$$

The higher this value, the higher the optical resolution power and imaging quality of a microoptical component or system.

#### 13.2.3.4 Point Spread Function

The PSF describes the accumulated impact of diffraction phenomena and optical aberrations, for example, spherical aberration, which gives rise to a diffusion of the focal point along and lateral to the direction of propagation of focused light. It thus indicates how an idealized punctiform object is imaged by an optical component or system. The PSF can be referred to as the spatial equivalent to the modulation transfer function and is mathematically determined via the Fourier-transform of the wave front where the convolution of the PSF and the wave front gives the real image point on a detector.

#### 13.2.3.5 Rayleigh-Criterion

According to the *Rayleigh-criterion*, which was postulated by the English physicist *John William Rayleigh*[2] (1842–1919), a wave front that is transmitted by an optical component or system can be regarded as diffraction-limited if its maximum wave front deviation $\Psi_{max}$, expressed by the peak-to-valley value, is equal to or smaller than a fourth of the wavelength according to

$$|\Psi_{max}(x)| \leq \frac{\lambda}{4}. \quad (13.9)$$

However, the distribution of light in an image point depends not only on the maximum wave front deviation, but also on its actual shape and local deformations. This fact is considered by the Maréchal-criterion.

#### 13.2.3.6 Maréchal-Criterion

The *Maréchal-criterion*, named after the French researcher *Robert Gaston André Maréchal* (1916–2007), represents an essential expansion of the above-described Rayleigh-criterion (Ottevaere et al., 2006). According to Maréchal, a wave front transmitted by an optical component or system is diffraction-limited if the root-mean-squared phase error $\Psi_{RMS}$ is equal to or smaller than a fourteenth of the wavelength:

$$\Psi_{RMS} = \sqrt{\int |\Psi(x)|^2\, dx - \left[\int |\Psi(x)|\, dx\right]^2} \leq \frac{\lambda}{14}. \quad (13.10)$$

In contrast to the Rayleigh-criterion, both the deformation of the wave front by optical aberrations and optical noise are thus considered in addition to the maximum wave front deviation.

#### 13.2.3.7 Geometrical Quality Scores

Microoptical components can also be specified on the basis of geometrical parameters such as contour inaccuracies or surface textures. According to DIN 4760, contour inaccuracies are classified into different orders as listed in Table 13.1.

For the evaluation of optics surfaces, merely the first four orders are of importance since the size of fifth- or sixth-order defects is much smaller than the wavelength and thus not optically active. The first order, the form deviation, corresponds to the contour accuracy of optics surfaces, which is specified by DIN ISO 10110 as presented in more detail in Section 6.3.1. Waviness, the second-order contour inaccuracy, is an

[2] Officially, his full name was *John William Strutt*, 3. Baron Rayleigh.

**TABLE 13.1**
**Classification of Contour Inaccuracies according to DIN 4760**

| Order of Contour Inaccuracy | Denomination of Contour Inaccuracy |
|---|---|
| First order | Form deviation |
| Second order | Waviness |
| Third and fourth order | Surface roughness (grooves or bucklings) |
| Fifth order | Roughness of the material structure |
| Sixth order | Point defects of the material |

intermediate surface error found between the form deviation and the surface roughness. It is distinguished from the latter parameter by defining a cut-off-frequency that separates waviness from surface roughness. Metaphorically speaking, surface waviness mirrors vibrations of tools and machines and consequently process instabilities due to such vibrations. Finally, the surface roughness is considered because it may cause diffuse reflection and scattering as expressed by the total integrated scatter function (see Section 6.3.4).

Moreover, the ratio of the root mean squared surface roughness $Rq$ and the sagitta $S$ (for definition of the sagitta see Section 7.7) is sometimes specified, where a low $Rq/S$-ratio indicates a high surface quality.

## 13.3 MICROOPTICAL COMPONENTS AND SYSTEMS

### 13.3.1 Microlenses

Basically, *microlenses* are miniaturized classical lenses; plano-convex lenses are mainly used in microoptics due to the fact that this lens type is comparatively simple to produce. The diameters and radii of curvature of such microlenses amount to some hundreds of microns, whereas the center thickness ranges from some tens to hundreds of microns. However, the basic lensmaker's equations for describing the relationship between the radii of curvature and the focal length can be applied (compare Section 5.2).

In order to reduce spherical aberration, the curved surface of a microlens is usually orientated toward the incident bundle of light as shown in Figure 13.4a. However, in some cases, the lens may be orientated the other way around as shown in Figure 13.4b. As a result, a microlens (and any plano-convex lens, respectively) features two characteristic focal lengths: the effective back focal length $EFL_{\text{back}}$ and the effective front focal length $EFL_{\text{front}}$. The first is given by

$$EFL_{\text{back}} = \frac{R}{n-1} - \frac{t_{\text{c}}}{n}, \tag{13.11}$$

where $R$ is the radius of curvature, $n$ is the index of refraction, and $t_{\text{c}}$ is the thickness of the lens. The effective front focal length follows from

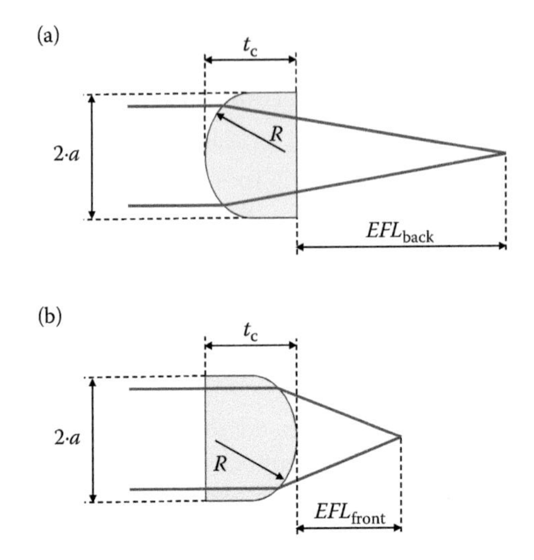


**FIGURE 13.4** Visualization of important parameters of microlenses, given by the lens aperture diameter $2 \cdot a$, the radius of curvature $R$, the lens center thickness $t_c$, and the effective back focal length $EFL_{back}$ (a) and front focal length $EFL_{front}$ (b), depending on the direction of light passing through such a microlens.

$$EFL_{front} = \frac{R}{n-1} = \frac{\left(S + \frac{a^2}{S}\right)}{2 \cdot (n-1)} \approx \frac{a^2}{S}. \tag{13.12}$$

Here, $S$ is the lens sagitta and $a$ half the lens diameter. The radius of curvature of a microlens can be calculated by

$$R = \frac{S \cdot (e+1)}{2} + \frac{a^2}{2 \cdot S}. \tag{13.13}$$

The parameter $e$ is the conus constant, which describes the basic shape of the lens surface and amounts to 1 for spherical surfaces, as presented in Table 4.2. The lens sagitta can consequently be calculated on the basis of the radius of curvature and half the lens diameter according to

$$S = R - \sqrt{R^2 - a^2}. \tag{13.14}$$

As described in more detail in Section 13.2.3.1, microoptical components can be characterized on the basis of their Fresnel number $F$. For a microlens, this value follows from half the lens diameter and the effective back focal length according to

$$F = \frac{a^2}{\lambda \cdot EFL_{back}}. \tag{13.15}$$

Microlenses can be used in different fields of application. For example, mounting an appropriate microlens on fiber entrance faces allows the fiber's numerical aperture and thus the coupling of light to increase. Further, light can be collimated or focused when placing microlenses at the exit face of an optical fiber as applied for optical telecommunication devices, optoacoustic microphones, fiber lasers, PillCams,[3] or endoscopes. Here, even microoptical systems consisting of several single microlenses are used in order to realize aberration-free imaging.

A special type of microlens is based on gradient index (GRIN) materials as introduced in Section 3.4. As shown in Figure 13.5, the use of such materials allows realizing focusing optical elements without any classical surface machining.

GRIN lenses typically feature low diameters in the range from some hundreds of microns to some millimeters, due to the manufacturing method, which is based on diffusion as presented in Section 13.4.7. The focal length $f$ of a GRIN lens can be determined according to

$$f = \frac{1}{n_0 \cdot g \cdot \sin(g \cdot l)}, \tag{13.16}$$

where $l$ is its geometric length, given by

$$l = \frac{2 \cdot \pi \cdot P}{g}. \tag{13.17}$$

The parameters $n_0$ and $g$ in Equation 13.16 are the index of refraction at the lens center and the geometrical gradient constant, respectively. The symbol $P$ in Equation 13.17 denotes the lens pitch as defined as follows: within a GRIN lens, a light beam propagates in the form of a sinus curve due to the radial distribution of the refractive index. One full sinusoidal path is defined as one pitch. The behavior of a GRIN lens is directly related to this value. For instance, a collimated bundle of light rays that passes through a converging GRIN lens is focused if

$$l = 0.23 \cdot P. \tag{13.18}$$

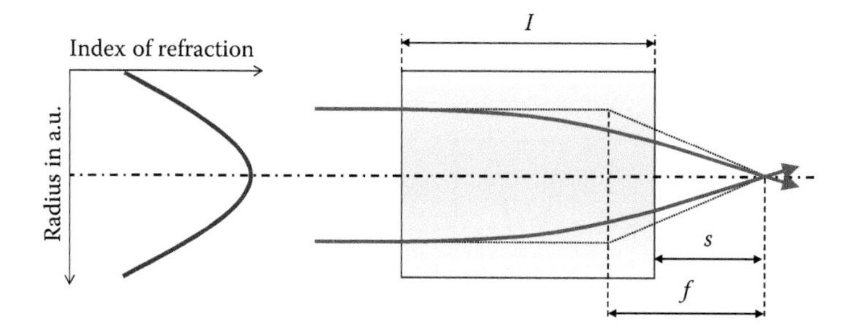


**FIGURE 13.5** Distribution of index of refraction in a rod GRIN lens (left) and impact of such distribution on light passing through such a lens with the geometrical length $l$, resulting in focusing of incident collimated light at a certain focal length $f$ and working distance $s$, respectively (right).

[3] PillCams are self-sufficient electro-optic microsystems that feature the shape and size of a pill. This PillCam is swallowed and passes through the entire alimentary tract, filming the gastric and intestinal walls (e.g., for detecting and analyzing stomach or bowel cancer).

The distance between the lens exit face and the focus, also referred to as working distance $s$ or $BFL$, is then given by

$$s = \frac{1}{n_0 \cdot g \cdot \tan(g \cdot l)}. \tag{13.19}$$

### 13.3.2 Microlens Arrays

The periodic arrangement of microlenses on a substrate is referred to as *microlens array* as schematically shown in Figure 13.6. In addition to the basic lens parameters as described in Section 13.3.1, such microlens arrays are characterized by the pitch $P$ (i.e., the lateral distance between the particular microlenses). Microlens arrays are further classified on the basis of the lens arrangement, for example, rectangular or hexagonal packed. As shown in Figure 13.7, the particular arrangement can result in different pitches in x and y directions. Another classification refers to the surface shape of the involved micro-lenses as expressed by the conus constant $e$.

There is a number of different applications for microlens arrays such as the generation of light sections (a.k.a. line foci),[4] the generation of spot patterns, or the homogenization of laser light. The latter application is performed in order to convert the distribution of light intensity of an initial raw laser beam (e.g., Gaussian beam profile) to a laser beam with a flat distribution of light intensity. This is achieved by

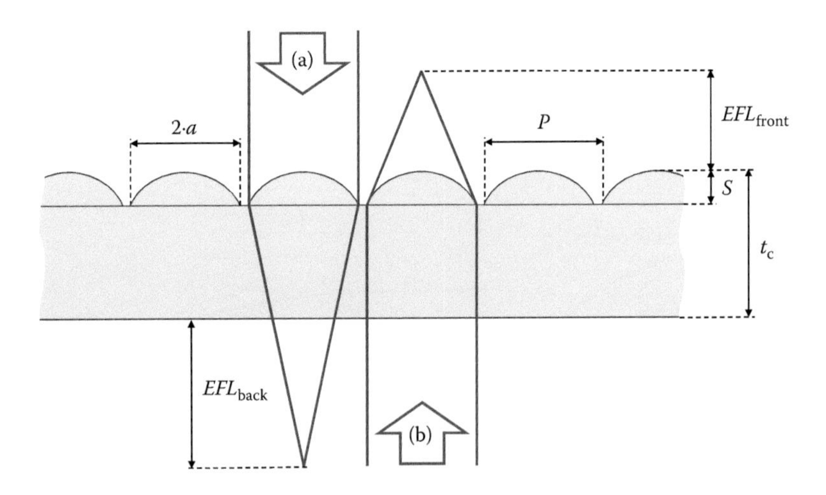


**FIGURE 13.6** Visualization of important parameters of microlens arrays, given by the lens aperture diameter $2 \cdot a$ of the involved microlenses, the lens pitch $P$, the lens center thickness $t_c$, and sagitta $S$, as well as the effective back focal length $EFL_{back}$ (a) and front focal length $EFL_{front}$ (b), depending on the direction of light passing through such a microlens array.

[4] For instance, line foci of high energy density are required for laser annealing of amorphous coatings. This process is usually performed using excimer laser irradiation and is thus referred to as excimer laser annealing. It is applied for the production of thin film transistors, for example, for flat screens.

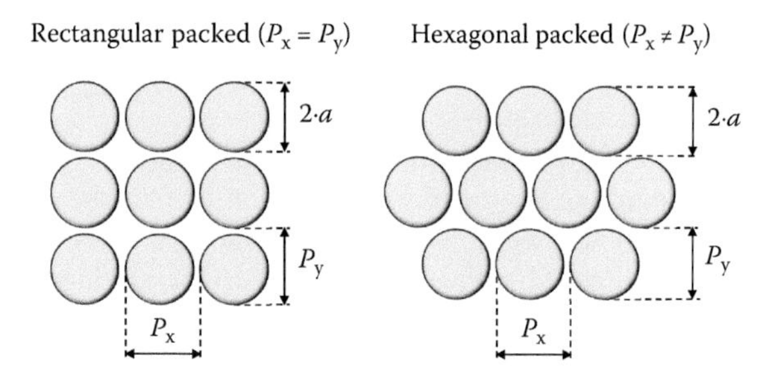


**FIGURE 13.7** Examples for different arrangements of microlenses on a microlens array resulting in different pitches. (Adapted from Nussbaum, P. et al., *Pure and Applied Optics: Journal of the European Optical Society Part A*, 6, 617–636, 1997.)

*homogenizers*, consisting of the combination of at least one microlens array and a Fourier lens as shown in Figure 13.8.

This type of laser beam homogenizer is referred to as nonimaging homogenizer. Here, an incoming raw laser beam is divided into several subaperture beams (one per microlens). These subaperture beams are subsequently superposed in an image plane by the Fourier lens. This superposition leads to the formation of a homogenous field, where the distribution of laser light can be described as a flat top as shown in Figure 13.8. The size of the homogeneous field $D_{hf}$ is given by

$$D_{hf} = \frac{P_{la} \cdot f_{Fl}}{f_{la}}. \tag{13.20}$$

Here, $P_{la}$ is the pitch of the lens array, $f_{la}$ its focal length, and $f_{Fl}$ the focal length of the used Fourier lens. Laser beam homogenizers can also be set up using two microlens arrays. Such homogenizers are known as imaging homogenizers, where the size of the homogeneous field follows from

$$D_{hf} = P_{la1} \cdot \frac{f_{Fl}}{f_{la1} \cdot f_{la2}} \cdot \left[ \left( f_{la1} + f_{la2} \right) - d \right]. \tag{13.21}$$

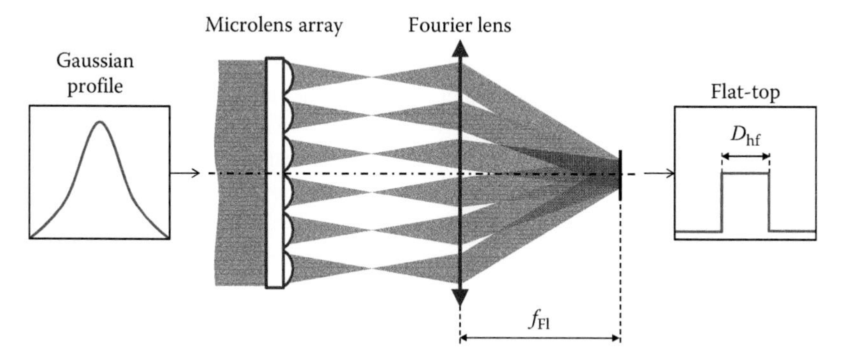


**FIGURE 13.8** Working principle of a nonimaging laser beam homogenizer for the conversion of an initial beam profile (here: Gaussian) to a flat-top. (Adapted from Bich, A. et al., *Proceedings of SPIE*, 6879, 68790Q-12008.)

It thus depends on the pitch of the first microlens array $P_{la1}$, the focal length of the Fourier lens $f_{Fl}$, the focal lengths of both microlens arrays, $f_{la1}$ and $f_{la2}$, and the distance $d$ between the microlens arrays.

Another application of microlens arrays is the improvement of light harvesting in photovoltaic cells, where the degree of efficiency can be increased by using microstructured cover glasses that allow the incoming sunlight to be focused onto the active semiconductor chips. Microlens arrays are further essential elements of so-called Hartmann-Shack wave front sensors. Here, an incident light wave front is divided into several subapertures by a microlens array, acting as spot generator, and the position of each focused subaperture is determined on a detector. A tilted or deformed wave front gives rise to a lateral displacement of the particular focus on the detector. This allows the calculation of the shape of the incoming wave front and its deformation from an idealized plane or sphere.

### 13.3.3 Micromirrors

*Micromirrors* are thin plates with high planeness and typical lateral dimensions from approximately $4 \cdot 4\,\mu m^2$ to $10 \cdot 10\,\mu m^2$. These plates are usually made of silicon and are used for setting up micromirror arrays (i.e., an arrangement of thin micromirrors on actuators). A micromirror array can thus be shaped and deformed dynamically by controlling these actuators and each particular micromirror, respectively. This allows the realization of thin film digital micromirror devices or spatial light modulators. Such active devices are known as adaptive optics and can be applied for the shaping of laser beams (Stephen and Vollertsen, 2013), the correction of wave fronts, the realization of bar code scanners, or light modulation, for example, the dynamic generation of diffraction patterns.

### 13.3.4 Gratings

*Gratings* are periodic arrangements of two-dimensional structures, for example, lines or rectangular or triangular grooves. Depending on the substrate material and the optical coatings, these devices can be operated in transmission (→ refracting grating) or reflection (→ reflecting grating). The characterization of optical gratings is based on the density of structures, which is usually specified by the grating constant $g$, that is, the groove or line density per millimeter (l/mm) or the number of line pairs per millimeter (lp/mm) where one line pair is given by one groove and one strip (given by the strip between grooves). Alternatively, the absolute lateral size $\Lambda$ (i.e., the length of a line pair given by the distance between neighboring grooves taking the strip between into account) can be specified. A selection of standard grating constants including the accompanying size of the particular structure and grooves is listed in Table 13.2.

Even though gratings feature comparatively simple geometries, high production accuracy is required as shown by the following example. Assuming an acceptable manufacturing tolerance of 5% for the distance of grooves, the required production accuracy for a grating with a grating constant of 1800 lp/mm amounts to 14 nm. Thus, sophisticated methods need to be applied for the production of optical gratings and microoptical elements in general as described in Section 13.4.

**TABLE 13.2**
**Selection of Gratings with Standard Grating Constants Including the Size of the Particular Structure and Width of Grooves**

| Grating Constant (lp/mm) | Size of Structure (μm) | Width of Groove (μm) |
|---|---|---|
| 300 | 3.33 | 1.66 |
| 600 | 1.67 | 0.84 |
| 1200 | 0.83 | 0.42 |
| 1600 | 0.63 | 0.32 |
| 1800 | 0.56 | 0.28 |

*Note:* The given sizes of structure and widths of grooves are valid for equidistant rectangular gratings.

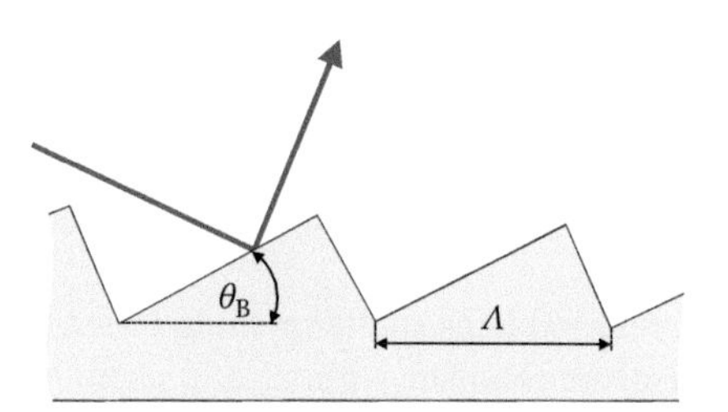


**FIGURE 13.9** Echelle grating with blaze angle $\theta_B$ and lateral size of grating structures $\Lambda$.

High-production accuracy gains further importance for gratings with defined three-dimensional structures. For example, *echelle gratings* (a.k.a. blazed gratings) feature a prismatic texture with a defined angle, the so-called *blaze angle* as visualized in Figure 13.9. The blaze angle is given by

$$\theta_B = \arcsin\frac{m \cdot \lambda}{2 \cdot g}, \tag{13.22}$$

with $m$ being the diffraction order, $\lambda$ the wavelength, and $g$ the grating constant. The realization of defined blaze angles allows optimizing the efficiency of such gratings for a defined diffraction order.

Blazed gratings and optical gratings in general are used for the analysis of spectra, spectral filtering, and monochromization, fiber coupling, and multiplexing or demultiplexing in optical telecommunication. Special types of gratings with nonperiodic free-form surface textures, so-called diffractive optical elements (DOEs) are, for example, used to generate holograms. Finally, gratings or arrays with prismatic or "moth eye" structures can be used as antireflective elements (Wilson and Hutley, 1982).

## 13.4 MANUFACTURING OF MICROOPTICS

Single microlenses can be produced by classical optical manufacturing methods as described in Chapters 7 through 10 using tools of adequate sizes. However, sophisticated manufacturing methods are required in order to manufacture complex microoptics devices, for example, gratings or microlens arrays. In 1981, the realization of microlenses via ion exchange and the resulting formation of GRIN materials were presented (Oikawa et al., 1981), and 7 years later, the production of microlenses by thermal reflow process was reported (Popovic et al., 1988). The suitability of laser-based methods (Haruna et al., 1990; Mihailov and Lazare, 1993), deep lithography processes (Frank et al., 1991), microjet printing (MacFarlane et al., 1994), and ion etching (Stern and Rubico Jay, 1994) for producing microoptical devices was then intensively investigated and shown in the 1990s (Ottevaere et al., 2006). Now, some of these approaches have become established in modern microoptics manufacturing. In this section, these main processes and methods and their underlying mechanisms are introduced.

### 13.4.1 Dry Etching

Dry etching techniques can be classified into three different sub-categories: physical etching, chemical etching, and chemical-physical etching, as described hereafter.

#### 13.4.1.1 Physical Etching

*Physical etching* is also referred to as sputter etching or ion beam etching (IBE) and represents a pure physical mechanism without any chemical reactions. As shown in Figure 13.10, material removal is achieved by ion bombardment and the kinetic energy of ions where, usually, ions of inert gases such as argon or cesium are used.

For this purpose, ions of a certain mass $m$ are generated and accelerated by an ion cannon. The ions thus propagate with the velocity $v$, given by

$$v = \sqrt{\frac{2 \cdot z \cdot e \cdot U}{m}}. \tag{13.23}$$

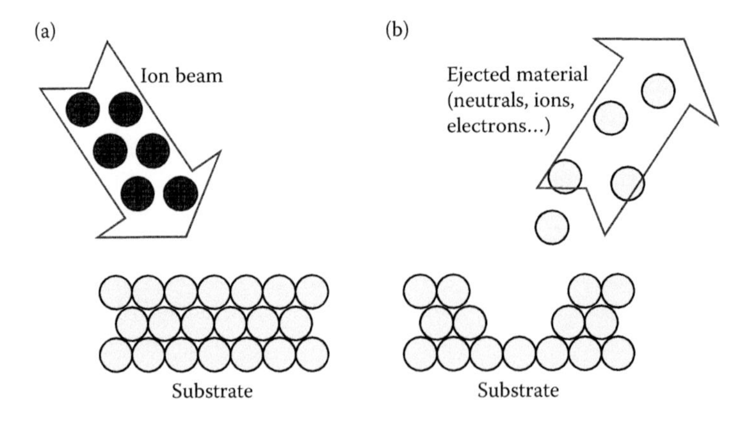


**FIGURE 13.10** Principle of ion etching: an incident primary ion beam (a) is guided onto the substrate surface, leading to the ejection or removal of substrate material such as secondary ions, neutrals, electrons, etc. (b).

Here, $z$ is the ion charge, $e$ the elementary electric charge,[5] and $U$ the acceleration voltage. The kinetic energy of an accelerated ion finally amounts to

$$E_{\text{kin}} = \frac{1}{2} \cdot m \cdot v^2 = z \cdot e \cdot U. \tag{13.24}$$

In order to maintain an ion's kinetic energy after its generation, this method is performed in a high vacuum or ultra-high vacuum with a high free length of path[6] and a low number of collision partners, respectively. During physical etching, the removal rate $N(t)$ (i.e., the number of particles removed from the substrate surface) can be calculated by

$$N(t) = N_0 \cdot \left(1 - e^{\left(-\frac{Y \cdot I_p \cdot t}{e \cdot N_0}\right)}\right). \tag{13.25}$$

It consequently depends on the number of particles on the substrate surface[7] $N_0$, the sputter yield $Y$, the primary current $I_p$, and the sputter time $t$. The primary current directly follows from the acceleration voltage, whereas the sputter yield results from the substrate material and its atomic number, the mass number and kinetic energy of the used ions, and the angle of incidence of the ion beam on the substrate surface.

Even though this method represents a powerful and state-of-the-art approach for the manufacture of optics surfaces of high precision, material removal by ion beam sputtering can result in a disturbing and unwanted effect, surface rippling. Such surface rippling is described by the Bradley-Harper-model (Bradley and Harper, 1988) which describes the preferential material removal at roughness or waviness valleys with respect to peaks.

However, this effect can be overcome by appropriate process control. Physical etching such as IBE has thus been established as one standard technique for the microstructuring of glasses and semiconductor wafers as performed during photolithography (see Section 13.4.2) as well as for polishing (Pearson, 1972; Stognij et al., 2002), precision correction, and aspherization of optics surfaces.

#### 13.4.1.2 Chemical Etching

In contrast to physical etching, material removal during *chemical etching* is achieved by chemical reactions. The required reactive species, such as radicals, are usually provided by plasma, applying suitable process gases such as fluorochemical compounds as described in more detail in Section 8.4.2. Using this method, two preconditions have to be fulfilled: the reaction product should be volatile, and the lifetime of reactive plasma species should be sufficiently long to cover the distance from the plasma volume to the work piece surface.

---

[5] The elementary electric charge amounts to $e = 1.6021766208 \cdot 10^{-19}$ C.

[6] The free length of path in ambient pressure (1013 hPa) is approximately 70 nm. In contrast, it amounts to approximately 10 cm to 1 km in high vacuum ($p \approx 10^{-3}$ to $10^{-7}$ hPa) and approximately 1 to $10^5$ km in ultra-high vacuum (where $p \approx 10^{-7}$ to $10^{-12}$ hPa).

[7] The number of particles on any surface amounts to approximately $10^{15}$ per $cm^2$.

#### 13.4.1.3 Chemical-Physical Etching

*Chemical-physical etching* is also known as *reactive ion etching.* It can be described as a combination of physical and chemical etching. As shown in Figure 13.11, both reactive species (R) and ions (+) are generated in a plasma volume that is in direct contact with the work piece surface (Figure 13.11a). As a first step, shown in Figure 13.11b, the reactive species traverse the plasma and adhere to the work piece surface. Second, ions are accelerated toward the surface and consequently collide with it (Figure 13.11c). The actual etching process is then induced by the kinetic energy provided by the collision of the incident ion and the surface. Finally, the reaction products, given by the compounds of the reactive plasma species and surface atoms, are dissolved and volatilized (Figure 13.11d).

The acceleration of ions is due to the formation of a so-called plasma sheath region at the substrate surface. Within such a plasma sheath, the quasineutrality of a plasma is not valid, leading to a potential well with a lower number of electrons close to the substrate surface with respect to the plasma volume. The plasma sheath thickness roughly corresponds to the Debye length $\lambda_D$, named after the Dutch-American physical chemist *Peter Joseph Debye* (1884–1966). It is generally given by

$$\lambda_D = \sqrt{\frac{\varepsilon_0 \cdot k_B \cdot T}{n_e \cdot e^2}}. \tag{13.26}$$

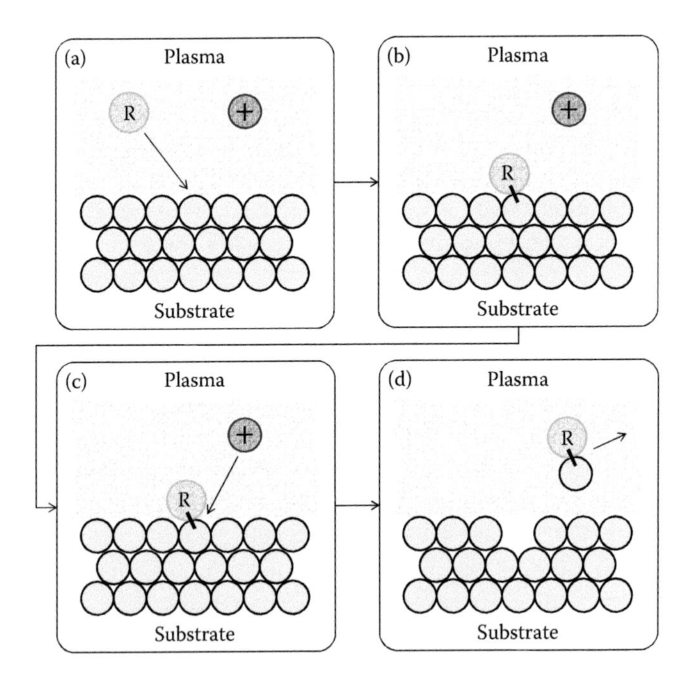


**FIGURE 13.11** Principle of chemical-physical etching by the formation of reactive species (R) and ions (+) in a plasma discharge. For detailed description, see running text.

Here, $\varepsilon_0$ is the vacuum permittivity,[8] $k_B$ is the Boltzmann constant,[9] $T$ is the temperature,[10] $n_e$ is the electron density, and $e$ is the elementary electric charge. The potential well within such plasma sheath regions represents the driving force for the acceleration of ions.

### 13.4.2 Photolithography

The above-described dry etching techniques can be applied for microstructuring of substrates via *photolithography* (a.k.a. optical lithography or UV lithography). This method consists of two main steps: the transfer of any pattern or structure on a substrate surface and the etching of this pattern. For the first step, the substrate is coated with a photoresist (i.e., a light-sensitive medium) via spin coating or similar coating procedures. As shown in Figure 13.12, this resist is subsequently exposed to light; usually UV light is used in order to realize small structure sizes due to the short wavelength. In doing so, any two-dimensional pattern can be transferred onto the resist surface by mask imaging.

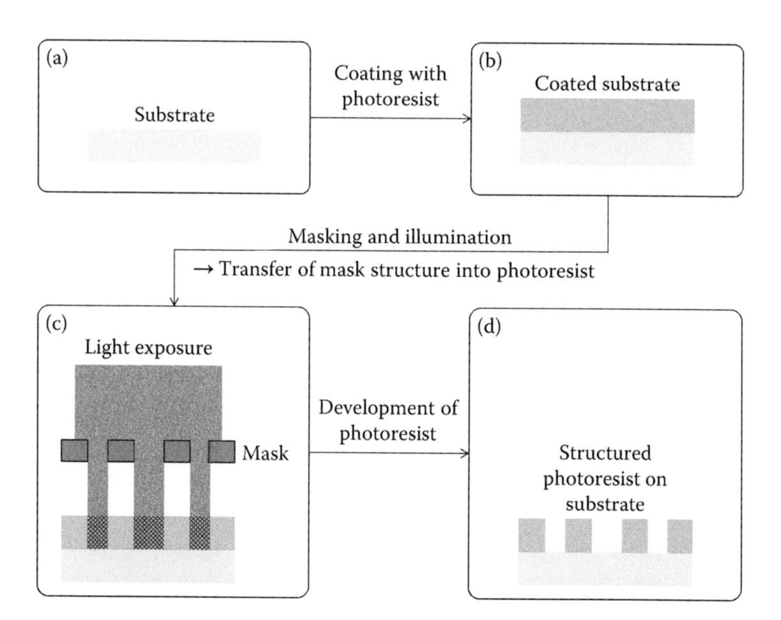


**FIGURE 13.12** First step of photolithography: work piece preparation by coating of a substrate with photoresist (a–b) and structuring of the photoresist by mask imaging (c) and final development (d).

[8] The vacuum permittivity amounts to $\varepsilon_0 = 8.854187817 \cdot 10^{-12}$ A s/V m.

[9] The Boltzmann constant is $k_B = 1.38064852 \cdot 10^{-23}$ J/K.

[10] The Debye length can be calculated for both ions and electrons. This is considered by inserting either the ion temperature $T_i$ or the electron temperature $T_e$ in Equation 13.26.

The irradiated areas of the resist are consequently chemically modified by the incident light and can be dissolved when applying appropriate developers.[11] The mask structure is then transferred onto the resist, where the choice of appropriate imaging systems allows a miniaturization of the actual mask dimension and pattern dimension, respectively. Once the photoresist is structured, the surface is further processed by dry etching as shown in Figure 13.13. The resist structure is thus transferred into the substrate material. Finally, residues of photoresist are stripped off, either by wet chemical cleaning or by further dry etching. This method, photolithography, allows the generation of complex microoptics structures, such as microlens arrays, DOEs, or optical gratings.

Instead of dry etching techniques, wet etching may be applied for material removal in the course of a photolithography process in some cases. Here, only select acids can be used, since glass features a high acid resistance, and it is thus commonly referred to as quasichemically inert. Suitable acids for wet etching of glass are listed in Table 13.3.

As an example, fused silica, which exclusively consists of the network former silicon dioxide ($SiO_2$), can be decomposed when exposed to hydrofluoric acid (HF) according to

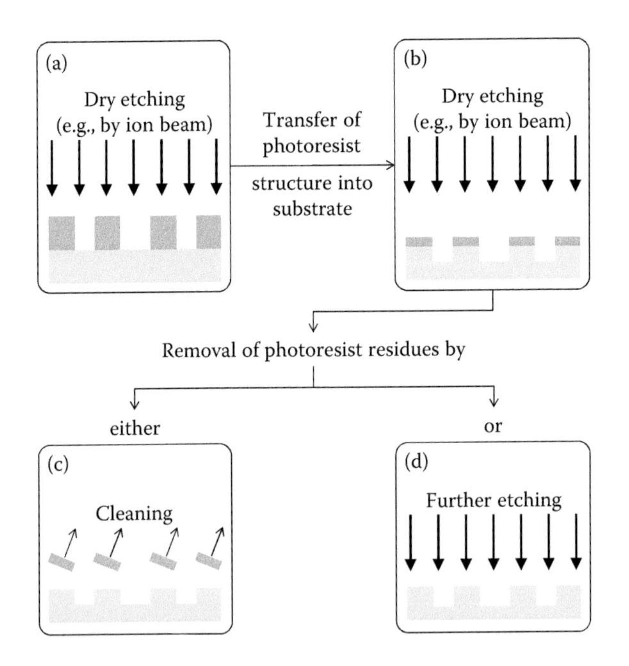


**FIGURE 13.13** Second step of photolithography; actual surface patterning by transferring the photoresist structure into the substrate by dry etching (a–b) and final removal of residues from the photoresist by cleaning (c) or further etching (d).

[11] This applies to so-called positive photoresists. In contrast, the irradiated areas become nonsoluble when using negative photoresists where the nonirradiated areas are dissolved by the developer. The first type of photoresist thus produces a positive copy of the mask, whereas the use of negative photoresists results in the formation of a negative copy of the mask.

**TABLE 13.3**
**Suitable Acids for Wet Etching of Glasses**

| Acid | Total Formula |
|---|---|
| Hydrofluoric acid | HF |
| Phosphoric acid | $H_3PO_4$ |
| Sulfuric acid | $H_2SO_4$ |

$$SiO_2 + 4HF \rightarrow SiF_4 + 2H_2O. \tag{13.27}$$

The volatile reaction products are thus gaseous silicon tetrafluoride ($SiF_4$) and gaseous water ($H_2O$). Decomposition of the network modifier sodium oxide ($Na_2O$) as existent in a number of multicomponent glasses (compare Table 3.5) can be achieved using sulfuric acid ($H_2SO_4$), where sodium sulfate ($Na_2SO_4$) and water are formed in the course of the chemical reaction,

$$Na_2O + H_2SO_4 \rightarrow Na_2SO_4 + H_2O. \tag{13.28}$$

In addition to the acids listed in Table 13.3, potassium hydroxide (KOH) allows a certain material removal since this compound attacks oxides. For instance, soluble potassium silicates are formed from the reaction of silicon dioxide ($SiO_2$) and potassium hydroxide.

### 13.4.3 Laser-Based Methods

Applying focused laser irradiation for materials processing has several advantages. First, this approach allows the generation of smallest free-form surface structures in the size range of some microns, without the use of masks. Second, laser irradiation can be referred to as a contactless tool without any wearing, thus featuring a high "form stability" as long as the process is well known and controlled. However, direct laser machining of glasses is still a challenging task, since the coupling of incoming laser energy into the glass surface is significantly inhibited due to the high transmission of glasses, resulting in poor near-surface absorption and high optical depth of penetration $d_{opt}$, respectively. The latter parameter is given by

$$d_{opt} = \frac{1}{\alpha}, \tag{13.29}$$

where $\alpha$ is the absorption coefficient of the irradiated medium (compare Section 2.4). Actually, most established laser sources emit light within the transmission range of optical glasses and especially of quartz glass and fused silica. This restriction can be overcome by different strategies: (1) the use of lasers with laser wavelengths in the ultraviolet or infrared wavelength region, (2) the application of ultrashort laser pulses, or (3) the modification of the glass surface transmission or absorption characteristics. In the latter case, different approaches that are mostly

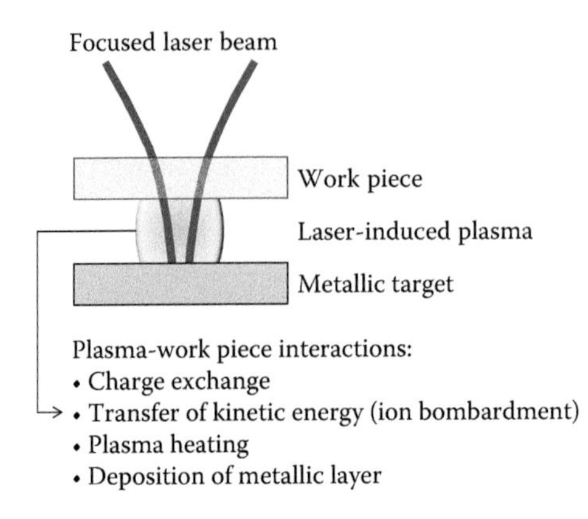


**FIGURE 13.14** Working principle of laser-induced plasma-assited ablation (LIPAA). (Adapted from Gerhard, C., *Atmospheric Pressure Plasma-Assisted Laser Ablation of Optical Glasses*, Cuvillier Verlag, Göttingen, Germany, 2014.)

based on the application of absorbing layers were developed in the last decades as presented hereafter. These approaches allow reducing the required laser energy for material removal and thus mitigating thermally induced disturbing effects, resulting in high machining quality.

#### 13.4.3.1 Laser-Induced Plasma-Assisted Ablation

*Laser-induced plasma-assisted ablation* (LIPAA) is based on the utilization of several plasma-matter interactions and plasma-induced effects: (1) a charge exchange amongst ions and electrons of the plasma plume and the surface of the glass work piece, (2) a transfer of kinetic energy, provided by radicals and ions, to the surface, (3) plasma heating of the work piece, and (4) the successive deposition of a metallic thin layer on the work piece surface. This thin layer gives rise to an increase in absorption of incoming laser irradiation at this interface (i.e., the work piece rear side). This is achieved by the working principle of LIPAA as shown in Figure 13.14.

Here, a laser beam passes the actual glass work piece and is focused onto a metallic target (e.g., stainless steel, tin, aluminum, or copper), which is placed at a certain distance from the work piece rear side. Due to the high laser fluence, a plasma is ignited on the metal surface, leading to the above-described effects and mechanisms. Further information on this promising technique can be found in the literature (Zhang et al., 1998a,b, 1999; Sugioka and Midorikawa, 2001; Hanada et al., 2006).

#### 13.4.3.2 Laser-Induced Backside Wet Etching

For *laser-induced backside wet etching* (LIBWE), the rear side of a glass work piece is brought in direct contact with an absorbing liquid as shown in Figure 13.15. The incoming focused laser irradiation is thus absorbed at this interface, and photo–thermal ablation is achieved due to the deposition of laser energy within the liquid close to the work piece rear side. Ablation is further supported by laser-induced shock waves, cavities, and thermo-elastic pressure within the used liquid and the

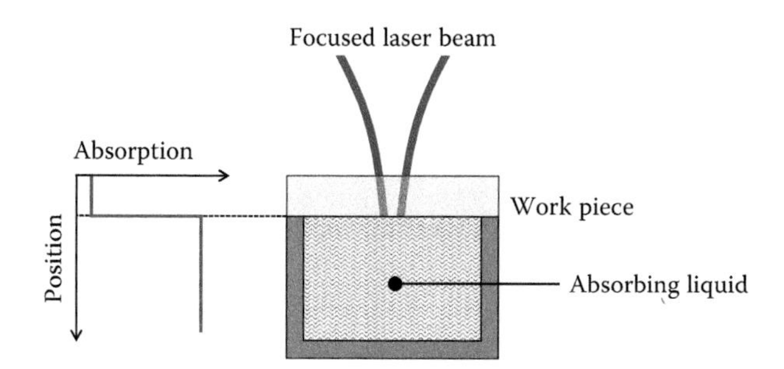


**FIGURE 13.15** Working principle of laser-induced backside wet etching (LIBWE). (Adapted from Gerhard, C., *Atmospheric Pressure Plasma-Assisted Laser Ablation of Optical Glasses*, Cuvillier Verlag, Göttingen, Germany, 2014.)

accompanying mechanical impact on the glass surface[12] (Zimmer and Böhme, 2008). A selection of suitable liquids for LIBWE as reported in the literature is listed in Table 13.4.

Applying LIBWE allows a significant reduction in laser ablation threshold $F_{th}$ (i.e., merely some hundreds of mJ/cm² in the case of fused silica)[13] (Wang et al., 1999). The laser ablation threshold depends on several material-specific parameters of both the particular glass and the used liquid according to

$$F_{th} = \left(\rho_g \cdot L_{T,g} \cdot c_g + \rho_l \cdot L_{T,l} \cdot c_l\right) \cdot \left(T_m - T_0\right). \tag{13.30}$$

Here, $\rho$ is the density, $L_T$ is the thermal diffusion length (a.k.a. thermal depth of penetration), and $c$ is the specific heat capacity where the indices "g" and "l" indicate these parameters for the machined glass ("g") and the used liquid ("l"). Further, $T_m$ is the melting temperature of the glass, and $T_0$ is its initial temperature (Hopp et al.,

**TABLE 13.4**
**Selection of Liquids Suitable for LIBWE as Reported in the Literature**

| Liquid | Molecular Formula | Reference |
|---|---|---|
| Solutions of acetone | $(CH_3)_2CO$ | Wang et al., 1999 |
| Pyrene | $C_{16}H_{10}$ | Wang et al., 1999 |
| Aqueous solutions of naphthalene-1,3,6-trisulfonic acid trisodium salt | $Np(SO_3Na)_3$ | Ding et al., 2002 |
| Solutions of pyrene admixture with acetone, tetrachloroethylene, and toluene | $C_2Cl_4$, $C_7H_8$ | Zimmer et al., 2003 |
| Liquid gallium | Ga | Zimmer et al., 2006a |
| Pure toluene | $C_7H_8$ | Huang et al., 2007 |

[12] This effect is comparable to the working principle of ultrasonic cleaning devices.
[13] For comparison, the laser ablation threshold amounts to approximately 6–10 J/cm² for pure fused silica.

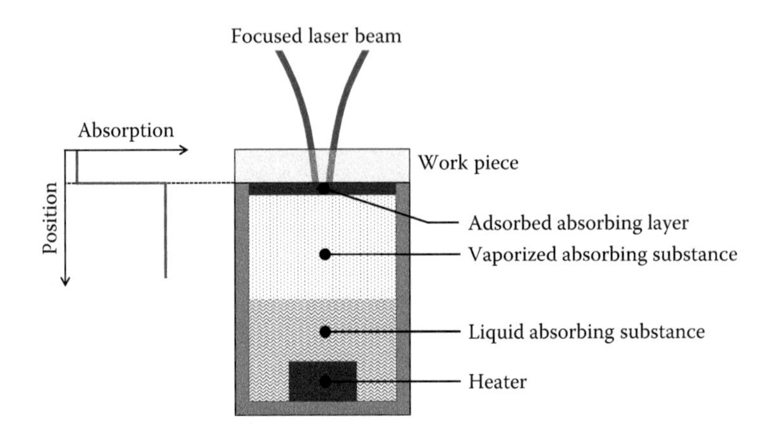


**FIGURE 13.16** Working principle of laser etching at a surface adsorbed layer (LESAL). (Adapted from Gerhard, C., *Atmospheric Pressure Plasma-Assisted Laser Ablation of Optical Glasses*, Cuvillier Verlag, Göttingen, Germany, 2014.)

2009). The laser ablation threshold can thus be determined on the basis of material properties where the thermal diffusion length in Equation 13.30 is given by

$$L_T = 2 \cdot \sqrt{\frac{\lambda}{\rho \cdot c} \cdot t}. \tag{13.31}$$

Here, $\lambda$ is the thermal conductivity of the glass or liquid, respectively, $\rho$ is its density, $c$ is its specific heat capacity, and $t$ is the time, in the present case the duration of laser irradiation.

#### 13.4.3.3 Laser Etching at a Surface Adsorbed Layer

*Laser etching at a surface adsorbed layer* (LESAL) is quite connatural to LIBWE, since near-surface increase in absorption at the work piece rear side is achieved by the adsorption of absorbing substances. As shown in Figure 13.16, these substances are filled into an open-topped vessel in the liquid state and the work piece is placed on top of this vessel.

Then, the substance is vaporized by heating and consequently deposits at the glass work piece rear side, leading to the formation of an absorbing layer. Material removal is then initiated by the incoming laser beam due to a fast heating of the adsorbed layer and the near-surface region of the work piece by local absorption of laser irradiation, the initialization of shock waves at the interface, and the desorption and decomposition of the absorbing layer. This method is mainly performed using pure toluene (Böhme and Zimmer, 2004; Zimmer et al., 2004) or carbon (Zimmer et al., 2006b) as an absorbing substance.

#### 13.4.3.4 Laser-Induced Backside Dry Etching

Similar to LIBWE and LESAL, *laser-induced backside dry etching* (LIBDE) is based on an increase in absorption at a glass work piece rear side by the deposition of absorbing layers where solid coatings such as metals are applied. Material removal is thus initiated by the coupling of incoming laser irradiation at the coated interface because of the formation of an absorbing laser-induced plasma (Zimmer et al., 2009). Suitable metals for this laser-based method include silver (Hopp et al., 2006), aluminum (Hopp et al., 2007), and tin (Hopp et al., 2008). LIBDE can also be performed using nonmetallic layer materials such as silicon monoxide (Ihlemann, 2008) or silicon suboxide ($SiO_x$, where $1 < x < 2$) (Klein-Wiele et al., 2006). This approach allows both laser backside structuring and laser front side structuring.

Another novel approach for increasing the near-surface absorption of glasses is the use of chemically active plasmas at atmospheric pressure. Such plasmas are fed by hydrogenous process gases. Within the plasma volume, hydrogen molecules are dissociated by electron impact, resulting in the formation of atomic hydrogen, which enters the glass due to its high diffusivity. As a result of the accompanying hydrolytic scission of the glass network and the chemical reduction of glass-forming oxides, optically active defects such as so-called 'E' centers (i.e., unpaired electrons within silicon dioxide tetrahedrons), nonbridging oxygen, and hydrogen centers (H-centers) are induced (Skuja, 1998). These defects give rise to an increase in absorption in the wavelength range from approximately 170–350 nm and thus support the coupling of incoming UV-laser irradiation (Gerhard et al., 2012, 2013a). The laser ablation threshold is consequently reduced, resulting in a reduction in thermal impact on the glass and an improved machining quality in terms of enhanced contour accuracy (Gerhard et al., 2014), reduced surface roughness of the laser-ablated area (Brückner et al., 2012; Hoffmeister et al., 2012), and a reduction of laser-induced debris on the glass surface (Gerhard et al., 2013b). Moreover, the ablation rate and the process efficiency, respectively, are increased in comparison to the abovementioned methods, which are based on the application of absorbing layers on the surface. The comparatively high ablation rate is due to the high penetration depth of hydrogen implanted by the plasma (Tasche et al., 2014).

#### 13.4.3.5 Laser Interference Patterning

Besides classical direct laser structuring, where either the focused laser beam is guided and controlled by so-called galvo-scanners or the work piece is moved on motorized xyz-stages, a number of different strategies are available for laser microstructuring. One approach, *laser interference patterning*, takes advantage of the wave character of light and the resulting interference phenomena (Bieda et al., 2015; Valle et al., 2015). For this purpose, a raw laser beam is split into two laser beams. As shown in Figure 13.17, both beams are then superposed on the work piece surface, where the optical axes of the beams are tilted by a certain angle $\theta$.

Due to constructive and destructive interference, the resulting laser beam intensity profile on the surface features a periodic striped pattern, where the distance between stripes of high and low intensity $\Lambda$ depends on the laser wavelength $\lambda$ and can further be adjusted by the tilt angle $\theta$ according to

$$\Lambda = \frac{\lambda}{2 \cdot \sin\theta}. \quad (13.32)$$

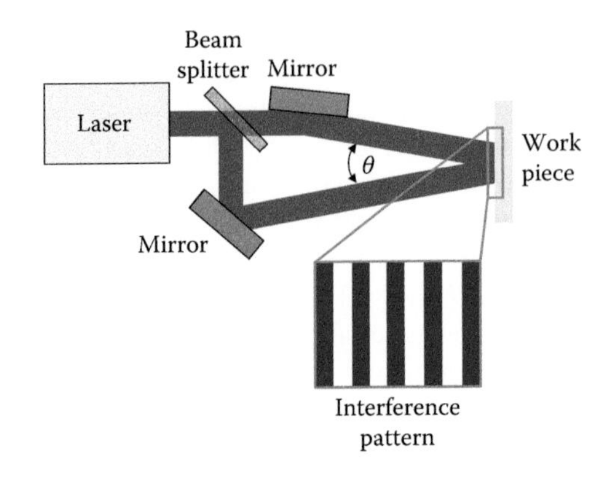


**FIGURE 13.17** Working principle of laser-interference patterning based on the superposition of laser beams at a tilt angle $\theta$. (Adapted from Bieda, M. et al., *Advanced Engineering Materials*, 17, 102–108, 2015.)

Once material removal by such superposed laser beams has occurred, the striped intensity pattern is transferred into the work piece surface, allowing the realization of surface structures, such as gratings, where the absolute lateral structure size corresponds to $\Lambda$ (i.e., the length of a line pair).

### 13.4.4 Embossing Methods

As presented in Section 7.2.1, molding is applied for shaping or preforming of optical components. However, this technique is also suitable for the production of finished and polished optics, where surface polishing is achieved by flame polishing due to the high temperature of the molds, compression molding dies, and pressing tools. In this case, the process is referred to as hot embossing or precision molding. This allows for the realization of complex-shaped (micro-)optical components such as aspherical lenses or lens arrays, Fresnel lenses, gratings, and DOEs at high quantities. The surface quality of such elements is lower than for classically manufactured optics.

#### 13.4.4.1 Hot Embossing

For *hot embossing*, a certain volume of glass is heated up to its processing temperature (see Section 3.2.2.2) and shaped by pressing in a microstructured mold as shown in Figure 13.18. Subsequently, the glass work piece is cooled, and the mold is removed; shaping is thus due to material displacement and compression. The latter can cause a change of the optical properties of the used glass. Another challenging aspect is material-dependent shrinking of the glass during the cooling process, resulting in a deviation from the target geometry. This issue has to be considered during the design of the tool (i.e., the mold) in order to compensate for such shrinking by appropriate allowances.

Hot embossing represents an alternative method for the production of complex or even free-form surfaces, which cannot be produced by classical manufacturing processes. It is also an economic method for high-volume business, for example, the manufacture of optical components for consumer markets. Against this background,

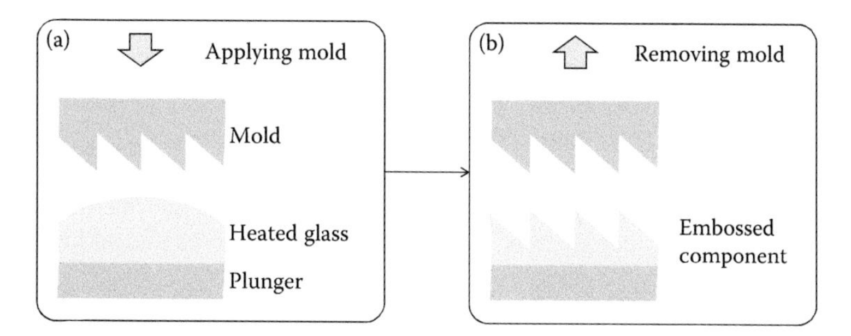


**FIGURE 13.18** Principle of hot embossing: a certain volume of heated glass is embossed by a structured mold (a), and the mold structure is transferred into the glass surface (b).

it also allows the use of optical plastics instead of glasses for low-cost lenses as for example for single-use cameras.

### 13.4.4.2 UV-Reactive Injection Molding

Microstructuring by embossing can also be applied to UV-curing optical liquids (i.e., resins of high viscosity). This approach is comparable to spray cast processes and is thus referred to as *UV-reactive injection molding* (UV-RIM). Here, at least one of the used tools, either the mold or the die, is transparent for the wavelength of light required for curing the optical liquid. As shown in Figure 13.19, the clear space

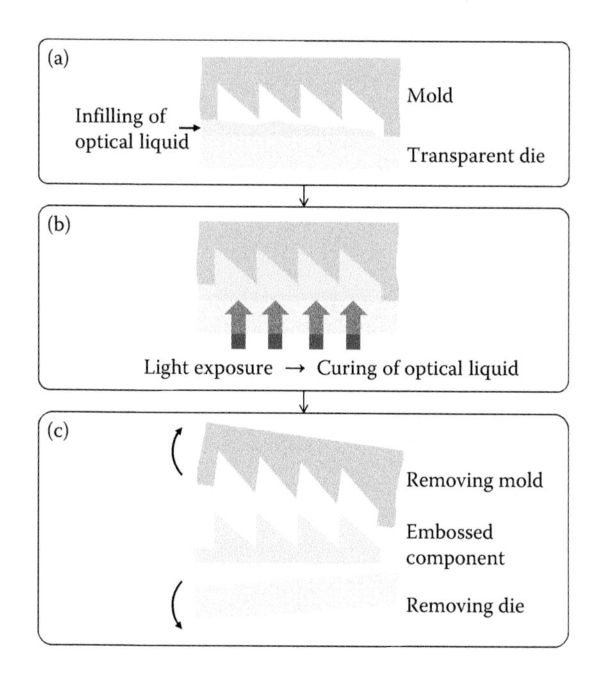


**FIGURE 13.19** Process steps of UV-reactive injection molding (UV-RIM): an optical liquid is filled between a mold and a transparent die (a) and subsequently cured by light exposure (b). The used mold and die are finally removed and the embossed component is taken out (c).

between the mold and the die is filled with the optical liquid, which consequently takes the shape of the clear space.

Curing and hardening of the work piece is then achieved by light exposure through the transparent tool component. Finally, the work piece is retrieved from the forming tool where, in some cases, mold release agents may be applied prior to infilling of the liquid in order to avoid adhesion to the tool after curing. Shrinking of the used liquid, usually copolymers, further has to be considered and addressed with appropriate allowances for the tools.

#### 13.4.4.3 Nanoimprint Lithography

For *nanoimprint lithography*, a substrate is coated with a UV-curing polymer such as photoresist by spin coating in order to realize a uniform layer thickness. Consequently, a microstructured die is pressed into this polymer, which is then exposed to UV-light and cured in this way. Finally, the die is removed.

Nanoimprint lithography allows the mechanical transfer of a structure into a photoresist; it could be applied as a preliminary preparation step for lithography by etching as presented in Section 13.4.2. Even though this method is fast and precise, the used microstructured dies wear after some thousands of cycles and have to be exchanged due to a successive degradation of surface geometry.

### 13.4.5 Microjet Printing

For *microjet printing*, droplets of optical liquids such as fine cement are deposited on a substrate by pouring, roller stamping, or printing. As long as the droplet volume of the applied liquid is sufficiently small, a spherical droplet shape is formed by self-organization due to the surface tension of the liquid. After the actual pouring or printing process, the droplets are cured by heat or UV irradiation and finally annealed by tempering, where necessary, depending on the type of liquid. A selection of suitable liquids as reported in the literature is listed in Table 13.5.

**TABLE 13.5**
**Selection of Liquids Suitable for Microjet Printing as Reported in the Literature**

| Liquid | Deposition Process | Reference |
|---|---|---|
| Thioether methacrylate (TEMA) | Pouring | Okamoto et al., 1999 |
| Polydimethylsiloxane (PDMS) | Roller stamping | Chang et al., 2007 |
| Polydimethylsiloxane (PDMS) | Inkjet printing | Sung et al., 2015 |
| (Dimethylsiloxane)epoxypropoxypropyl terminated (DMS-DGE), polymerized with diamine:1.3bis(aminomethyl) cyclohexane (BAC) | Printing by microcantilever spotter system | Bardinal et al., 2007 |
| Two-component epoxy resin | Printing by microplotter system | Zang et al., 2014 |

The radius of curvature and focal length, respectively, can be varied by controlling the contact time of the printer or by changing the contact angle between the liquid and the substrate. The latter can be carried out by applying appropriate coatings to the substrate prior to droplet deposition (Hartmann et al., 2000), controlling temperature during cooling (Sung et al., 2015), or adjusting the surface tension coefficient between the substrate and the liquid by varying the temperature and viscosity, respectively, of the liquid (Kim et al., 2003). Microjet printing finally allows the realization of aspherically shaped lens surfaces (Lee et al., 2014) when using the gravity-induced mechanism of self-organization of hanging droplets.

### 13.4.6 Thermal Reflow Method

The first step performed during the *thermal reflow process* for realizing microoptical patterns such as lens arrays is quite comparable to the first process step of photolithography (see Section 13.4.2). A substrate is coated with a photoresist, and the structure of a mask is transferred into the photoresist by light exposure, where the incident light gives rise to a chemical modification of irradiated photoresist. This modified resist is then removed with developers, resulting in a rectangular surface profile as shown in Figure 13.20.

Subsequent heating of the work piece results in melting of the photoresist, which forms a spherical surface due to its surface tension. After cooling and hardening, a microlens pattern is generated. Finally, the generated surface structure can be transferred into the substrate by etching processes in order to obtain a monolithic microoptical component.

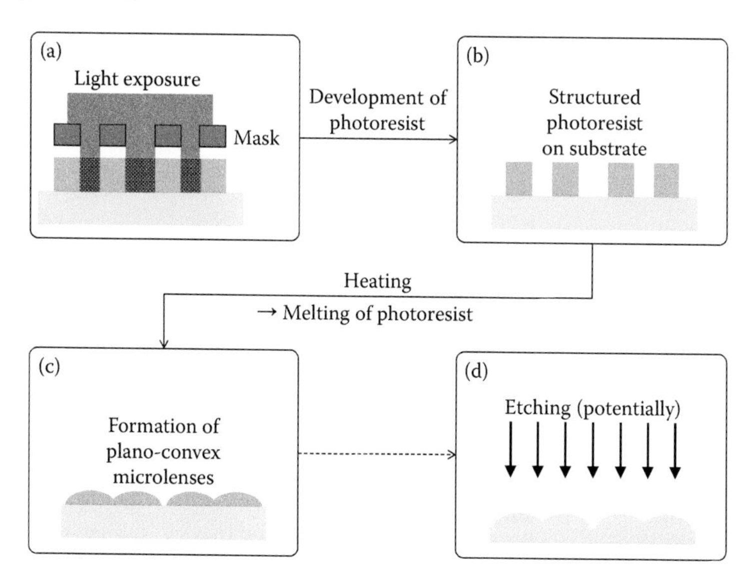


**FIGURE 13.20** Process steps of the thermal reflow method; a photoresist layer is patterned by light exposure and development (a-b) and subsequently molten, resulting in the formation of microlenses due to surface tension of the molten resist (c). If necessary, the photoresist structure can be transferred into the substrate by subsequent etching (d).

### 13.4.7 Ion Exchange

*Ion exchange* allows the realization of gradient index (GRIN) materials and lenses as already presented in Section 3.4. This method is based on diffusion (i.e., a nondirectional random walk of particles due to a difference in concentration). The mass transport of particles, ions in the present case, then occurs from zones of high concentration toward zones of low concentration. This behavior is generally described by Fick's first law, named after the German physicist *Adolf Eugen Fick* (1829–1901) and given by

$$J = \frac{1}{A} \cdot \frac{dN}{dt} = -D \cdot \frac{\partial c}{\partial x}. \tag{13.33}$$

Here, $J$ is the particle flux (or ion flux in ion diffusion), $A$ is the area where diffusion occurs, $N$ is the number of particles or ions, $t$ is time, $D$ is the diffusion coefficient, and the term $\partial c/\partial x$ denotes the concentration gradient. The intensity of diffusion can thus be determined on the basis of the diffusion coefficient. Some examples of ion exchanges typically used for the production of GRIN lenses and their particular diffusion coefficients are listed in Table 13.6.

In practice, such ion exchange is achieved by placing a glass in a dipping bath, usually a salt bath that contains the required ions. The gradient in index of refraction, following from the exchange of ions from the salt bath into the glass and from the glass into the salt bath, respectively, is controlled by adjusting the temperature[14] and the dwell time of the glass in the bath. The mean depth of penetration of ions $d_{dif}$ into the glass (a.k.a. characteristic diffusion depth) is thus both temperature- and time-dependent and can be estimated by

$$d_{\text{dif}} \approx 2 \cdot \sqrt{D \cdot t}. \tag{13.34}$$

Ion exchange in salt baths allows realizing GRIN lenses in the form of rotational-symmetric rods. For the production of microlens arrays, droplets of salt solution are applied on a substrate by inkjet printers, microcantilever spotter systems, or

**TABLE 13.6**
**Examples for Diffusion Coefficients during Ion Exchange in Glasses**

| Glass | Ion Exchange | Diffusion Coefficient in cm²/s |
|---|---|---|
| Alumina silicate | Silver for sodium ($Ag^+$ for $Na^+$) | $0.2–9 \cdot 10^{-7}$ |
| | Lithium for sodium ($Li^+$ for $Na^+$) | $0.59–1.1 \cdot 10^{-7}$ |
| | Sodium for lithium ($Na^+$ for $Li^+$) | $0.42–0.43 \cdot 10^{-7}$ |
| Soda alumina silicate | Thallium for sodium ($Tl^+$ for $Na^+$) | $0.3–8.6 \cdot 10^{-7}$ |
| | Copper for sodium ($Cu^+$ for $Na^+$) | $4.7–6.4 \cdot 10^{-7}$ |

*Source:* Visconti, A.J., and Bentley, J.L., *Optical Engineering*, 52, 112103.

[14] Controlling the temperature during ion exchange processes is of significant importance since the diffusion coefficient is a temperature-dependent parameter.

microplotters. Here, a further effect, swelling, is applied additionally. Such swelling is achieved by the choice of implanted ions, which feature a higher volume than the ions removed from the glass substrate. Swelling finally results in the formation of a curvature on the substrate surface, where the shape of curvature is given by the distribution of implanted ions. By this modification of the initially plane substrate surface to a slightly convex one, an additional lens effect is obtained. This procedure allows reducing the focal length and increasing the numerical aperture of GRIN microlens arrays.

### 13.4.8 3D Printing

*3D printing* represents the newest approach for the production of microlenses and microoptical components. This method is based on two-photon absorption of femtosecond laser irradiation within liquid photoresists, resulting in its curing and hardening. Due to the small laser spot sizes, complex microoptical systems with diameters of approximately 50 µm, and accuracies in the submicrometer range can be realized. At the time of printing for the present book, the feasibility of this technique was just reported in the literature (Gissibl et al., 2016a,b), but it turns out to be a powerful and potential approach for miniaturization of optical systems in the near future.

## 13.5 SUMMARY

Optical components with lateral dimensions in the order of magnitude of the wavelength of light are referred to as microoptical elements. Such elements can be evaluated and specified on the basis of different parameters such as the Strehl-ratio or the Maréchal-criterion, where diffraction effects are considered. For the manufacture, different techniques are available: dry etching can be classified into pure physical, pure chemical, or combined physical-chemical processes. These processes are applied for microstructuring via photolithography. Moreover, different laser-based techniques can be applied where the poor near-surface absorption of optical glasses is overcome by appropriate absorbing coatings. Moreover, embossing techniques such as hot embossing, reactive injection molding, or nanoimprint lithography, as well as microjet printing and the thermal reflow method, allow for the fabrication of microlens arrays or gratings. These techniques can be used for the realization or transfer of any structure into polymer coatings on glass substrates, for example, prior to etching. In contrast, GRIN materials are produced by a gradient-like modification of the glass bulk material due to diffusion and the exchange of ions from the glass and a surrounding medium, usually a salt bath. Finally, 3D printing based on two-photon absorption of femtosecond laser irradiation in optical liquids represents a novel and promising approach for the manufacture of complex microoptical systems.

## 13.6 FORMULARY AND MAIN SYMBOLS AND ABBREVIATIONS

**Airy disc diameter $D_{\text{Airy}}$:**

$$D_{\text{Airy}} = 2.4392 \cdot \frac{\lambda \cdot f}{2 \cdot a}$$

$\lambda$ wavelength
$f$ focal length
$a$ half diameter of aperture stop

**Talbot length $z_T$ (general description):**

$$z_T = \frac{\lambda}{1 - \sqrt{1 - \frac{\lambda^2}{(2 \cdot a)^2}}}$$

$\lambda$ wavelength
$a$ half diameter of aperture stop

**Talbot length $z_T$ (special case, a >> $\lambda$):**

$$z_T \approx \frac{2 \cdot (2 \cdot a)^2}{\lambda}$$

$\lambda$ wavelength
$a$ half diameter of aperture stop

**Fresnel number $F$:**

$$F = \frac{a^2}{d \cdot \lambda}$$

$a$ half diameter of aperture stop
$d$ distance between optical element and detector
$\lambda$ wavelength

**Strehl-ratio $S$:**

$$S = \frac{I_{\text{real}}}{I_{\text{theo}}}$$

$I_{\text{real}}$ maximum intensity of real image disc
$I_{\text{theo}}$ maximum intensity of Airy disc intensity pattern

**Michelson contrast $C_M$ (a.k.a. modulation $M$):**

$$C_M = M = \frac{I_{\max} - I_{\min}}{I_{\max} + I_{\min}}$$

$I_{\max}$ intensity of bright object areas
$I_{\min}$ intensity of dark object areas

**Modulation transfer function $MTF$:**

$$MTF(R) = \frac{M_{\text{image}}(R)}{M_{\text{object}}(R)}$$

$M_{\text{image}}$ modulation of image
$M_{\text{object}}$ modulation of object
$R$ spatial frequency

**Cut-off-frequency $f_{\text{cut-off}}$ (where $MTF(R)=0$):**

$$f_{\text{cut-off}} = \frac{1}{\arctan\left(\frac{\lambda}{2 \cdot a}\right)}$$

$\lambda$ wavelength
$a$ half diameter of aperture stop

**Rayleigh-criterion (definition of wave front diffraction limit):**

$$|\Psi_{\max}(x)| \leq \frac{\lambda}{4}$$

$\Psi_{\max}$ peak-to-valley wave front deviation
$\lambda$ wavelength

**Maréchal-criterion (definition of wave front diffraction limit):**

$$\Psi_{\text{RMS}} = \sqrt{\int |\Psi(x)|^2 \, dx - \left[\int |\Psi(x)| \, dx\right]^2} \leq \frac{\lambda}{14}$$

$\Psi_{\text{RMS}}$ the root-mean-squared phase error
$\Psi(x)$ position-dependent phase error
$\lambda$ wavelength

**Effective back focal length $EFL_{\text{back}}$ of a spherical microlens in air:**

$$EFL_{\text{back}} = \frac{R}{n-1} - \frac{t_c}{n}$$

$R$ radius of curvature
$n$ index of refraction
$t_c$ lens center thickness

**Effective front focal length $EFL_{\text{front}}$ of a spherical microlens in air:**

$$EFL_{\text{front}} = \frac{R}{n-1} = \frac{\left(S + \frac{a^2}{S}\right)}{2 \cdot (n-1)} \approx \frac{a^2}{S}$$

$R$ radius of curvature
$n$ index of refraction
$S$ lens sagitta
$a$ half diameter of lens

**Radius of curvature $R$ of a microlens in air:**

$$R = \frac{S \cdot (e+1)}{2} + \frac{a^2}{2 \cdot S}$$

$S$ lens sagitta
$e$ conus constant
$a$ half diameter of lens

**Sagitta of a microlens in air:**

$$S = R - \sqrt{R^2 - a^2}$$

$R$ radius of curvature
$a$ half diameter of lens

**Fresnel number $F$ of a microlens:**

$$F = \frac{a^2}{\lambda \cdot EFL_{\text{back}}}$$

$a$ half diameter of lens
$\lambda$ wavelength
$EFL_{\text{back}}$ effective back focal length

**Focal length $f$ of a GRIN lens:**

$$f = \frac{1}{n_0 \cdot g \cdot \sin(g \cdot l)}$$

$n_0$ index of refraction at center
$g$ geometrical gradient constant
$l$ geometric length of lens

**Geometric length of a GRIN lens:**

$$l = \frac{2 \cdot \pi \cdot P}{g}$$

$P$ lens pitch
$g$ geometrical gradient constant

**Working distance $s$ of a GRIN lens (a.k.a. BFL):**

$$s = \frac{1}{n_0 \cdot g \cdot \tan(g \cdot l)}$$

$n_0$ index of refraction at center
$g$ geometrical gradient constant
$l$ geometric length of lens

**Size of the homogeneous field $D_{\text{hf}}$ (nonimaging laser beam homogenizer):**

$$D_{\text{hf}} = \frac{P_{\text{la}} \cdot f_{\text{Fl}}}{f_{\text{la}}}$$

$P_{\text{la}}$ pitch of lens array
$f_{\text{la}}$ focal length of lens array
$f_{\text{Fl}}$ focal length of Fourier lens

**Size of the homogeneous field $D_{hf}$ (imaging laser beam homogenizer):**

$$D_{hf} = P_{la1} \cdot \frac{f_{Fl}}{f_{la1} \cdot f_{la2}} \cdot \left[ \left( f_{la1} + f_{la2} \right) - d \right]$$

$P_{la1}$ pitch of first lens array
$f_{Fl}$ focal length of Fourier lens
$f_{la1}$ focal length of first lens array
$f_{la2}$ focal length of second lens array
$d$ distance between lens arrays

**Blaze angle $\theta_B$ of an echelle grating:**

$$\theta_B = \arcsin \frac{m \cdot \lambda}{2 \cdot g}$$

$m$ diffraction order
$\lambda$ wavelength
$g$ grating constant

**Velocity of ions $v$ (during physical etching):**

$$v = \sqrt{\frac{2 \cdot z \cdot e \cdot U}{m}}$$

$m$ ion mass
$z$ ion charge
$e$ elementary electric charge
$U$ ion acceleration voltage

**Kinetic energy $E_{kin}$ of accelerated ions (during physical etching):**

$$E_{kin} = \frac{1}{2} \cdot m \cdot v^2 = z \cdot e \cdot U$$

$m$ ion mass
$v$ ion velocity
$z$ ion charge
$e$ elementary electric charge
$U$ ion acceleration voltage

**Removal rate $N(t)$ (during physical etching):**

$$N(t) = N_0 \cdot \left( 1 - e^{\left( -\frac{Y \cdot I_p \cdot t}{e \cdot N_0} \right)} \right)$$

$N_0$ number of particles on the substrate surface
$Y$ sputter yield
$I_p$ primary current
$t$ sputter time

**Debye length $\lambda_D$ of a plasma:**

$$\lambda_D = \sqrt{\frac{\varepsilon_0 \cdot k_B \cdot T}{n_e \cdot e^2}}$$

$\varepsilon_0$ vacuum permittivity
$k_B$ Boltzmann constant
$T$ temperature
$n_e$ electron density
$e$ elementary electric charge

**Optical depth of penetration $d_{opt}$ (e.g., for laser irradiation):**

$$d_{opt} = \frac{1}{\alpha}$$

$\alpha$ absorption coefficient

**Laser ablation threshold $F_{th}$ for LIBWE:**

$$F_{th} = \left(\rho_g \cdot L_{T,g} \cdot c_g + \rho_l \cdot L_{T,l} \cdot c_l\right) \cdot \left(T_m - T_0\right)$$

$\rho_g$ density of machined glass
$L_{T,g}$ thermal diffusion length of machined glass
$c_g$ specific heat capacity of machined glass
$\rho_l$ density of used absorbing liquid
$L_{T,l}$ thermal diffusion length of used absorbing liquid
$c_l$ specific heat capacity of used absorbing liquid
$T_m$ melting temperature of machined glass
$T_0$ initial process temperature (normally room ambient temperature)

**Thermal diffusion length $L_T$:**

$$L_T = 2 \cdot \sqrt{\frac{\lambda}{\rho \cdot c} \cdot t}$$

$\lambda$ thermal conductivity
$\rho$ density
$c$ specific heat capacity
$t$ time (e.g., duration of laser irradiation)

**Lateral structure size $\Lambda$ produced by laser interference patterning:**

$$\Lambda = \frac{\lambda}{2 \cdot \sin\theta}$$

$\lambda$ wavelength
$\theta$ tilt angle of superposed laser beams

**Fick's first law (of diffusion):**

$$J = \frac{1}{A} \cdot \frac{dN}{dt} = -D \cdot \frac{\partial c}{\partial x}$$

$J$ particle flux
$A$ area
$N$ number of particles
$t$ time
$D$ diffusion coefficient
$\partial c / \partial x$ concentration gradient

**Mean depth of penetration $d_{dif}$ during diffusion:**

$$d_{dif} \approx 2 \cdot \sqrt{D \cdot t}$$

$D$ diffusion coefficient
$t$ time

## REFERENCES


Bardinal, V., Daran, E., Leïchlé, T., Vergnenègre, C., Levallois, C., Camps, T., Conedera, V., Doucet, J.B., Carcenac, F., Ottevaere, H., and Thienpont, H. 2007. Fabrication and characterization of microlens arrays using a cantilever-based spotter. *Optics Express* 15:6900–6907.

Bich, A., Rieck, J., Dumouchel, C., Roth, S., Weible, K.J., Eisner, M., Voelkel, R., Zimmermann, M., Rank, M., Schmidt, M., Bitterli, R., Ramanan, N., Ruffieux, P., Scharf, T., Noell, W., Herzig, H.P., and de Rooij, N. 2008. Multifunctional micro-optical elements for laser beam homogenizing and beam shaping. *Proceedings of SPIE* 6879:68790Q-1.

Bieda, M., Schmädicke, C., Roch, T., and Lasagni, A. 2015. Ultra-low friction on 100Cr6-steel surfaces after direct laser interference patterning. *Advanced Engineering Materials* 17:102–108.

Böhme, R., and Zimmer, K. 2004. Low roughness laser etching of fused silica using an adsorbed layer. *Applied Surface Science* 239:109–116.

Bradley, R.M., and Harper, J.M.E. 1988. Theory of ripple topography induced by ion bombardment. *Journal of Vacuum Science and Technology A* 6:2390.

Brückner, S., Hoffmeister, J., Ihlemann, J., Gerhard, C., Wieneke, S., and Viöl, W. 2012. Hybrid laser-plasma micro-structuring of fused silica based on surface reduction by a low-temperature atmospheric pressure plasma. *Journal of Laser Micro/Nanoengineering* 7:73–76.

Chang, C.-Y., Yang, S.-Y., and Chu, M.-H. 2007. Rapid fabrication of ultraviolet-cured polymer microlens arrays by soft roller stamping process. *Microelectronic Engineering* 84:355–361.

Ding, X., Kawaguchi, Y., Niino, H., and Yabe, A. 2002. Laser-induced high-quality etching of fused silica using a novel aqueous medium. *Applied Physics A* 75:641–645.

Frank, M., Kufner, M., Kufner, S., and Testorf, M. 1991. Microlenses in polymethyl methacrylate with high relative aperture. *Applied Optics* 30:2666–2667.

Gerhard, C. 2014. *Atmospheric Pressure Plasma-Assisted Laser Ablation of Optical Glasses.* Göttingen, Germany: Cuvillier Verlag.

Gerhard, C., Dammann, M., Wieneke, S., and Viöl, W. 2014. Sequential atmospheric pressure plasma-assisted laser ablation of photovoltaic cover glass for improved contour accuracy. *Micromachines* 5:408–419.

Gerhard, C., Heine, J., Brückner, S., Wieneke, S., and Viöl, W. 2013a. A hybrid laser-plasma ablation method for improved nanosecond laser machining of heavy flint glass. *Lasers in Engineering* 24:391–403.

Gerhard, C., Tasche, D., Brückner, S., Wieneke, S., and Viöl, W. 2012. Near-surface modification of optical properties of fused silica by low-temperature hydrogenous atmospheric pressure plasma. *Optics Letters* 37:566–568.

Gerhard, C., Weihs, T., Tasche, D., Brückner, S., Wieneke, S., and Viöl, W. 2013b. Atmospheric pressure plasma treatment of fused silica, related surface and near-surface effects and applications. *Plasma Chemistry and Plasma Processing* 33:895–905.

Gissibl, T., Thiele, S., Herkommer, A., and Giessen, H. 2016a. Sub-micrometre accurate free-form optics by three-dimensional printing on single-mode fibres. *Nature Communications* 7:11763.

Gissibl, T., Thiele, S., Herkommer, A., and Giessen, H. 2016b. Two-photon direct laser writing of ultracompact multi-lens objectives. *Nature Photonics* 10:554–560.

Hanada, Y., Sugioka, K., Obata, K., Garnov, S.V., Miyamoto, I., and Midorikawa, K. 2006. Transient electron excitation in laser-induced plasma-assisted ablation of transparent materials. *Journal of Applied Physics* 99:043301.

Hartmann, D.M., Kibar, O., and Esener, S.C. 2000. Characterization of a polymer microlens fabricated by use of the hydrophobic effect. *Optics Letters* 25:975–977.

Haruna, M., Takahashi, M., Wakahayashi, K., and Nishihara, H. 1990. Laser beam lithographed micro-Fresnel lenses. *Applied Optics* 29:5120–5126.

Hoffmeister, J., Gerhard, C., Brückner, S., Ihlemann, J., Wieneke, S., and Viöl, W. 2012. Laser micro-structuring of fused silica subsequent to plasma-induced silicon suboxide generation and hydrogen implantation. *Physics Procedia* 39:613–620.

Hopp, B., Smausz, T., and Bereznai, M. 2007. Processing of transparent materials using visible nanosecond laser pulses. *Applied Physics A* 87:77–79.

Hopp, B., Smausz, T., Csizmadia, T., Budai, J., Oszkó, A., and Szabó, G. 2008. Laser-induced back-side dry etching: wavelength dependence. *Journal of Physics D* 41:175501.

Hopp, B., Smausz, T., Vass, C., Szabó, G., Böhme, R., Hirsch, D., and Zimmer, K. 2009. Laser-induced backside dry and wet etching of transparent materials using solid and molten tin as absorbers. *Applied Physics A* 94:899–904.

Hopp, B., Vass, C., Smausz, T., and Bor, Z. 2006. Production of submicrometre fused silica gratings using laser-induced backside dry etching technique. *Journal of Physics D* 39:4843–4847.

Huang, Z.Q., Hong, M.H., Tiaw, K.S., and Lin, Q.Y. 2007. Quality glass processing by laser induced backside wet etching. *Journal of Laser Micro/Nanoengineering* 2:194–199.

Ihlemann, J. 2008. Micro patterning of fused silica by laser ablation mediated by solid coating absorption. *Applied Physics A* 93:65–68.

Kim, K.-R., Chang, S., and Oh, K. 2003. Refractive microlens on fiber using UV-curable fluorinated acrylate polymer by surface-tension. *IEEE Photonics Technology Letters* 15:1100–1102.

Klein-Wiele, J.-H., Békési, J., Simon, P., and Ihlemann, J. 2006. Fabrication of $SiO_2$ phase gratings by UV laser patterning of silicon suboxide layers and subsequent oxidation. *Journal of Laser Micro/Nanoengineering* 1:211–214.

Lee, W.M., Upadhya, A., Reece, P.J., and Phan, T.G. 2014. Fabricating low cost and high performance elastomer lenses using hanging droplets. *Biomedical Optics Express* 5:1626–1635.

MacFarlane, D.L., Narayan, V., Tatum, J.A., Cox, W.R., Chen, T., and Hayes, D.J. 1994. Microjet fabrication of microlens arrays. *IEEE Photonics Technology Letters* 6:1112–1114.

Mihailov, S., and Lazare, S. 1993. Fabrication of refractive microlens arrays by excimer laser ablation of amorphous Teflon. *Applied Optics* 32:6211–6218.

Nussbaum, P., Völkel, R., Herzig, H.P., Eisner, M., and Haselbeck, S. 1997. Design, fabrication and testing of microlens arrays for sensors and microsystems. *Pure and Applied Optics: Journal of the European Optical Society Part A* 6:617–636.

Oikawa, M., Iga, K., Sanada, T., Yamamoto, N., and Nishizawa, K. 1981. Array of distributed-index planar micro-lenses prepared from ion exchange technique. *Japanese Journal of Applied Physics* 20:L296–L298.

Okamoto, T., Mori, M., Karasawa, T., Hayakawa, S., Seo, I., and Sato, H. 1999. Ultraviolet-cured polymer microlens arrays. *Applied Optics* 38:2991–2996.

Ottevaere, H., Cox, R., Herzig, H.P., Miyashita, T., Naessens, K., Taghizadeh, M., Völkel, R., Woo, H.J., and Thienpont, H. 2006. Comparing glass and plastic refractive microlenses fabricated with different technologies. *Journal of Optics A: Pure and Applied Optics* 8:S407–S429.

Pearson, A.D. 1972. Ion beam polishing of glass. *Materials Research Bulletin* 7:567–571.

Popovic, Z.D., Sprague, R.A., and Connell, G.A.N. 1988. Technique for monolithic fabrication of microlens arrays. *Applied Optics* 27:1281–1284.

Skuja, L. 1998. Optically active oxygen-deficiency-related centers in amorphous silicon dioxide. *Journal of Non-Crystalline Solids* 239:16–48.

Stephen, A., and Vollertsen, F. 2013. Compact machining module for laser chemical manufacturing. *Production Engineering Research and Development* 7:541–545.

Stern, M.B., and Rubico Jay, T. 1994. Dry etching for coherent refractive microlens arrays. *Optical Engineering* 33:3547–3551.

Stognij, A.I., Novitskii, N.N., and Stukalov, O.M. 2002. Nanoscale ion beam polishing of optical materials. *Technical Physics Letters* 28:17–20.

Sugioka, K., and Midorikawa, K. 2001. Novel technology for laser precision microfabrication of hard materials. *RIKEN Review* 32:36–42.

Sung, Y.-L., Jeang, J., Lee, C.-H., and Shih, W.-C. 2015. Fabricating optical lenses by ink-jet printing and heat-assisted in situ curing of polydimethylsiloxane for smartphone microscopy. *Journal of Biomedical Optics* 20:047005.

Talbot, H.F. 1836. Facts relating to optical science. *The London and Edinburgh Philosophical Magazine and Journal of Science* 9:401–407.

Tasche, D., Gerhard, C., Ihlemann, J., Wieneke, S., and Viöl, W. 2014. The impact of O/Si ratio and hydrogen content on ArF excimer laser ablation of fused silica. *Journal of the European Optical Society—Rapid Publications* 9:14026.

Valle, J., Burgui, S., Langheinrich, D., Gil, C., Solano, C., Toledo-Arana, A., Helbig, R., Lasagni, A., and Lasa, I. 2015. Evaluation of surface microtopography engineered by direct laser interference for bacterial anti-biofouling. *Macromolecular Bioscience* 15:1060–1069.

Visconti, A.J., and Bentley, J.L. 2013. Fabrication of large-diameter radial gradient-index lenses by ion exchange of Na+ for Li+ in titania silicate glass. *Optical Engineering* 52:112103.

Wang, J., Niino, H., and Yabe, A. 1999. One-step microfabrication of fused silica by laser ablation of an organic solution. *Applied Physics A* 68:111–113.

Wilson, S.J., and Hutley, M.C. 1982. The optical properties of "moth eye" antireflection surfaces. *Optica Acta* 29:993–1009.

Zang, Z., Tang, X., Liu, X., Lei, X., and Chen, W. 2014. Fabrication of high quality and low cost microlenses on a glass substrate by direct printing technique. *Applied Optics* 53:7868–7871.

Zhang, J., Sugioka, K., and Midorikawa, K. 1998a. Direct fabrication of microgratings in fused quartz by laser-induced plasma-assisted ablation with a KrF excimer laser. *Optics Letters* 23:1486–1488.

Zhang, J., Sugioka, K., and Midorikawa, K. 1998b. Laser-induced plasma-assisted ablation of fused quartz using the fourth harmonic of a Nd+:YAG laser. *Applied Physics A* 67:545–549.

Zhang, J., Sugioka, K., and Midorikawa, K. 1999. High-quality and high-efficiency machining of glass materials by laser-induced plasma-assisted ablation using conventional nanosecond UV, visible, and infrared lasers. *Applied Physics A* 69:S879–S882.

Zimmer, K., and Böhme, R. 2008. Laser-induced backside wet etching of transparent materials with organic and metallic absorbers. *Laser Chemistry* 2008:170632.

Zimmer, K., Böhme, R., and Rauschenbach, B. 2004. Laser etching of fused silica using an adsorbed toluene layer. *Applied Physics A* 79:1883–1885.

Zimmer, K., Böhme, R., and Rauschenbach, B. 2006a. Enhancing the etch rate at backside etching of fused silica. *Journal of Laser Micro/Nanoengineering* 1:292–296.

Zimmer, K., Böhme, R., and Rauschenbach, B. 2006b. Laser backside etching of fused silica due to carbon layer ablation. *Applied Physics A* 82:325–328.

Zimmer, K., Böhme, R., Vass, C., and Hopp, B. 2009. Time-resolved measurements during backside dry etching of fused silica. *Applied Surface Science* 255:9617–9621.

Zimmer, K., Braun, A., and Böhme, R. 2003. Etching of fused silica and glass with excimer laser at 351 nm. *Applied Surface Science* 208–209:199–204.

# 14 Cleaning

## 14.1 INTRODUCTION

Cleaning represents the final step of optics manufacturing but is also of essential importance after each manufacturing step. As an example, insufficient cleaning after roughing or lapping may cause severe scratches during subsequent fine lapping or polishing due to remaining comparatively large abrasive grains. Moreover, residues of working materials such as raw cement, oil-based coolants and lubricants, abrasive grains, tool debris, polishing agents, and defoamers[1] may cause stains at the surface and consequently impact the further processing or final function of an optical component. For instance, contamination can give rise to increased surface absorption and reduced laser-induced damage threshold, respectively (Génin et al., 1997; Neauport et al., 2005; Bude et al., 2014). In order to avoid such effects, surface cleanliness is critical and described in more detail in Section 6.3.3, where not only scratches and digs, but also stains are covered by DIN ISO 10110 (index number "5").

Even after cleaning, optics surfaces can feature carbonaceous contamination by residues from cleaning agents such as acetone or ethanol (Gerhard et al., 2013). Moreover, storing may lead to further surface contamination by adsorption of organic contaminations from the ambient air in common environmental conditions (Langmuir, 1918). Such contaminations might cause a reduction of the adhesion strength and long-term stability of coatings and fine cement applied to contaminated surfaces or induce glass corrosion.

In this chapter, different methods and approaches for classical cleaning and precision cleaning are introduced.

## 14.2 CLEANING METHODS

### 14.2.1 Classical Cleaning

Classical cleaning is usually performed by wet chemical methods. First, coarse residues and contaminants from the manufacturing process are removed by soaking the work piece in adequate solvents such as ethanol or acetone. Second, the work piece is wiped manually with cloths or lens-cleaning paper several times until no visible contaminations can be observed. Third, the work piece is usually finally cleaned successively in different *ultrasonic baths*, where different solvents as listed in Table 14.1 and mixtures of solvents are used as cleaning fluids. The composition of the cleaning fluids used here is adapted to the type of glass in terms of chemical resistance (see Section 6.2.4.1), and actual cleaning is achieved by both a chemical and mechanical impact of the fluid on the surface. In the latter case, ultrasonic pulses

[1] The contaminations mentioned here may also accumulate within scratches or the silica gel layer formed during polishing, making removal difficult or even impossible.

**TABLE 14.1**
**Solvents Used in Optical Manufacturing and Cleaning of Optics**

| Solvent | Total Formula |
|---|---|
| Acetone | $(CH_3)_2CO$ |
| Ethanol | $C_2H_5OH$ |
| Ethyl acetate | $C_2H_5COOCH_3$ |
| Isopropyl alcohol | $(CH_3)_2CHOH$ |
| Methyl ethyl ketone | $CH_3COC_2H_5$ |
| Methylene chloride | $CH_2Cl_2$ |
| Toluene | $C_6H_5CH_3$ |
| Trichloroethane | $Cl_3$:$CH_3$ |
| Trichlorotrifluoroethane | $Cl_2F$.$ClF_2$ |
| Xylenes | $C_6H_4(CH_3)_2$ |

are generated by an ultrasonic transmitter and propagate in the form of a longitudinal wave within the liquid. The negative pressure regions of these waves give rise to the formation of small cavities on the work piece surface, which are then condensed by subsequently incident high-pressure amplitudes of the ultrasonic wave. As a result, the surface is affected by a pressure impulse, and contaminants are removed mechanically (Posth et al., 2012).

### 14.2.2 Precision Cleaning

For some applications such as laser or ultraviolet optics, higher grades of surface cleanliness than specified by DIN ISO 10110 may be required, and even residues of cleaning agents can directly impact the laser-induced damage threshold or long-term stability of the used optical components. High surface cleanliness can be obtained by different approaches as introduced hereafter. As listed in Table 14.1, cleaning agents are carbonaceous compounds. Such compounds can be efficiently removed by plasma cleaning.

#### 14.2.2.1 Plasma Cleaning

*Plasma cleaning* processes have turned out to be suitable for the removal of carbonaceous surface contaminants such as hydrocarbons ($-CH_2-$), represented by residues from cleaning agents or adsorbed from the ambient air. Commonly, plasma cleaning is performed at low pressure applying oxygenic working gases (Fischer, 2012). Cleaning is then realized by several interacting mechanisms. On one hand, plasma-induced extreme ultraviolet irradiation allows the decomposition of contaminants by photo-desorption, and on the other hand, hydrocarbons are decomposed by reactive plasma species such as oxygen radicals ($O^*$) or ozone ($O_3$) (Hansen et al., 1993). According to

$$(-CH_2-CH_2-)+6O^* \rightarrow 2CO_2+2H_2O, \tag{14.1}$$

hydrocarbons are oxidized to volatile carbon dioxide ($CO_2$) and water ($H_2O$) in the course of the plasma-chemical process. Finally, a certain sputtering by ions contributes to cleaning but could also give rise to a degradation of surface accuracy due to ion-induced wrinkling (see Section 13.4.1.1).

In order to overcome the impact of ion bombardment on surface accuracy and to substitute the required vacuum equipment, the suitability of atmospheric pressure plasmas for cleaning glass surfaces was investigated in the last years (Shun'ko and Belkin, 2007; Buček et al., 2008; Iwasaki et al., 2008a,b). This approach is mainly used for cleaning glass containers, for example, the sterilization of medical devices (Cheruthazhekatt et al., 2010) but allows increasing the laser-induced damage threshold of optical media (Gerhard et al., 2017) due to the removal of surface-adherent hydrocarbons and residues from UV-absorbing polishing agents and will thus most likely gain importance in optics manufacturing in the future.

#### 14.2.2.2 Dry-Ice Blasting

Precision cleaning of glass surfaces can also be performed by *dry-ice blasting* (a.k.a. carbon dioxide snow cleaning). Here, either solid carbon dioxide snow pellets or liquid carbon dioxide is used. In the first case, removal of contaminants is a purely mechanical process, whereas the latter procedure can be described as follows: liquid carbon dioxide is mixed with compressed air and sprayed onto the work piece surface. Due to an abrupt relaxation during mixing, the liquid carbon dioxide becomes solid, resulting in the formation of small dry-ice crystals. These crystals then sublime at the work piece surface; cleaning is thus due to both thermal and mechanical effects, where the process temperature is given by the dry-ice temperature of approximately −79°C. The mechanical impact can be adjusted by the choice of the geometry of the used spray nozzles. Since dry ice features a comparatively low hardness of approximately 2 Mohs,[2] this method is also suitable for cleaning sensitive surfaces.

#### 14.2.2.3 High-Precision Cleaning by Surface Modification

High-precision cleaning is, for example, applied in order to increase the laser-induced damage threshold of laser optics. Damage of such components may lead to downtime and interruption of production processes in laser-based manufacturing process chains for bulk articles and can thus provoke notable costs. For UV-laser optics such as protection windows, sapphire has been established as a commonly used optical medium, and several approaches for increasing the surface quality were especially developed for this medium. For instance, the removal of chemically modified near-surface layers can be realized by ion beam polishing (Giuliano, 1972) or laser polishing (Wei et al., 2012). Subsurface damage such as microcracks that can act as a seam for contaminants can be removed by annealing (Pinkas et al., 2010), and even pristine surface can be generated by the deposition and subsequent annealing of appropriate thin layers on sapphire surfaces (Park and Chan, 2002).

---

[2] According to the Mohs scale of hardness, named after the German-Austrian mineralogist *Friedrich Mohs* (1773–1839), the hardness of minerals is classified into 10 classes of hardness, where 1 represents the softest and 10 the hardest material (see Table 7.2 in Section 7.2.2.1).

## 14.3 SUMMARY

Surface and subsurface contaminations may have a severe impact on the final functionality and further processing of optical components. Against this background, cleaning is of great importance. In classical optics manufacturing, surface cleaning is performed manually and with the aid of ultrasonic baths. The underlying mechanisms during ultrasonic cleaning are a chemical decomposition of surface-adherent contaminants and a mechanical impact by locally induced pressure. Precision cleaning can be realized by plasmas, where hydrocarbons are oxidized by oxygen species, or by means of dry-ice blasting on the basis of mechanical and thermal processes. Finally, surface modification techniques are suitable for cleaning selected optical media.

## REFERENCES

Buček, A., Homola, T., Aranyosiová, M., Velič, D., Plecenik, T., Havel, J., Sťahel, P., and Zahoranová, A. 2008. Atmospheric pressure nonequilibrium plasma treatment of glass surface. *Chemické Listy* 102:S1459–S1462.

Bude, J., Miller, P., Baxamusa, S., Shen, N., Laurence, T., Steele, W., Suratwala, T., Wong, L., Carr, W., Cross, D., and Monticelli, M. 2014. High fluence laser damage precursors and their mitigation in fused silica. *Optics Express* 22:5839–5851.

Cheruthazhekatt, S., Černák, M., Slavíček, P., and Havel, J. 2010. Gas plasmas and plasma modified materials in medicine. *Journal of Applied Biomedicine* 8:55–66.

Fischer, S.M. 2012. Plasma und Glas—ein starkes Team, effektive Reinigung mit Niederdruckplasma, *Magazin für Oberflächentechnik* 66:1–2 (in German).

Génin, F.Y., Kozlowski, M.R., and Brusasco, R.M. 1997. Catastrophic failure of contaminated fused silica optics at 355 nm. *Proceedings of the Annual International Conference on Solid-State Lasers for Applications to Inertial Confinement Fusion* 2:978–986.

Gerhard, C., Tasche, D., Munser, N., and Dyck, H. 2017. Increase in nanosecond laser-induced damage threshold of sapphire windows by means of direct dielectric barrier discharge plasma treatment. *Optics Letters* 42:49–52.

Gerhard, C., Weihs, T., Tasche, D., Brückner, S., Wieneke, S., and Viöl, W. 2013. Atmospheric pressure plasma treatment of fused silica, related surface and near-surface effects and applications. *Plasma Chemistry and Plasma Processing* 33:895–905.

Giuliano, C.R. 1972. Laser-induced damage in transparent dielectrics: Ion beam polishing as a means of increasing surface damage thresholds. *Applied Physics Letters* 21:39–41.

Hansen, R.W.C., Bissen, M., Wallace, D., Wolske, J., and Miller, T. 1993. Ultraviolet/ozone cleaning of carbon-contaminated optics. *Applied Optics* 32:4114–4116.

Iwasaki, M., Inui, H., Matsudaira, Y., Kano, H., Yoshida, N., Ito, M., and Hori, M. 2008a. Nonequilibrium atmospheric pressure plasma with ultrahigh electron density and high performance for glass surface cleaning. *Applied Physics Letters* 92:081503.

Iwasaki, M., Matsudaira, Y., Takeda, K., Ito, M., Miyamoto, E., Yara, T., Uehara, T., and Hori, M. 2008b. Roles of oxidizing species in a nonequilibrium atmospheric-pressure pulsed remote $O_2/N_2$ plasma glass cleaning process. *Journal of Applied Physics* 103:023303.

Langmuir, I. 1918. The adsorption of gases on plane surfaces of glass, mica and platinum. *Journal of the American Chemical Society* 40:1361–1403.

Neauport, J., Lamaignere, L., Bercegol, H., Pilon, F., and Birolleau, J.-C. 2005. Polishing-induced contamination of fused silica optics and laser induced damage density at 351 nm. *Optics Express* 13:10163–10171.

Park, H., and Chan, H.M. 2002. A novel process for the generation of pristine sapphire surfaces. *Thin Solid Films* 422:135–140.

Pinkas, M., Lotem, H., Golan, Y., Einav, Y., Golan, R., Chakotay, E., Haim, A., Sinai, E., Vaknin, M., Hershkovitz, Y., and Horowitz, A. 2010. Thermal healing of the subsurface damage layer in sapphire. *Materials Chemistry and Physics* 124:323–329.

Posth, O., Wunderlich, H., Flämmich, M., and Bergner, U. 2012. Reinigung von Vakuumbauteilen für UHV- und UCV-Anwendungen. *Vakuum in Forschung und Praxis* 24:18–25 (in German).

Shun'ko, E.V., and Belkin, V.S. 2007. Cleaning properties of atomic oxygen excited to metastable state $2s^2 2p^4$ ($^1S_0$). *Journal of Applied Physics* 102:083304.

Wei, X., Xie, X.Z., Hu, W., and Huang, J.F. 2012. Polishing sapphire substrates by 355 nm ultraviolet laser. *International Journal of Optics* 2012:238367.

# Appendix

## A.1 TABLE OF SYMBOLS AND ABBREVIATIONS

Some symbols listed here may have several different meanings and may be extended by appropriate indices, prefixes, or suffixes in the running text.

| Symbol/Abbreviation | Meaning |
|---|---|
| $A$ | Absorbance, area, coefficient of asphericity, or Seidel coefficient |
| $a$ | Half diameter of aperture stop/lens surface or object distance |
| $A, B$ | Material-specific Cauchy parameters |
| $a'$ | Image distance |
| $AOI$ | Angle of incidence |
| $B, C$ | Material-specific Sellmeier coefficients |
| $BFL$ | Back focal length |
| $c$ | Specific heat capacity or speed of light |
| $C$ | Concentration |
| $CE$ | Centering error |
| $C_{\mathrm{M}}$ | Michelson contrast |
| $C_{\mathrm{p}}$ | Degree of coverage or Preston's coefficient |
| $D$ | Diffusion coefficient, distortion, diameter, or size |
| $d$ | Distance, depth, or defect value |
| $D_{\mathrm{Airy}}$ | Airy disc diameter |
| $D_{\mathrm{max}}$ | Maximum distortion |
| $d_{\mathrm{opt}}$ | Optical depth of penetration |
| $d_{\mathrm{tol}}$ | Acceptable fault tolerance |
| $e$ | Conus constant or elementary electric charge |
| $EFL$ | Effective focal length |
| $E_{\mathrm{kin}}$ | Kinetic energy |
| $f$ | Focal length or frequency |
| $F$ | Force, Fresnel number, or finesse |
| $F_{\mathrm{th}}$ | Laser ablation threshold |
| $g$ | Geometrical gradient constant or grating constant |
| $G$ | Grindability |
| $h$ | Height |
| $HK$ | Knoop hardness |
| $I$ | Intensity or current |
| $J$ | Particle flux |
| $K$ | Extinction coefficient or photo-elastic coefficient |
| $k_{\mathrm{B}}$ | Boltzmann constant |
| $l$ | Length |
| $L$ | Total load |
| $L_{\mathrm{T}}$ | Thermal diffusion length |
| $m$ | Diffraction order, magnification, or mass |

| | |
|---|---|
| $M$ | Modulation |
| $MF$ | Merit function |
| $MRR$ | Material removal rate |
| $MTF$ | Modulation transfer function |
| $n$ | Index of refraction |
| $N$ | Number or complex index of refraction |
| $N(t)$ | Removal rate |
| $n_e$ | Electron density |
| $n_t$ | Grinding tool drive |
| $OD$ | Optical density |
| $O_l$ | Longitudinal offset |
| $O_p$ | Parallel offset |
| $p$ | Pressure |
| $P$ | Lens pitch |
| $P_{x,y}$ | Partial dispersion |
| $Q$ | Heat input |
| $R$ | Radius or reflectance |
| $r$ | Radius or reflectivity |
| $R_{Petzval}$ | Petzval field curvature |
| $Rq$ | Root mean squared surface roughness |
| $S$ | Sagitta or Strehl-ratio |
| $s$ | Working distance |
| $S_I$ | Seidel sum for spherical aberration |
| $S_{II}$ | Seidel sum for coma |
| $S_{III}$ | Seidel sum for astigmatism |
| $S_{IV}$ | Seidel sum for Petzval field curvature |
| $S_V$ | Seidel sum for distortion |
| $T$ | Temperature or transmittance |
| $t$ | Time or thickness |
| $TIS$ | Total integrated scatter |
| $U$ | Ion acceleration voltage |
| $u$ | Object height |
| $u'$ | Image height |
| $v$ | Speed or velocity |
| $V$ | Abbe number or volume |
| $W$ | Work |
| $w$ | Object/aperture angle |
| $w'$ | Image angle |
| $Y$ | Sputter yield |
| $z$ | Ion charge |
| $z_T$ | Talbot length |
| $\alpha$ | Absorption coefficient, coefficient of thermal expansion, or angle |
| $\beta$ | Magnification |
| $\delta$ | Deviation or optical path difference |
| $\Delta f$ | Form deviation |
| $\Delta h$ | Change in height |
| $\Delta l$ | Change in length |

| | |
|---|---|
| $\Delta m$ | Change in work piece mass |
| $\Delta n$ | Birefringence or deviation in index of refraction |
| $\Delta s$ | Relative travel |
| $\delta_{sag}$ | Sagittal coma |
| $\delta_{tan}$ | Tangential coma |
| $\Delta w$ | Wave front deformation |
| $\Delta z$ | Grinding tool feed motion or runout |
| $\varepsilon$ | Angle of incidence |
| $\varepsilon'$ | Angle of refraction |
| $\varepsilon_0$ | Vacuum permittivity |
| $\varepsilon_B$ | Brewster's angle |
| $\varepsilon_{crit}$ | Critical angle of total internal reflection |
| $\theta$ | Tilt angle |
| $\theta_B$ | Blaze angle |
| $\Lambda$ | Lateral structure size |
| $\lambda$ | Thermal conductivity or wavelength |
| $\lambda_D$ | Debye length of a plasma |
| $\mu$ | Coefficient of friction |
| $\rho$ | Density |
| $\sigma$ | Collision diameter |
| $\sigma_m$ | Mechanical tension |
| $\varphi$ | Defocus |
| $\varphi_{max}$ | Maximum distance between focal planes (astigmatism) |
| $\Psi$ | Position-dependent phase error or wave front deviation |
| $\omega$ | Angular frequency of light |

## A.2 ENGLISH-GERMAN DICTIONARY OF TECHNICAL TERMS

| English | German |
|---|---|
| Abbe diagram (glass map) | Abbe-Diagramm |
| Abbe number (*V*-number) | Abbe-Zahl |
| Ablation | Ablation |
| Absorption | Absorption |
| Absorption coefficient | Absorptionskoeffizient |
| Achromatic doublet | Achromat |
| Acid resistance | Säurebeständigkeit |
| Airy disc | Beugungsscheibchen |
| Alkali resistance | Alkalibeständigkeit |
| Angle of incidence | Einfallswinkel |
| Angle of refraction | Brechungswinkel |
| Antireflective coating | Antireflexschicht |
| Asphere | Asphäre |
| Assembly | Zusammenbau |
| Astigmatism | Astigmatismus |
| Batch | Gemenge |

| | |
|---|---|
| Beveling | Fasen |
| Birefringence | Doppelbrechung |
| Bound abrasive grinding | Schleifen mit gebundenem Korn |
| Brewster's angle | Brewsterwinkel |
| Bubble | Blase |
| Canada balsam | Kanadabalsam |
| Carrier | Träger, Tragkörper |
| Cement | Kitt |
| Cement wedge | Kittkeil |
| Cementing | Kitten |
| Cementing error | Kittfehler |
| Cementing work station | Kittzentrierplatz |
| Centering | Zentrieren |
| Centering error | Zentrierfehler |
| Centering machine | Zentriermaschine |
| Centering runout | Zentrierschlag |
| Centering tool | Zentrierwerkzeug |
| Center thickness | Mittendicke |
| Chemical hypothesis | Chemische Abtragshypothese |
| Chemical vapor deposition | Chemische Gasphasenabscheidung |
| Chromatic aberration | Chromatische Aberration, Farbfehler |
| Circular saw bench | Tischkreissäge |
| Clamping | Klemmen, Spannen |
| Clamping bell | Spannglocke |
| Cleaning | Reinigung |
| Climatic resistance | Klimabeständigkeit |
| Cloth | Tuch |
| Coating | Beschichtung |
| Coefficient of thermal expansion | Thermischer Ausdehnungskoeffizient |
| Colored glass | Farbglas |
| Coma | Koma |
| Compression molding | Formpressen |
| Concave | Konkav |
| Condition for achromatism | Achromasiebedingung |
| Constructive interference | Konstruktive Interferenz |
| Contour accuracy | Formtreue, Passung, Passe |
| Conus constant | Konuskonstante |
| Converging lens | Sammellinse |
| Convex | Konvex |
| Cooling lubricant | Kühlschmiermittel |
| Critical angle of total internal reflection | Grenzwinkel der Totalreflexion |
| Crown glass | Kronglas |
| Crystal | Kristall |
| Cut-off-frequency | Grenzfrequenz |
| Cutting | Trennschleifen |
| Cylindrical cup wheel | Topfschleifwerkzeug |

| | |
|---|---|
| Cylindrical lens | Zylinderlinse |
| Decenter | Dezentrierung |
| Deflection prism | Umlenkprisma |
| Destructive interference | Destruktive Interferenz |
| Deviation | Ablenkung |
| Diameter | Durchmesser |
| Diffraction | Beugung |
| Dig | Loch |
| Dispersion | Dispersion |
| Dispersion angle | Dispersionswinkel |
| Dispersion formula | Dispersionsrelation |
| Dispersion prism | Dispersionsprisma |
| Distortion | Verzeichnung |
| Diverging lens | Zerstreuungslinse |
| Double-sided grinding | Doppelläppen |
| Dry etching | Trockenätzen |
| Dry-ice blasting | Trockeneisstrahlen |
| Eccentric | Exzenter |
| Edge thickness | Randdicke |
| Effective focal length | Brennweite |
| Embossing | Prägen |
| Extinction coefficient | Extinktionskoeffizient |
| Felt | Filz |
| Fine cement | Feinkitt |
| Fine contour error | Feinpassfehler |
| Finish grinding | Feinschleifen |
| Flame pyrolysis | Flammenpyrolyse |
| Flat grinding | Planschleifen |
| Flint glass | Flintglas |
| Flow hypothesis | Fließhypothese |
| Fluid jet polishing | Flüssigkeitsstrahlpolieren |
| Focal point | Brennpunkt |
| Free length of path | Freie Weglänge |
| Fresnel equation | Fresnelgleichung |
| Fretting hypothesis | Reibverschleißhypothese |
| Fused silica | Synthetisches Quarzglas |
| Gauge glass | Probeglas |
| Glass ceramic | Glaskeramik |
| Glass transition temperature | Glasübergangstemperatur |
| Gluing | Kleben |
| Gob | Klumpen |
| Grade of grinding | Schleifgrad |
| Grade of polishing | Poliergrad |
| Gradient index material | Gradientenindexmaterial |
| Grain | Korn |
| Grating | Gitter |

| | |
|---|---|
| Grindability | Schleifbarkeit |
| Grinding | Schleifen |
| Hardness | Härte |
| Hollow drilling | Hohlschleifen |
| Image distance | Bildweite |
| Image height | Bildhöhe |
| Image space | Bildraum |
| Imaging equation | Abbildungsgleichung |
| Inclusion | Einschluss |
| Index of refraction | Brechungsindex |
| Inhomogeneity | Inhomogenität |
| Interferometer | Interferometer |
| Internal transmittance | Reintransmissionsgrad |
| Ion exchange | Ionenaustausch |
| Lapping | Läppen |
| Laser polishing | Laserpolieren |
| Laser welding | Laserschweißen |
| Laser-induced damage threshold | Laserzerstörschwelle |
| Layer | Schicht |
| Layer growth | Schichtwachstum |
| Lens | Linse |
| Lens clock | Feinzeiger |
| Lensmaker's equation | Linsenmacherformel |
| Lever arm machine | Hebelarmmaschine |
| Longitudinal offset | Längsversatz |
| Loose abrasive grinding | Schleifen mit losem Korn |
| Magnification | Vergrößerung |
| Manufacturing drawing | Fertigungszeichnung |
| Measuring bell | Messglocke |
| Meltdown | Niederschmelzen |
| Merit function | Zielfunktion |
| Microlens | Mikrolinse |
| Mirror | Spiegel |
| Mirror coating | Spiegelschicht |
| Modulation transfer function | Modulationsübertragungsfunktion |
| Molding | Formpressen |
| Mount | Fassung |
| Mounting | Montieren |
| Mounting error | Monatgefehler |
| Multicomponent glass | Mehrkomponentenglas |
| Network former | Netzwerkbildner |
| Network modifier | Netzwerkwandler |
| Object distance | Objektweite |
| Object height | Objekthöhe |
| Object space | Objektraum |
| Optical contact bonding | Ansprengen |
| Optical density | Optische Dichte |
| Optical path difference | Optische Weglängendifferenz |

| | |
|---|---|
| Optical retardation | Optische Verzögerung |
| Parallel offset | Parallelversatz |
| Partial dispersion | Teildispersion |
| Petzval field curvature | Bildfeldwölbung |
| Phosphate resistance | Phosphatbeständigkeit |
| Photolithography | Photolithographie |
| Physical vapor deposition | Physikalische Gasphasenabscheidung |
| Pitch | Pech |
| Plaining | Läutern |
| Plasma cleaning | Plasmareinigen |
| Plasma polishing | Plasmapolieren |
| Plate | Platte |
| Polarization prism | Polarisationsprisma |
| Polishing | Polieren |
| Polishing pad | Poliermittelträger |
| Polishing suspension | Poliermittelsuspension |
| Position tolerance | Lagetoleranz |
| Precision centering | Zentrierdrehen |
| Preshaping | Vorformen |
| Principal dispersion | Hauptdispersion |
| Prism | Prisma |
| Prism wedge angle | Prismenkeilwinkel |
| Quartz glass | Quarzglas |
| Radius of curvature | Krümmungsradius |
| Raw cement | Grobkitt |
| Ray entrance height | Strahleinfallshöhe |
| Reactive ion etching | Reaktives Ionenätzen |
| Refining agent | Läuterungsmittel |
| Reflectance | Reflexionsgrad |
| Reflection | Reflexion |
| Reflective coating | Reflexschicht |
| Refraction | Brechung |
| Removal hypothesis | Abtragshypothese |
| Rough grinding | Vorschleifen |
| Roughing | Schruppen |
| Roughness | Rauheit |
| Rounding | Rundieren |
| Sagitta | Pfeilhöhe |
| Scratch | Kratzer |
| Screw connecting | Schraubverbindung |
| Setting angle | Anstellwinkel |
| Shape | Form |
| Silica gel layer | Kieselgelschicht |
| Single-component glasses | Einkomponentenglas |
| Sliding angle | Gleitwinkel |
| Soldering | Löten |
| Spherical aberration | Sphärische Aberration, Kugelgestaltsfehler |
| Spherical cup wheel | Schleifschale |

| | |
|---|---|
| Spherometer | Sphärometer |
| Spindle | Spindel |
| Sputtering | Sputtern |
| Stabilizer | Stabilisator |
| Stacking | Stapeln |
| Stain | Fleck |
| Stain resistance | Fleckenbeständigkeit |
| Stress birefringence | Spannungsdoppelbrechung |
| Striae | Schlieren |
| Striae class | Schlierenklasse |
| Surface cleanliness | Oberflächensauberkeit |
| Tappet | Zapfen |
| Thick lens | Dicke Linse |
| Thin lens | Dünne Linse |
| Tilt | Verkippung |
| Tolerance | Toleranz |
| Tool | Werkzeug |
| Toric lens | Torische Linse |
| Total transmittance | Gesamttransmissionsgrad |
| Transmission | Transmission |
| Turning lathe | Drehbank, Rundiermaschine |
| Ultrasonic bath | Ultraschallbad |
| Wavelength | Wellenlänge |
| Wedge | Keil |
| Window | Fenster |
| Wobble circle | Taumelkreis |

## A.3 FURTHER READING (NOT CITED IN THE PRESENT WORK)


Bäumer, S. (ed.). 2005. *Handbook of Plastic Optics*. Weinheim, Germany: Wiley-VCH Verlag.

Bray, C. 2001. *Dictionary of Glass: Materials and Techniques*. Baltimore, MD: Johns Hopkins University Press.

Bureau of Naval Personnel. 1969. *Basic Optics and Optical Instruments*. New York: Dover Publications.

Herzig, H.P. (ed.). 1997. *Micro-Optics—Elements, Systems and Applications*. London/Bristol: Taylor & Francis.

Kufner, M., and Kufner, S. 1997. *Micro-Optics and Lithography*. Brussels, Belgium: VUB University Press.

Malacara, D. 1992. *Optical Shop Testing*. Hoboken, NJ: John Wiley & Sons.

Scholze, H. 1988. *Glas*. Berlin and Heidelberg: Springer (in German).

Stenzel, O. 2014. *Optical Coatings—Material Aspects in Theory and Practice*. Berlin and Heidelberg: Springer Verlag.

Twyman, F. 1989. *Prism and Lens Making*. Bristol and New York: Adam Hilger.

Vogel, W. 2012. *Glaschemie*. Berlin and Heidelberg: Springer (in German).

Williamson, R. 2011. *Field Guide to Optical Fabrication*. Bellingham, WA: SPIE Press.
Yoder, P.R. Jr. 2008. *Mounting of Optics in Optical Instruments*. Bellingham, WA: SPIE Press.

## A.4 EXERCISES

### A.4.1 Basics of Light Propagation

#### Exercise A.4.1.1

A light beam coming from vacuum with $n=1$ enters an optical medium with an index of refraction of $n'=1.5$, where the angle of incidence is $\varepsilon=40°$. Determine the angle of refraction $\varepsilon'$ and the deviation of the light beam from its original direction of propagation $\delta$.

#### Exercise A.4.1.2

Determine the reflectance $R_s$ for s-polarized and $R_p$ p-polarized light as well as the total reflectance $R_{tot}$ at the interface of an optical medium with an index of refraction of $n'=1.6$, where the index of refraction of the ambient medium is $n=1$. The angle of incidence of incoming light is 50°.

#### Exercise A.4.1.3

A light ray enters an optical interface at normal incidence. The indices of refraction before and behind this interface are $n=1.33$ and $n'=1.78$, respectively. Determine the total reflectance for this case.

#### Exercise A.4.1.4

The Brewster's angle at a glass surface amounts to $\varepsilon_B=57°$, where the index of refraction of the ambient medium is $n=1$. Determine the index of refraction of the glass.

#### Exercise A.4.1.5

A deflection prism is made of a glass with an index of refraction of $n_{glass}=1.48$, where the ambient medium is air with an index of refraction of $n_{air}=1$. Determine the critical angle of total internal reflection $\varepsilon_{crit}$ for this case. How does $\varepsilon_{crit}$ change (absolute value) if a drop of water with an index of refraction of $n_{water}=1.33$ is put on the prism surface?

#### Exercise A.4.1.6

A light beam with an initial intensity of $I_0=1\,W/cm^2$ passes a glass plate, where the absorption coefficient of the glass material is $\alpha=0.0024072\,cm^{-1}$. After passing the glass plate, the light beam intensity amounts to $I_t=0.98\,W$. Determine the internal transmission $T_i$ and the thickness $t$ of this glass plate.

#### Exercise A.4.1.7

Determine the absorbance $A$ of a glass sample with a thickness of 25 mm, where the absorption coefficient of the glass is $\alpha=0.0056346\,cm^{-1}$.

## A.4.2 Optical Materials

### Exercise A.4.2.1

Within an optical medium, the speed of s-polarized light is $c_s=2 \cdot 10^8$ m/s, whereas the speed of p-polarized light is $c_p=1.99 \cdot 10^8$ m/s. Determine the birefringence of this medium, assuming the speed of light in vacuum to amount to $c_0=3 \cdot 10^8$ m/s.

### Exercise A.4.2.2

An optical glass is described by the following Sellmeier coefficients: $B_1=1.039$, $B_2=0.232$, $B_3=1.011$, $C_1=0.006\,\mu m^2$, $C_2=0.021\,\mu m^2$, and $C_3=103.561\,\mu m^2$. Determine the total reflectance $R_{tot}$ for perpendicular incidence at the interface from air, with $n=1$ to this glass at a wavelength of $\lambda=500$ nm.

### Exercise A.4.2.3

The surface reflectivity of an unknown glass is measured at normal incidence ($\varepsilon=0°$) for three different wavelengths, 480, 546, and 644 nm. The particularly measured reflectivities are $R_{480}$nm=8.3%, $R_{546}$nm=8.1%, and $R_{644}$nm=7.8%. The ambient medium is air with $n=1$. Define the type of glass by calculating its Abbe number.

### Exercise A.4.2.4

The homogeneity of a glass sample with a thickness of $t=1$ cm is evaluated via wave front measurement (in transmission). The detected wave front deformation amounts to $\lambda/10$ at a test wavelength of 633 nm. Determine the inhomogeneity class that represents this glass sample.

### Exercise A.4.2.5

A gradient index lens features a geometrical gradient constant of $g=0.653$/mm. The index of refraction at the center of this lens is $n_0=1.616$. Determine the index of refraction at the position $r=500\,\mu m$.

### Exercise A.4.2.6

Determine the volume removed during grinding of a glass with a grindability of $G=165$, where grinding was performed for 30 s. The removed volume of the reference glass is $\Delta V_{ref}=0.4$ mm³.

## A.4.3 Optical Components

### Exercise A.4.3.1

A thin lens is made of a glass with an index of refraction of 1.64. Its radii of curvature are $R_1=53$ mm and $R_2=-146$ mm. Determine the effective focal length *EFL* of this lens. What is the EFL of an equivalent thick lens with a center thickness of $t_c=6$ mm?

### Exercise A.4.3.2

The angle of incidence of light on a prism with a wedge angle of $\alpha=50°$ is $\varepsilon_1=30°$. The prism material has an index of refraction of 1.55. The ambient medium is air with $n_{air}=1$. Determine the exit angle of light $\varepsilon_2'$ at the prism's exit surface.

### Exercise A.4.3.3

As determined experimentally, the minimum deviation $\delta_{min}$ of light passing through a prism with a wedge angle of $\alpha=45°$ is 23°. Determine the index of refraction of the prism material.

### Exercise A.4.3.4

A dispersion prism with a wedge angle of $\alpha=40°$ is made of a glass with a main dispersion of $n_F-n_C=0.0442$; its index of refraction is 1.9546 at a wavelength of 486 nm. Determine the dispersion angle $\delta_d$ of incident white light after passing this prism, where the angle of incidence on the first prism surface is 70° and the wavelengths of interest are $\lambda_1=656$ nm and $\lambda_2=486$ nm.

### Exercise A.4.3.5

A light beam is sent through a wedge made of a material with an index of refraction of $n=1.68$ and is deviated by 3° from its original direction of propagation. Determine the wedge angle $\alpha$ of the wedge.

### Exercise A.4.3.6

A light beam is sent through a plane-parallel plate made of a material with an index of refraction of $n=1.57$. The thickness of this plate is $t=20$ mm and the angle of incidence of the light beam at the front face of the plate is $\varepsilon = 45°$. Determine the parallel and longitudinal offset $O_p$ and $O_l$ of the light beam after passing through the plate.

## A.4.4 Design of Optical Components

### Exercise A.4.4.1

An object with a height of $u=60$ mm is imaged by a lens, where the object distance is $a=1000$ mm, and the image distance is $a'=50$ mm. Determine the effective focal length *EFL* of this lens and the image height $u'$.

### Exercise A.4.4.2

A thin lens and a thick biconvex lens with a center thickness of $t_c=10$ mm are made of a material with an index of refraction of $n=1.6$. The radii of curvature of both lenses are $R_1=75$ mm and $R_2=100$ mm, respectively. Determine the particular effective focal length *EFL* of these lenses.

### Exercise A.4.4.3

Determine the longitudinal spherical aberration for an optical interface with a radius of curvature of $R=100$ mm, a maximum ray entrance height of $h_{max}=15$ mm, and a minimum one of $h_{min}=1$ mm. The index of refraction behind this optical interface is $n=1.75$.

### Exercise A.4.4.4

An achromatic doublet is made of two lenses: a converging and a diverging one. The effective focal length of the converging lens is $EFL_1=8.95$ mm. It is made of a glass with an Abbe number of $V_1=50$. The EFL of the diverging lens is $EFL_2=-17$ mm. Determine the Abbe number of the material of this lens.

#### Exercise A.4.4.5

The actual lateral image coordinate of an image is $u_a'=9$ mm, whereas this value should theoretically amount to $u_t'=9.6$ mm according to the paraxial imaging model. Determine the type of distortion.

#### Exercise A.4.4.6

As determined by basic considerations and analysis on an imaging task, the Seidel sum for spherical aberration of an optical system should amount to $S_I=0.005$, where the upper and the lower limit should be $d_{max}=0.004$ and $d_{min}=0.002$, respectively. For the design of such an optical system, an existing start system with a Seidel sum of $S_I=0.01$ is chosen. This system should now be optimized. Determine the absolute value of the merit function $MF$ for this case where merely spherical aberration is considered as defect.

### A.4.5 Tolerancing of Optical Components and Systems

#### Exercise A.4.5.1

An optical glass is specified by the identification marks "0/40," "1/2·0.25," and "2/4, 1." What do these identification marks mean in practice?

#### Exercise A.4.5.2

The quality of an optical glass sample with a nominal index of refraction of $n=1.5154$ and a thickness of $t=10$ mm is evaluated via wave front measurements, where a wave front deformation of $\Delta w=40$ nm is determined. Determine the inhomogeneity class of this glass.

#### Exercise A.4.5.3

The Knoop hardness of an optical glass is measured according to DIN EN ISO 4545, where the force applied to the used diamond tip is $F=10$ N, and the length of the resulting imprint of the diamond tip on the glass surface is $l=0.17$ mm. Determine the class of hardness of this glass.

#### Exercise A.4.5.4

A glass rod with a length of $l_0=300$ mm is heated from 300 to 350 K. After such heating, the length of this glass rod has increased and amounts to 300.12 mm. Determine the coefficient of thermal expansion $\alpha$ of the glass material.

#### Exercise A.4.5.5

A lens surface features an interference pattern as shown in Figure A.1, where the given value for the observable deviation of the Newton fringe from its basic shape $d$ was measured by an interferometer. The test wavelength of the used interferometer is 600 nm.

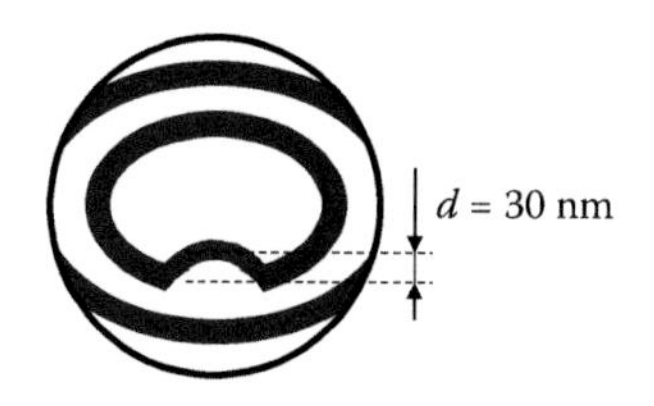


**FIGURE A.1** Measured interference pattern.

Which specification in a manufacturing drawing describes this pattern according to DIN ISO 10110?

#### Exercise A.4.5.6

A convex lens surface with a nominal radius of curvature of 6 mm is marked with "3/50(44)" in a manufacturing drawing. This lens is destined for an imaging task where the wavelength of interest is 1064 nm. Discuss the possible impact of the given specification of surface accuracy on the imaging quality.

#### Exercise A.4.5.7

The centering error of a lens made of a glass with an index of refraction of 1.65 is measured by determining the radius of the wobble circle resulting from rotation of the lens. For observation, an optical setup with a magnification of $m=5$ is used. The distance from the lens principal plane to the detector plane is $d=350$ mm, and the measured wobble circle radius amounts to $r_w=800\,\mu$m. Determine the centering error of this lens.

#### Exercise A.4.5.8

The polished surface of an optical component with a diameter of 25.4 mm is specified according to DIN ISO 10110 and marked with "5/2·0.16" and "P4." Determine the required surface quality in terms of roughness and defect size/area.

#### Exercise A.4.5.9

The root mean squared roughness of a lens surface is $Rq=33$ nm. Determine the fraction of specular (i.e., directed) reflected light at a wavelength of 380 nm, assuming perpendicular incidence of light.

### A.4.6 Shape Forming

#### Exercise A.4.6.1

A lens surface with a target radius of curvature of $R_c=50$ mm shall be rough ground using a cylindrical cup wheel with a diameter of $D_{cw}=30$ mm and a cutting edge radius of $r=2$ mm. Which setting angle $\alpha$ has to be chosen for a convex lens surface and a concave one, respectively?

### Exercise A.4.6.2

A lens surface is lapped with loose abrasives made of silicon carbide, where the type designation of the lapping grain material is "F240." Estimate the depth of micro cracks $d_{mc}$ resulting from the lapping procedure.

### Exercise A.4.6.3

A lens is fine ground using a spherical cup wheel marked with the type designation "D3." Estimate the thickness of material to be removed in order to obtain a microcrack-free surface.

### Exercise A.4.6.4

A spherical cup wheel with abrasive pellets and a diameter of $D_{cw}=40\,mm$ features a degree of coverage of $C_p=16\%$ (i.e., the pellet density). The pellets have a diameter of $D_p=4\,mm$. Determine the number of pellets $N$ on the cup wheel's surface.

### Exercise A.4.6.5

A plane lens surface with a diameter of $D_{wp}=25.4\,mm$ is ground, where the drive of the grinding tool amounts to $n_t=1000\,rpm$. Determine the resulting cutting velocity $v_c$.

### Exercise A.4.6.6

A plane glass window surface with a diameter of 22 mm is ground. The initial thickness of this window is 6 mm, and the target thickness after grinding is 5 mm. For the given process conditions, the material removal rate is $MRR=6.3\,mm^3/min$. Determine the required duration of the grinding process $t$ for achieving the target thickness.

### Exercise A.4.6.7

A lens with a diameter of $D_l=25.4\,mm$ shall be beveled by grinding its edge in a spherical cup wheel. Which cup wheel should be chosen?

### Exercise A.4.6.8

A convex lens shall be rough ground, where the radius of curvature after rough grinding should amount to $R_c=101\,mm$. Calculate the sagitta of this rough ground lens (with respect to a plane surface), which is measured using a spherometer with a ring-shaped bell with a diameter of 40 mm.

## A.4.7 Polishing

### Exercise A.4.7.1

A fused silica surface is polished for 6 h, where the mean process temperature amounts to 90°C. The diffusion coefficient of water into fused silica amounts to $D=10^{-18}\,cm^2/s$ at this temperature. Estimate the mean diffusion depth $d_{dif}$ of water from the used aqueous polishing suspension into the fused silica surface after the polishing process.

**Exercise A.4.7.2**

A glass surface with a diameter of 2.5 cm is polished, where the weight of the polishing tool is 2 kg. The relative velocity between the work piece and the polishing tool is $v=4$ m/s, and the Preston coefficient is $C_p=10^{-7}$ cm²/N. Determine the material removal rate *MRR*.

**Exercise A.4.7.3**

A lens made of fused silica is polished for 1 h. The area of the lens surface is $A=507$ mm². The total load during polishing is 20 N, and the relative velocity between the work piece and the polishing tool amounts to 2 m/s. Determine the time-dependent work *W* required for material removal for three different polishing pad materials: pitch with a medium hardness of 26, synthetic felt, and polyurethane.

**Exercise A.4.7.4**

The mass of a polishing suspension with a concentration of $C_s=17\%$ is $m_s=1.2$ kg. Determine the mass of the polishing agent $m_{pa}$ within this suspension.

### A.4.8 Cementing

**Exercise A.4.8.1**

Two prisms are cemented to a prism group using fine cement with an index of refraction of 1.58. Due to a cementing error, the cement layer between the two cemented plane prism surfaces features a wedge angle of 2°. Determine the deviation $\delta$ caused by this cement layer.

**Exercise A.4.8.2**

An achromatic doublet is cemented where the radii of curvature of the surfaces to be cemented are $R_1=29$ mm and $R_2=30$ mm. The used fine cement has an index of refraction of 1.5. Due to the difference in radii of curvature, a concave-convex lens (i.e., a positive meniscus) is formed between the actual lenses by the cement layer. Determine the effective focal length *EFL* of this cement layer assuming its thickness to be negligible.

### A.4.9 Centering

**Exercise A.4.9.1**

The centering error of a lens amounts to 1.38′. The lens is made of a glass with an index of refraction of 1.5. For the determination of the centering error, an optical setup with a magnification of $m=6$ was used, where the distance from the lens principal plane to the detector plane was $d=500$ mm. Determine the measured wobble circle radius $r_w$.

**Exercise A.4.9.2**

A lens with the radii of curvature $R_1=50$ mm and $R_2=100$ mm features a centering error of 0.008 rad. Determine the runout $\Delta z$.

### Exercise A.4.9.3

A symmetric biconvex lens with the radii of curvature $R=R_1=R_2=75\,\text{mm}$ is centered where the diameters of the used clamping bells is $D=D_1=D_2=20\,\text{mm}$. Is this lens self-aligned by the clamping bells?

## A.4.10 Coating

### Exercise A.4.10.1

Determine the reflectance of a glass surface with and without a single-layer antireflective coating made of silicon dioxide with an index of refraction of $n_c=1.46$. The index of refraction of the glass is $n_g=1.57$, and the index of refraction of the ambient medium is $n_a=1$. The angle of incidence of light is 0° (i.e., normal incidence).

### Exercise A.4.10.2

Compare the transmittances of an air-gapped lens doublet without any coating and with an antireflective coating with a residual reflectance of $R_r=0.5\%$. Assumptions: normal incidence ($\varepsilon=0°$) at all surfaces, index of refraction of all lenses $n'=1.5$, index of refraction of ambient medium $n=1$; bulk absorption and scattering are not considered.

## A.4.11 Assembly of Optomechanical Systems

### Exercise A.4.11.1

A beveled lens with a radius of curvature of the beveled surface of $R=75\,\text{mm}$ is fixed in a mount by screw connecting where the lens bevel acts as mounting surface. Due to inaccuracies in manufacturing, the bevel is not rotation-symmetric but features differences in bevel leg lengths. The maximum bevel leg length is $l_{max}=800\,\mu\text{m}$, and the minimum one is $l_{min}=50\,\mu\text{m}$. Determine the tilt angle $\alpha$ of the lens with respect to the mount cylinder axis, which results in this case, and indicate the corresponding specification of the centering error according to DIN ISO 10110.

## A.4.12 Microoptics

### Exercise A.4.12.1

An object consisting of bright and dark areas is imaged by a microlens. The intensity of light within the bright object areas amounts to $0.49\,\text{W/m}^2$, whereas the dark object areas feature an intensity of $0.28\,\text{W/cm}^2$. After imaging, the light intensity within the bright image areas is $0.31\,\text{W/cm}^2$. The intensity of the dark areas amounts to $0.19\,\text{W/cm}^2$. Determine the modulation transfer function *MTF*. The spatial frequency shall not be considered.

### Exercise A.4.12.2

A nonimaging homogenizer consists of a Fourier lens with a focal length of $f_{Fl}=25\,\text{mm}$ and a microlens array with a pitch of $P_{la}=400\,\mu\text{m}$. The focal length of the involved microlenses is $f_{la}=500\,\mu m$. Determine the size $D_{hf}$ of the resulting homogeneous field. In order to realize an imaging homogenizer, a second microlens array (focal length $f_{la2}=600\,\mu m$) is added to the abovementioned setup. Which distance $d$ between the microlens arrays should be chosen in order to obtain the same size of the resulting homogenous field as determined above?

### Exercise A.4.12.3

An echelle grating shall be used at a wavelength of $\lambda=633\,nm$ and be optimized for the first diffraction order. Its grating constant is $g=1800$ line pairs per millimeter. Determine the required blaze angle of $\theta_B$ of this grating.

### Exercise A.4.12.4

An optics surface is precision polished by ion beam etching. The kinetic energy $E_{kin}$ of the incident ions is 200 eV (electron volts), and the ion mass is $m=6.6 \cdot 10^{-26}\,kg$. Determine the velocity $v$ of the ions.

### Exercise A.4.12.5

An optical medium (plastic) with a heat deflection temperature of $T_h=200°C$ shall be microstructured via reactive ion etching. The Debye length of the plasma sheath within the used process chamber amounts to $\lambda_D=1.58\,\mu m$, and the electron density within the plasma is $n_e=10^{18}$ per cubic meter. Is the approach of reactive ion etching suitable for this task?

### Exercise A.4.12.6

A planar gradient index (GRIN) lens is produced by applying a droplet of salt solution on a glass substrate in order to replace sodium ions ($Na^+$) by lithium ions ($Li^+$). The target center thickness of the planar GRIN lens is $t_c=500\,\mu m$, and the diffusion coefficient is $D=0.85 \cdot 10^{-7}\,cm^2/s$. Estimate the duration $t$ of the ion exchange process.

## A.5 SOLUTION OF EXERCISES

### A.4.1 Basics of Light Propagation

### Exercise A.4.1.1

The angle of refraction can be calculated on the basis of Snell's law and amounts to

$$\varepsilon' = \arcsin\left(\frac{n \cdot \sin\varepsilon}{n'}\right) = \arcsin\left(\frac{1 \cdot \sin 40°}{1.5}\right) = 25.37°.$$

The deviation of the light beam from its original direction of propagation is then

$$\delta = \varepsilon - \varepsilon' = 40° - 25.37° = 14.63°.$$

### Exercise A.4.1.2

The partial reflectance for s-polarized and p-polarized light follows from the Fresnel equations. Applying the basic Fresnel equations, the angle of refraction has to be determined initially:

$$\varepsilon' = \arcsin\left(\frac{n \cdot \sin\varepsilon}{n'}\right) = \arcsin\left(\frac{1 \cdot \sin 50°}{1.6}\right) = 28.61°.$$

The reflectance then amounts to

$$R_s = \left|\frac{n \cdot \cos\varepsilon - n' \cdot \cos\varepsilon'}{n \cdot \cos\varepsilon + n \cdot \cos\varepsilon'}\right|^2 = \left|\frac{1 \cdot \cos 50° - 1.6 \cdot \cos 28.61°}{1 \cdot \cos 50° + 1.6 \cdot \cos 28.61°}\right|^2 = \left|\frac{0.64 - 1.41}{0.64 + 1.41}\right|^2$$

$$= 0.14 = 14.11\%$$

for s-polarized light and to

$$R_p = \left|\frac{n' \cdot \cos\varepsilon - n \cdot \cos\varepsilon'}{n' \cdot \cos\varepsilon + n \cdot \cos\varepsilon'}\right|^2 = \left|\frac{1.6 \cdot \cos 50° - 1 \cdot \cos 28.61°}{1.6 \cdot \cos 50° + 1 \cdot \cos 28.61°}\right|^2 = \left|\frac{1.03 - 0.88}{1.03 + 0.88}\right|^2 = 6 \cdot 10^{-3}$$

$$= 0.62\%$$

for p-polarized light. The total reflectance is thus

$$R_{tot} = \frac{R_s + R_p}{2} = \frac{0.14 + 6 \cdot 10^{-3}}{2} = 0.073 = 7.3\%.$$

### Exercise A.4.1.3

At normal incidence, the total reflection can be determined by the simplified Fresnel equation. It thus amounts to

$$R_{tot} = \left(\frac{n' - n}{n' + n}\right)^2 = \left(\frac{1.78 - 1.33}{1.78 + 1.33}\right)^2 = \left(\frac{0.45}{3.11}\right)^2 = 0.021 = 2.1\%.$$

### Exercise A.4.1.4

A:

The Brewster's angle is given by

$$\varepsilon_B = \arctan\left(\frac{n'}{n}\right).$$

Solving this equation for the index of refraction of the glass, $n'$, and inserting the given values gives

$$n' = n \cdot \tan\varepsilon_B = 1 \cdot \tan 57° = 1.53987.$$

### Exercise A.4.1.5

The critical angle of total internal reflection is generally given by

$$\varepsilon_{\text{crit}} = \arcsin\left(\frac{n}{n'}\right).$$

At the interface glass–air, it consequently amounts to

$$\varepsilon_{\text{crit}} = \arcsin\left(\frac{1}{1.48}\right) = 42.51°$$

and to

$$\varepsilon_{\text{crit}} = \arcsin\left(\frac{1.33}{1.48}\right) = 63.98°$$

at the interface glass–water. The absolute value of the change is thus

$$|\Delta\varepsilon_{\text{crit}}| = |42.51° - 63.98°| = 21.47°.$$

### Exercise A.4.1.6

The thickness $t$ of the glass plate can be determined on the basis of the Beer-Lambert law,

$$I_{\text{t}} = I_0 \cdot e^{-\alpha \cdot t}.$$

The ratio of the transmitted intensity to $I_{\text{t}}$ and the initial intensity $I_0$ gives the internal transmission $T_{\text{i}}$ (reflection losses are not considered) according to

$$\frac{I_{\text{t}}}{I_0} = T_i = e^{-\alpha \cdot t} = 0.98 = 98\%.$$

Solving this equation for the thickness where the inverse function of the exponential function is the Napierian logarithm gives the thickness $t$ of the glass plate according to

$$t = \frac{\ln\left(\frac{I_t}{I_0}\right)}{-\alpha} = \frac{\ln T_i}{-\alpha} = \frac{\ln\left(\frac{0.98\ \text{W}/\text{cm}^2}{1\ \text{W}/\text{cm}^2}\right)}{-\frac{0.0024072}{\text{cm}}} = 8.4\ \text{cm}.$$

### Exercise A.4.1.7

The absorbance is given by

$$A = 1 - T_i$$

with

$$T_i = e^{-\alpha \cdot t}.$$

It finally amounts to

$$A = 1 - e^{-\alpha \cdot t} = e^{-\frac{0.0056346}{\text{cm}} \cdot 2.5\,\text{cm}} = 1 - 0.986 = 0.014 = 1.4\%.$$

## A.4.2 Optical Materials

### Exercise A.4.2.1

For the determination of birefringence, the ordinary and extraordinary indices of refraction have to be calculated in advance. The ordinary index of refraction amounts to

$$n_o = \frac{c_0}{c_s} = \frac{3 \cdot 10^8\ \text{m/s}}{2 \cdot 10^8\ \text{m/s}} = 1.5.$$

and the extraordinary one is

$$n_{eo} = \frac{c_0}{c_p} = \frac{3 \cdot 10^8\ \text{m/s}}{1.99 \cdot 10^8\ \text{m/s}} = 1.50754.$$

Finally, the birefringence of the optical medium is

$$\Delta n = n_{eo} - n_o = 1.50754 - 1.5 = 0.00754.$$

### Exercise A.4.2.2

For the determination of the total reflectance $R$ at perpendicular incidence according to

$$R_{tot} = \left( \frac{n' - n}{n' + n} \right)^2,$$

the index of refraction of the glass $n'$ has to be calculated initially using the Sellmeier equation,

$$n' = \sqrt{1 + \frac{B_1 \cdot \lambda^2}{\lambda^2 - C_1} + \frac{B_2 \cdot \lambda^2}{\lambda^2 - C_2} + \frac{B_3 \cdot \lambda^2}{\lambda^2 - C_3}}.$$

It thus amounts to

$$n' = \sqrt{1 + \frac{1.03 \cdot (0.5\ \mu\text{m})^2}{(0.5\ \mu\text{m})^2 - 0.006\ \mu\text{m}^2} + \frac{0.232 \cdot (0.5\ \mu\text{m})^2}{(0.5\ \mu\text{m})^2 - 0.021\ \mu\text{m}^2} + \frac{1.011 \cdot (0.5\ \mu\text{m})^2}{(0.5\ \mu\text{m})^2 - 103.561\ \mu\text{m}^2}}$$

$$= 1.51855.$$

The total reflectance at normal incidence and at a wavelength of 500 nm is thus

$$R_{tot} = \left( \frac{1.51855 - 1}{1.51855 + 1} \right)^2 = 0.0424 = 4.24\%.$$

### Exercise A.4.2.3

Since the center wavelength is 546 nm in the present case, the Abbe number $V_e$, referring to the Fraunhofer line $e$, needs to be calculated according to

$$V_e = \frac{n_e - 1}{n_{F'} - n_{C'}}.$$

For this purpose, the indices of refraction at the Fraunhofer lines $e$, $F'$, and $C'$ have to be determined. This can be realized on the basis of the equation for the total reflectance at normal incidence,

$$R_{tot} = \left(\frac{n' - n}{n' + n}\right)^2.$$

Solving this equation for the index of refraction of the glass $n'$ gives

$$n' = \frac{n \cdot \left(\sqrt{R_{tot}} + 1\right)}{1 - \sqrt{R_{tot}}}.$$

The particular indices of refraction of the glass are thus

$$n'(480\text{ nm}) = \frac{1 \cdot \left(\sqrt{0.083} + 1\right)}{1 - \sqrt{0.083}} = 1.80899 = n_{F'},$$

$$n'(546\text{ nm}) = \frac{1 \cdot \left(\sqrt{0.081} + 1\right)}{1 - \sqrt{0.081}} = 1.79564 = n_e,$$

and

$$n'(644\text{ nm}) = \frac{n \cdot \left(\sqrt{0.078} + 1\right)}{1 - \sqrt{0.078}} = 1.78823 = n_{C'}.$$

The Abbe number is then

$$V_e = \frac{1.79564 - 1}{1.80899 - 1.78823} = 38.33.$$

The unknown glass is a flint glass since the Abbe number is (much) lower than 50.

### Exercise A.4.2.4

The inhomogeneity class is defined by the deviation in index of refraction $\Delta n$, generally given by

$$\Delta n = \frac{\Delta w}{2 \cdot t}.$$

The wave front deformation amounts to $\Delta w = \lambda/10 = 633\,\text{nm}/10 = 63.3\,\text{nm}$ and the glass sample thickness is $1\,\text{cm} = 10\,\text{mm} = 10^7\,\text{nm}$. Hence, the deviation in index of refraction is

$$\Delta n = \frac{63.3\text{ nm}}{2 \cdot 10^7\text{ nm}} = 3.165 \cdot 10^{-6}.$$

The inhomogeneity class is thus 3, since

$$4 \cdot 10^{-6}(\text{class } 3) > 3.165 \cdot 10^{-6} > 2 \cdot 10^{-6}(\text{class } 4).$$

### Exercise A.4.2.5

The radial distribution of the refractive index $n(r)$ within a gradient index lens is given by a hyperbolic secant distribution according to

$$n(r) = n_0 \cdot \sec h(g \cdot r).$$

Generally, the hyperbolic secant can also be rewritten as

$$\sec h(x \cdot y) = \frac{1}{\cosh(x \cdot y)} = \frac{2}{e^{(x \cdot y)} + e^{-(x \cdot y)}}.$$

The radial distribution of the index of refraction within a gradient index lens is thus

$$n(r) = n_0 \cdot \frac{2}{e^{(g \cdot r)} + e^{-(g \cdot r)}}.$$

For the given parameters, the index of refraction at the position $r = 500\,\mu\text{m}$ is then

$$n(0.5\text{ mm}) = 1.616 \cdot \frac{2}{e^{\left(\frac{0.653}{\text{mm}} \cdot 0.5\text{ mm}\right)} + e^{-\left(\frac{0.653}{\text{mm}} \cdot 0.5\text{ mm}\right)}} = 1.616 \cdot \frac{2}{1.3861 + 0.7214} = 1.53357.$$

### Exercise A.4.2.6

The removed volume amounts to

$$\Delta V_{\text{glass}} = G \cdot \Delta V_{\text{ref}} = 1.65 \cdot 0.4\text{ mm}^3 = 0.66\text{ mm}^3.$$

The factor of 1.65 results from dividing the particular glass grindability by the grindability of the reference glass (165/100 = 1.65).

## A.4.3 Optical Components

### Exercise A.4.3.1

The effective focal length of the thin lens amounts to

$$EFL = \frac{1}{1.64 - 1} \cdot \left( \frac{53\text{ mm} \cdot (-146\text{ mm})}{(-146\text{ mm}) - 53\text{ mm}} \right) = 60.76\text{ mm}.$$

Taking the given center thickness of $t_c$=6 mm into account, the effective focal length of the equivalent thick lens is then

$$EFL = \frac{1}{1.64-1} \cdot \frac{1.64 \cdot 53\,\text{mm} \cdot (-146\,\text{mm})}{(1.64-1) \cdot 6\,\text{mm} + 1.64 \cdot (-146\,\text{mm} - 53\,\text{mm})} = 61.48\,\text{mm}.$$

**Exercise A.4.3.2**

First, the angle of incidence of light at the interface glass–air within the prism has to be determined. It amounts to

$$\varepsilon_2 = \alpha - \arcsin\left(\frac{\sin\varepsilon_1}{n}\right) = 50° - \arcsin\left(\frac{\sin 30°}{1.55}\right) = 31.18°.$$

The exit angle of light is thus

$$\varepsilon_2' = \arcsin(n \cdot \sin\varepsilon_2) = \arcsin(1.55 \cdot \sin 31.18°) = 53.37°.$$

**Exercise A.4.3.3**

The index of refraction of the prism material amounts to

$$n = \frac{\sin\left(\frac{\delta_{\min} + \alpha}{2}\right)}{\sin\left(\frac{\alpha}{2}\right)} = \frac{\sin\left(\frac{23° + 45°}{2}\right)}{\sin\left(\frac{45°}{2}\right)} = \frac{0.559}{0.383} = 1.56397.$$

**Exercise A.4.3.4**

The given index of refraction is valid at the second wavelength of interest, $\lambda_2$=486 nm (i.e., the Fraunhofer line *F*). Since the main dispersion is known, the index of refraction at the first wavelength of interest, $\lambda_1$=656 nm (Fraunhofer line *C*) can be determined as follows:

$$n_F - n_C = 0.0442 \rightarrow n_C = n_F - 0.0442 = 1.9104.$$

In order to calculate the dispersion angle according to

$$\delta_d = \delta(\lambda_2) - \delta(\lambda_1),$$

the resulting deviation $\delta$ for each particular wavelength and index of refraction has to be determined in advance. At $\lambda_2$=486 nm, it amounts to

$$\delta(\lambda_2) = 70° + 22.44° - 40° = 52.44°.$$

with

$$\varepsilon_2' = \arcsin(n \cdot \sin\varepsilon_2) = \arcsin(1.9546 \cdot \sin 11.26°) = 22.44°.$$

and

$$\varepsilon_2 = \alpha - \arcsin\left(\frac{\sin\varepsilon_1}{n}\right) = 40° - \arcsin\left(\frac{\sin 70°}{1.9546}\right) = 40° - 28.74° = 11.26°.$$

At $\lambda_1 = 656\,\text{nm}$, the deviation is calculated in the same way:

$$\delta(\lambda_1) = \varepsilon_1 + \varepsilon_2' - \alpha = 70° + 20.45° - 40° = 50.45°.$$

with

$$\varepsilon_2' = \arcsin(n \cdot \sin\varepsilon_2) = \arcsin(1.9104 \cdot \sin 10.54°) = 20.45°.$$

and

$$\varepsilon_2 = \alpha - \arcsin\left(\frac{\sin\varepsilon_1}{n}\right) = 40° - \arcsin\left(\frac{\sin 70°}{1.9104}\right) = 40° - 29.46° = 10.54°.$$

Finally, the dispersion angle amounts to

$$\delta_\text{d} = \delta(\lambda_2) - \delta(\lambda_1) = 52.44° - 50.45° = 1.99°.$$

**Exercise A.4.3.5**

The wedge angle $\alpha$ of wedges is generally given by

$$\delta = \alpha \cdot (n-1),$$

with $\delta$ being the deviation of light. In the present case, the wedge angle thus amounts to

$$\alpha = \frac{\delta}{(n-1)} = \frac{3°}{(1.68-1)} = 4.41°.$$

**Exercise A.4.3.6**

The parallel offset of the light beam can be calculated according to

$$O_\text{p} = t \cdot \sin\varepsilon \cdot \left(1 - \frac{\cos\varepsilon}{\sqrt{n^2 - \sin^2\varepsilon}}\right).$$

It is thus

$$O_\text{p} = 20\text{ mm} \cdot \sin 45° \cdot \left(1 - \frac{\cos 45°}{\sqrt{1.57^2 - \sin^2 45°}}\right) = 14.14\text{ mm} \cdot \left(1 - \frac{0.71}{\sqrt{2.47 - 0.5}}\right)$$

$$= 6.93\text{ mm}.$$

Note that the term $\sin^2\varepsilon$ can be expressed as

$$\sin^2 \varepsilon = \sin(\varepsilon)^2 = \frac{1}{2} \cdot \left[1 - \cos(2 \cdot \varepsilon)\right].$$

The longitudinal offset is

$$O_l = t \cdot \frac{n-1}{n} = 20 \text{ mm} \cdot \frac{1.57-1}{1.57} = 7.26 \text{ mm}.$$

## A.4.4 Design of Optical Components

### Exercise A.4.4.1

The effective focal length is generally given by

$$\frac{1}{EFL} = \frac{1}{a} + \frac{1}{a'}.$$

It thus amounts to 47.62 mm. For the determination of the wanted image height, the magnification $\beta$ has to be calculated in advance. It is

$$\beta = \frac{a'}{a} = \frac{50 \text{ mm}}{1000 \text{ mm}} = 0.05.$$

The image height is then

$$\beta = \frac{u'}{u} \rightarrow u' = \beta \cdot u = 0.05 \cdot 60 \text{ mm} = 3 \text{ mm}.$$

### Exercise A.4.4.2

The effective focal length of a thin lens is generally given by

$$\frac{1}{EFL} = (n-1) \cdot \left(\frac{1}{R_1} - \frac{1}{R_2}\right).$$

In the present case, it is thus

$$\frac{1}{EFL} = (1.6-1) \cdot \left(\frac{1}{75 \text{ mm}} - \frac{1}{100 \text{ mm}}\right) \rightarrow EFL = 555.6 \text{ mm}.$$

For an equivalent thick lens, the effective focal length follows from

$$\frac{1}{EFL} = (n-1) \cdot \left(\frac{1}{R_1} - \frac{1}{R_2}\right) + \frac{(n-1)^2 \cdot t_c}{n \cdot R_1 \cdot R_2}$$

and amounts to

$$\frac{1}{EFL} = (1.6-1) \cdot \left(\frac{1}{75 \text{ mm}} - \frac{1}{100 \text{ mm}}\right) + \frac{(1.6-1)^2 \cdot 10 \text{ mm}}{1.6 \cdot 75 \text{ mm} \cdot 100 \text{ mm}} \rightarrow EFL = 476.2 \text{ mm}.$$

### Exercise A.4.4.3

The longitudinal spherical aberration is given by the difference in back focal length according to

$$\Delta BFL = BFL(h_{\min}) - BFL(h_{\max}),$$

where the particular back focal length for a given ray entrance height follows from

$$BFL = R + \frac{h}{n \cdot \sin\left[\arcsin\left(\frac{h}{R}\right) - \arcsin\left(\frac{h}{n \cdot R}\right)\right]}.$$

For the given maximum ray entrance height, the back focal length is

$$BFL(h_{\max}) = 100\text{ mm} + \frac{15\text{ mm}}{1.75 \cdot \sin\left[\arcsin\left(\frac{15\text{ mm}}{100\text{ mm}}\right) - \arcsin\left(\frac{15\text{ mm}}{1.75 \cdot 100\text{ mm}}\right)\right]}$$

$$= 236.36\text{ mm}$$

and

$$BFL(h_{\min}) = 100\text{ mm} + \frac{1\text{ mm}}{1.75 \cdot \sin\left[\arcsin\left(\frac{1\text{ mm}}{100\text{ mm}}\right) - \arcsin\left(\frac{1\text{ mm}}{1.75 \cdot 100\text{ mm}}\right)\right]}$$

$$= 236.98\text{ m}.$$

for the given minimum ray entrance height. The longitudinal spherical aberration finally amounts to

$$\Delta BFL = 236.98\text{ mm} - 236.36\text{ mm} = 0.62\text{ mm} = 620\ \mu\text{m}.$$

### Exercise A.4.4.4

For achromatic doublets, the products of the effective focal length and the Abbe number of the involved single lenses generally equal according to an amount expressed by

$$EFL_1 \cdot V_1 = -EFL_2 \cdot V_2.$$

The Abbe number of the diverging lens is thus

$$V_2 = \frac{EFL_1 \cdot V_1}{-EFL_2} = \frac{8.95\text{ mm} \cdot 50}{-(-17\text{ mm})} = 26.32.$$

### Exercise A.4.4.5

The type of distortion can be identified via the determination of percentaged distortion $D_{\text{per}}$ according to

$$D_{\text{per}} = \frac{u'_{\text{a}} - u'_{\text{t}}}{u'_{\text{t}}} \cdot 100\%.$$

In the present case, it amounts to

$$D_{\text{per}} = \frac{9\text{ mm} - 9.6\text{ mm}}{9.6\text{ mm}} \cdot 100\% = -6.25\%.$$

Since this value is negative, the type of distortion is barrel distortion.

### Exercise A.4.4.6

Generally, the merit function $MF$ is given by

$$MF = \sum_i d_{i,\text{rel}}^2 = \sum_i \left( \frac{d_{i,\text{a}} - d_{i,\text{t}}}{d_{i,\text{tol}}} \right)^2.$$

Since, in the present case, the only defect of interest is spherical aberration, this expression can be rewritten as

$$MF = \left| \frac{d_{\text{a}} - d_{\text{t}}}{d_{\text{tol}}} \right|^2.$$

The actual defect value $d_{\text{a}}$ is given by the Seidel sum $S_{\text{I}}=0.01$ of the used start system, and the target defect value $d_{\text{t}}$ is $S_{\text{I}}=0.005$ as determined in the course of the analysis of the imaging task. The acceptable fault tolerance $d_{\text{tol}}$ follows from the given upper and lower limit and amounts to

$$d_{\text{tol}} = \frac{0.004 - 0.002}{2} = 0.001.$$

The absolute value of the MF finally is

$$MF = \left| \frac{0.01 - 0.005}{0.001} \right|^2 = 25.$$

## A.4.5 Tolerancing of Optical Components and Systems

### Exercise A.4.5.1

The code number "0" identifies stress birefringence. In the present case, a maximum difference in optical path length of 40 nm per 10 mm reference optical path length due to stress birefringence is acceptable. Further, the amount and size of bubbles and inclusions is specified by the code number "1." A maximum of two bulk defects with a maximum cross-sectional area of 0.25 mm² per bulk defect is thus acceptable. Finally, inhomogeneity and striae are indicated by the code number "2." The number "4" is the inhomogeneity class, and the number "1" represents the striae class. In the present case, the acceptable maximum deviation in index of refraction amounts to $\pm 2 \cdot 10^{-6}$ (inhomogeneity class 4) and the maximum share of striae is 10% (striae class 1).

### Exercise A.4.5.2

For the determination of the inhomogeneity class, the deviation in index of refraction $\Delta n$ has to be calculated. It amounts to

$$\Delta n = \frac{\Delta w}{2 \cdot t} = \frac{40 \text{ nm}}{2 \cdot 10 \text{ mm}} = 2 \cdot 10^{-6}.$$

This glass can thus be classified into inhomogeneity class 4.

### Exercise A.4.5.3

The class of hardness directly follows from the Knoop hardness $HK$ of the tested glass. This value amounts to

$$HK = 1.451 \cdot \frac{F}{l^2} = 1.451 \cdot \frac{10 \text{ N}}{(0.17 \text{ mm})^2} = 502.$$

The tested glass can thus be classified into the class of hardness 5 (where $HK = 450$–$550$).

### Exercise A.4.5.4

The coefficient of thermal expansion of the glass amounts to

$$\alpha = \frac{1}{\Delta T} \cdot \frac{\Delta l}{l_0} = \frac{1}{50 \text{ K}} \cdot \frac{0.12 \text{ mm}}{300 \text{ mm}} = 8 \cdot 10^{-6} \text{ K}^{-1}.$$

### Exercise A.4.5.5

Generally, surface accuracy is indicated by the expression "3/A(B/C)" according to DIN ISO 10110. The strategy for the specification of the given interference pattern is shown in Figure A.2 below.

**FIGURE A.2** Evaluation and specification of the measured interference pattern.

First, the maximum sagitta $A$ and the surface irregularity $B$ are determined by counting the observable Newton fringes in two directions orthogonal to each other. Moreover, the fine contour error $C$ follows from the distance between two interference fringes $a$ (i.e., half the test wavelength) and the deviation $d$. As shown in Figure A.2 above, two fringes are observed in y-direction (where the fringes are counted from the lens center to its edge). The parameter "A" is thus 2. In contrast, merely one fringe is found in x-direction perpendicular to the y-axis, resulting in a parameter "B" of 1 ($2-1$ fringes $=1$ fringe). Moreover, the fine contour error and parameter $C$ are given by $d/a = 30\,\text{nm}/300\,\text{nm} = 0.1$. The full specification of the observed interference pattern according to DIN ISO 10110 is thus "3/2(1/0.1)."

## Exercise A.4.5.6

As a result of the given specification of surface accuracy "3/50(44)," the maximum acceptable surface irregularity and deviation from the nominal radius of curvature of the lens surface are six fringes and three wavelengths, respectively. The acceptable tolerance of the radius of curvature $A$ in y-direction is

$$A = 25 \cdot 1064 \text{ nm} = 26.6\ \mu\text{m}.$$

The tolerance range for the radius of curvature in this direction is thus approximately 5.97–6.03 ($6\,\text{mm} \pm 26.6\,\mu\text{m}$). In contrast, the acceptable tolerance of the radius of curvature $B$ in x-direction amounts to

$$A = 3 \cdot 1064 \text{ nm} = 3.192\ \mu\text{m},$$

resulting in a tolerance range from 5.9968 to 6.0032 mm ($6\,\text{mm} \pm 3.192\,\mu\text{m}$). As a worst case scenario, the actual lens surface could thus feature a radius of curvature of 5.97 mm in y-direction and 6.0032 mm in x-direction, but fulfill the given specification. This difference in radii of curvature represents a toric lens surface, which directly results in the formation of astigmatism even for incident light propagating parallel to the optical axis of the lens. This is also visualized the by comparison of the simulated spot diagrams of the present case in Figure A.3 below.

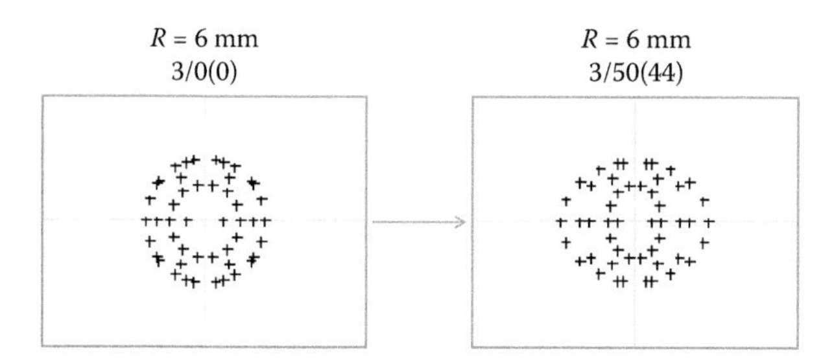


**FIGURE A.3** Visualization of the impact of manufacturing tolerances on imaging quality by the particular spot patterns of a lens without any manufacturing errors (left) and a lens with surface inaccuracy within the specified manufacturing tolerances (right).

### Exercise A.4.5.7

The centering error amounts to

$$CE = \frac{1720 \cdot r_w}{m \cdot d \cdot (n-1)} = \frac{1720 \cdot 0.8 \text{ mm}}{5 \cdot 350 \text{ mm} \cdot (1.65 - 1)} = 1.21'.$$

### Exercise A.4.5.8

The specification "5/2·0.16" defines the maximum number and area of surface defects. In the given case, a maximum of two quadratic defects with an edge length of 0.16 mm is acceptable. The resulting area per defect is thus $A = (0.16\,\text{mm})^2 = 0.0256\,\text{mm}^2$, and the maximum total defect area is $0.0512\,\text{mm}^2$. Moreover, the term "P4" defines the acceptable residual roughness of the polished surface, that is, 1.6–3.2 nm (see Section 6.3.4).

### Exercise A.4.5.9

The fraction of specular reflected light can be calculated with the aid of the total integrated scatter *TIS* function, given by

$$TIS = 1 - e^{-\left(\frac{4 \cdot \pi \cdot \cos AOI \cdot Rq}{\lambda}\right)^2}.$$

This interrelationship gives the fraction of diffusively reflected light. It amounts to

$$TIS = 1 - e^{-\left(\frac{4 \cdot \pi \cdot \cos 0^\circ \cdot 33 \text{ nm}}{380 \text{ nm}}\right)^2} = 0.696.$$

The fraction of specular reflected light is thus $1 - TIS = 0.304$ (i.e., 30.4%).

## A.4.6 Shape Forming

### Exercise A.4.6.1

For a convex lens surface, the required setting angle is

$$\alpha_{CX} = \arcsin \frac{D_{cw}}{2 \cdot (R_c + r)} = \arcsin \frac{30 \text{ mm}}{2 \cdot (50 \text{ mm} + 2 \text{ mm})} = 19.77^\circ,$$

whereas it amounts to

$$\alpha_{CC} = \arcsin \frac{D_{cw}}{2 \cdot (R_c - r)} \arcsin \frac{30 \text{ mm}}{2 \cdot (50 \text{ mm} - 2 \text{ mm})} = 18.21^\circ$$

for a concave lens surface.

### Exercise A.4.6.2

The type designation "F240" corresponds to a mean grain size of $D_g = 45\,\mu\text{m}$. (see Section 7.4.1.2). The depth of microcracks is thus approximately

$$d_{mc} \approx 0.3 \cdot D_g \approx 0.3 \cdot 45\ \mu\text{m} \approx 13.5\ \mu\text{m}.$$

### Exercise A.4.6.3

The grinding tool denomination "D3" indicates an abrasive grain size of $D_g = 3.5 \pm 1.5\,\mu m$ (see Section 7.3.1). Consequently, the minimum grain size is $D_{g,min} = 2\,\mu m$, and the maximum one is $D_{g,max} = 5\,\mu m$. In order to obtain a surface without any microcracks, the thickness of removed material should thus amount to at least $0.6\,\mu m$, but better to $1.5\,\mu m$ according to

$$d_{mc} \approx 0.3 \cdot D_g .$$

However, microcracks cannot be avoided in classical optical manufacturing, and even polishing induces microcracks, since the polishing suspension contains abrasive grains. Against this background, unconventional techniques without any mechanical impact by grains such as plasma polishing allow the removal or prevention of microcracks.

### Exercise A.4.6.4

The degree of coverage of a spherical cup wheel with abrasive pellets is given by

$$C_p = \frac{4 \cdot N \cdot \left(\frac{D_p}{2}\right)^2}{D_{cw}^2}.$$

The number of pellets in the present case is thus

$$N = \frac{C_p \cdot D_{cw}^2}{4 \cdot \left(\frac{D_p}{2}\right)^2} = \frac{0.16 \cdot (40\text{ mm})^2}{4 \cdot \left(\frac{4\text{ mm}}{2}\right)^2} = 16.$$

### Exercise A.4.6.5

The cutting velocity $v_c$ can be determined based on the assumption that the tool diameter $D_t$ is twice the lens surface diameter $D_{wp}$ (as usually valid in practice). It is thus 50.8 mm. In this case, the cutting velocity amounts to

$$v_c = D_t \cdot \pi \cdot n_t = 50.8\text{ mm} \cdot \pi \cdot 6000\text{ s}^{-1} = 957.6\text{ m/s}.$$

Note that the drive of the cutting tool was given in rounds per minute (rpm) and has to be converted to the SI-unit $s^{-1}$ ($1000\,rpm = 6000\,s^{-1}$).

### Exercise A.4.6.6

The initial thickness of the glass window is 6 mm, and its target thickness is 5 mm. Thus, a glass layer with a thickness of $t_1 = 1$ mm has to be removed. The volume of this layer follows from

$$\Delta V_1 = t_1 \cdot \pi \cdot r^2,$$

with $r$ being half the diameter of the glass window. The volume to be removed is thus

$$\Delta V_1 = 1\ \text{mm} \cdot \pi \cdot (11\ \text{mm})^2 = 380\ \text{mm}^3.$$

Finally, the duration or time of the grinding process required for removing this volume amounts to

$$t = \frac{\Delta V}{MRR} = \frac{380\ \text{mm}^3}{6.3\ \text{mm}^3/\text{min}} = 60\ \text{min}.$$

### Exercise A.4.6.7

For generating a bevel with a bevel angle of 45°, the radius of curvature $R_{cw}$ of the spherical cup wheel used for beveling is given by

$$R_{cw} = \frac{D_1}{\sqrt{2}}.$$

Thus, a (concave) spherical cup wheel with a radius of curvature of $R_{cw} = 17.96\,\text{mm}$ should be chosen in the present case.

### Exercise A.4.6.8

The diameter of the used ring-shaped bell is 40 mm; its radius is thus $r_s = 20\,\text{mm}$. The latter value is needed for the calculation of the sagitta as follows:

$$S = \frac{\left(\frac{r_s^2}{R_c}\right)}{1 + \sqrt{\left(1 - \frac{r_s^2}{R_c^2}\right)}} = \frac{\frac{(20\ \text{mm})^2}{101\ \text{mm}}}{1 + \sqrt{\left(1 - \frac{(20\ \text{mm})^2}{(101\ \text{mm})^2}\right)}} = 2\ \text{mm}.$$

## A.4.7 Polishing

### Exercise A.4.7.1

The mean diffusion depth can be estimated by

$$d_{dif} \approx 2 \cdot \sqrt{D \cdot t},$$

with $t$ being the polishing process duration of $6\,\text{h} = 21{,}600\,\text{s}$. The mean diffusion depth is thus

$$d_{dif} \approx 2 \cdot \sqrt{10^{-18}\ \text{cm}^2/\text{s} \cdot 21{,}600\ \text{s}} = 2.9 \cdot 10^{-7}\ \text{cm} = 2.9\ \text{nm}.$$

### Exercise A.4.7.2

The material removal rate is generally given by

$$MRR = C_p \cdot \frac{L}{A} \cdot \frac{\Delta s}{\Delta t},$$

where $L$ is the total load, $A$ is the work piece area, and the expression $\Delta s/\Delta t$ gives the relative velocity $v$. The total load can be determined on the basis of the weight or mass $m$ of the polishing tool and amounts to

$$L = m \cdot g = 2\text{ kg} \cdot 9.81\text{ m}/\text{s}^2 = 19.62\text{ kg} \cdot \text{m}/\text{s}^2 = 19.62\text{ N}.$$

Here, the parameter $g$ is the acceleration of gravity of approximately 9.81 m/s². Further, the work piece area is

$$A = \pi \cdot r^2 = \pi \cdot (1.25\text{ cm})^2 = 4.91\text{ cm}^2.$$

The material removal rate finally amounts to

$$MRR = C_\text{p} \cdot \frac{L}{A} \cdot \frac{\Delta s}{\Delta t} = 10^{-7}\text{ cm}^2/\text{N} \cdot \frac{19.62\text{ N}}{4.91\text{ cm}^2} \cdot 4\text{ m}/\text{s} = 1.6 \cdot 10^{-6}\text{ m}/\text{s} = 1.6\ \mu\text{m}/\text{s}.$$

**Exercise A.4.7.3**

The work required for material removal is generally given by

$$W = \mu \cdot A \cdot L \cdot v \cdot t,$$

where $\mu$ is the polishing pad material-specific coefficient of friction as listed in Table 8.2 in Section 8.3.1. ($\mu_{\text{pitch}}=0.735$, $\mu_{\text{felt}}=0.685$, $\mu_{\text{polyurethane}}=0.622$). After converting the given work piece area and the polishing time in the SI-units ($A=5.07\text{ m}^2$ and $t=3600\text{ s}$), the required work for the polishing pad material pitch can be calculated as follows:

$$W_{\text{pitch}} = 0.735 \cdot 5.07\text{ m}^2 \cdot 20\text{ N} \cdot 2\text{ m}/\text{s} \cdot 3600\ s = 536{,}609\text{ Nm} \approx 537\text{ kJ}.$$

Accordingly, the work required for material removal amounts to approximately 500 kJ, when using felt as polishing pad material, and to approximately 454 kJ in the case of polyurethane.

**Exercise A.4.7.4**

The polishing suspension concentration is given by

$$C_\text{s} = \frac{m_{\text{pa}}}{m_\text{s}} \cdot 100\%,$$

the mass of the polishing agent within the suspension is thus

$$m_{\text{pa}} = \frac{C_\text{s} \cdot m_\text{s}}{100\%} = \frac{17\% \cdot 1.2\text{ kg}}{100\%} = 0.204\text{ kg} = 204\text{ g}.$$

## A.4.8 Cementing

### Exercise A.4.8.1

In the present case, the cement layer can be treated as a wedge as introduced in Section 4.3.4. The deviation caused by such a wedge is generally given by

$$\delta = \alpha \cdot (n-1).$$

It thus amounts to

$$\delta = 2° \cdot (1.58-1) = 1.16°.$$

### Exercise A.4.8.2

Since the thickness of the cement layer can be neglected, the equation for calculating the effective focal length of a thin lens with different radii of curvature, generally given by

$$EFL = \frac{1}{n-1} \cdot \left( \frac{R_1 \cdot R_2}{R_2 - R_1} \right).$$

can be applied in the present case. The effective focal length of the cement layer thus amounts to

$$EFL = \frac{1}{1.5-1} \cdot \left( \frac{29\text{ mm} \cdot 30\text{ mm}}{30\text{ mm} - 29\text{ mm}} \right) = 1740\text{ mm} \approx 1.7\text{ m}.$$

## A.4.9 Centering

### Exercise A.4.9.1

The centering error is generally given by

$$CE = \frac{1720 \cdot r_w}{m \cdot d \cdot (n-1)}.$$

After solving this interrelationship for the wobble circle radius $r_w$ and inserting the given parameters, the wobble circle radius can be determined:

$$r_w = \frac{CE \cdot m \cdot d \cdot (n-1)}{1720} = \frac{1.38' \cdot 6 \cdot 500\text{ mm} \cdot (1.5-1)}{1720} = 1.2\text{ mm}.$$

### Exercise A.4.9.2

Generally, the centering error is given by

$$CE = 3434 \cdot \left( \frac{\Delta z}{R_1} + \frac{\Delta z}{R_2} \right),$$

where its unit is arc seconds. For a description of the centering error in radians, the factor 3434 is not considered. The runout thus amounts to

$$\Delta z = \frac{CE}{\left(\frac{1}{R_1} + \frac{1}{R_2}\right)} = \frac{0.008}{\left(\frac{1}{50\ \text{mm}} + \frac{1}{100\ \text{mm}}\right)} = 0.2\overline{6}\ \text{mm} = 266\ \mu\text{m}.$$

### Exercise A.4.9.3

Self-alignment of lenses during centering occurs if the sliding angle $\alpha$ is higher than 7°. In the present case, this angle amounts to

$$\alpha = \arcsin\left(\frac{D_1}{2 \cdot R_1}\right) + \arcsin\left(\frac{D_2}{2 \cdot R_2}\right) = \arcsin\left(\frac{20\ \text{mm}}{2 \cdot 75\ \text{mm}}\right) + \arcsin\left(\frac{20\ \text{mm}}{2 \cdot 75\ \text{mm}}\right) = 15.33°.$$

Consequently, the lens is self-aligned by the clamping bells.

## A.4.10 Coating

### Exercise A.4.10.1

For a single-layer antireflective coating, the reflectance at normal incidence is given by

$$R_{\text{coated}} = \left(\frac{n_{\text{g}} \cdot n_{\text{a}} - n_{\text{c}}^2}{n_{\text{g}} \cdot n_{\text{a}} + n_{\text{c}}^2}\right)^2.$$

It thus amounts to

$$R_{\text{coated}} = \left(\frac{1.57 \cdot 1 - (1.46)^2}{1.57 \cdot 1 + (1.46)^2}\right)^2 = 0.023 = 2.3\%.$$

In contrast, the uncoated glass surface features a reflectance of

$$R_{\text{uncoated}} = \left(\frac{1.57 - 1}{1.57 + 1}\right)^2 = 0.049 = 4.9\%.$$

The reflectance of the uncoated glass surface is thus approximately two times higher than the reflectance of the glass surface with single-layer antireflective coating.

### Exercise A.4.10.2

The transmittance $T_{\text{ls}}$ of each coated lens surface follows from the residual reflectance $R_{\text{r}}$ and amounts to

$$T_{\text{ls}} = 1 - R_{\text{r}} = 1 - 0.005 = 0.995 = 99.5\%,$$

where absorption within the glass bulk material is not considered. Since the transmittance is the same for all lens surfaces, the total transmittance of the coated doublet amounts to

$$T_{total} = T_{ls}^{n} = 0.996^{4} = 0.980 = 98\%.$$

Here, $n$ is the number of lens surfaces (four for two lenses).

The total trasnmittance of the doublet without any coating results from the reflectance of the uncoated glass surfaces as follows: based on the assumption of normal incidence, the total reflectance at each lens surface can be determined according to

$$R_{tot} = \left(\frac{n' - n}{n' + n}\right)^{2}$$

and amounts to

$$R_{tot} = \left(\frac{1.5 - 1}{1.5 + 1}\right)^{2} = 0.04 = 4\%.$$

In this case, the transmittance of each lens surface thus accounts for

$$T_{ls} = 1 - R_{r} = 1 - 0.04 = 0.96 = 96\%.$$

Finally, the total transmittance is

$$T_{total} = T_{ls}^{n} = 0.96^{4} = 0.85 = 85\%.$$

The transmittance of the lens doublet is consequently increased by 13% when applying the abovementioned antireflective coating with a residual reflectance of 0.5%.

## A.4.11 Assembly of Optomechanical Systems

### Exercise A.4.11.1

The tilt angle amounts to

$$\alpha = \frac{l_{max} - l_{min}}{2 \cdot R} = \frac{0.8\text{ mm} - 0.05\text{ mm}}{2 \cdot 75\text{ mm}} = 5 \cdot 10^{-3}\text{ rad}.$$

This corresponds to approximately 0.29° or 17 arc min. The corresponding specification of the centering error according to DIN ISO 10110 is thus "4/17′."

## A.4.12 Microoptics

### Exercise A.4.12.1

The modulation transfer function is given by the ratio of the modulations $M$ (or Michelson contrasts) of the image and the object according to

$$MTF = \frac{M_{\text{image}}}{M_{\text{object}}}.$$

The modulation is generally given by

$$M = \frac{I_{\max} - I_{\min}}{I_{\max} + I_{\min}}.$$

It amounts to

$$M_{\text{object}} = \frac{0.49\ \text{W/m}^2 - 0.28\ \text{W/m}^2}{0.49\ \text{W/m}^2 + 0.28\ \text{W/m}^2} = 0.27$$

for the object and to

$$M_{\text{image}} = \frac{0.31\ \text{W/m}^2 - 0.19\ \text{W/m}^2}{0.31\ \text{W/m}^2 + 0.19\ \text{W/m}^2} = 0.24$$

for the image. The modulation transfer function finally amounts to

$$MTF = \frac{0.24}{0.27} = 0.\bar{8}.$$

### Exercise A.4.12.2

The size of the homogeneous field of the nonimaging homogenizer amounts to

$$D_{\text{hf}} = \frac{P_{\text{la}} \cdot f_{\text{Fl}}}{f_{\text{la}}} = \frac{400\ \mu\text{m} \cdot 25\ \text{mm}}{500\ \mu\text{m}} = 20\ \text{mm}.$$

This size is the target value for the determination of the distance between the two microlens arrays of the imaging homogenizer based on

$$D_{\text{hf}} = P_{\text{la1}} \cdot \frac{f_{\text{Fl}}}{f_{\text{la1}} \cdot f_{\text{la2}}} \cdot \left[ \left( f_{\text{la1}} + f_{\text{la2}} \right) - d \right].$$

After solving this expression, the distance thus amounts to

$$d = -\frac{f_{\text{la1}} \cdot f_{\text{la2}} \cdot D_{\text{hf}}}{P_{\text{la1}} \cdot f_{\text{Fl}}} + f_{\text{la1}} + f_{\text{la2}} = -\frac{500\ \mu\text{m} \cdot 600\ \mu\text{m} \cdot 20\ \text{mm}}{400\ \mu\text{m} \cdot 25\ \text{mm}} + 500\ \mu\text{m} + 600\ \mu\text{m}$$

$$= 500\ \mu\text{m}.$$

### Exercise A.4.12.3

The grating constant amounts to 1800 line pairs per millimeter, corresponding to a lateral size of grating structures of 555.6 nm. The blaze angle is thus

$$\theta_{\text{B}} = \arcsin \frac{m \cdot \lambda}{2 \cdot g} = \arcsin \frac{1 \cdot 633\ \text{nm}}{2 \cdot 555.6\ \text{nm}} = 34.73°.$$

### Exercise A.4.12.4

First, the given kinetic energy of 200 eV has to be converted in a SI-conform unit (i.e., Joule $J$). 200 eV correspond to $3.2 \cdot 10^{-17}$ J. After solving the general expression for the kinetic energy,

$$E_{\text{kin}} = \frac{1}{2} \cdot m \cdot v^2,$$

for the wanted velocity $v$, this value can be calculated

$$v = \sqrt{\frac{2 \cdot E_{\text{kin}}}{m}} = \sqrt{\frac{2 \cdot 3.2 \cdot 10^{-17}\ \text{J}}{6.6 \cdot 10^{-26}\ \text{kg}}} = 31.14\ \text{m/s}.$$

### Exercise A.4.12.5

In order to identify the suitability of reactive ion etching for the microstructuring of the given optical medium, the temperature within the process chamber has to be determined. This can be realized by solving

$$\lambda_{\text{D}} = \sqrt{\frac{\varepsilon_0 \cdot k_{\text{B}} \cdot T}{n_{\text{e}} \cdot e^2}}$$

for the temperature $T$ according to

$$T = \frac{n_{\text{e}} \cdot \lambda_{\text{D}}^2}{\varepsilon_0 \cdot k_{\text{B}}}.$$

Here, $\varepsilon_0$ is the vacuum permittivity, $k_{\text{B}}$ is the Boltzmann constant, and $e$ is the elementary electric charge. The temperature is thus

$$T = \frac{10^{18}\ \text{m}^{-3} \cdot \left(1.6 \cdot 10^{-19}\ \text{As}\right)^2 \cdot \left(1.58 \cdot 10^{-6}\ \text{m}\right)^2}{8.85 \cdot 10^{-12}\ \text{As/Vm} \cdot 1.38 \cdot 10^{-23}\ \text{J/K}} = 523.77\ \text{K}.$$

This value corresponds to 250.67°C, which is higher than the heat deflection temperature of $T_{\text{h}}$=200°C. Reactive ion etching at the given conditions is thus not suitable for microstructuring of this comparatively temperature-sensitive optical medium.

### Exercise A.4.12.6

In the present case, the center thickness of the planar GRIN lens corresponds to the mean depth of penetration of ions $d_{\text{dif}}$ into the glass. This value is given by

$$t_{\text{c}} = d_{\text{dif}} \approx 2 \cdot \sqrt{D \cdot t}.$$

The duration of the ion exchange process is thus

$$t = \frac{t_{\text{c}}^2}{4 \cdot D} = \frac{\left(0.05\ \text{cm}\right)^2}{4 \cdot 0.85 \cdot 10^{-7}\ \text{cm}^2/\text{s}} = 7353\ \text{s} \approx 2\ \text{h}.$$

# Index

Note: Italic page numbers refer to figures and tables.

## D

## P



## Q



## R